The Cellular Response to the Genotoxic Insult
The Question of Threshold for Genotoxic Carcinogens

Issues in Toxicology

Series Editors:

Professor Diana Anderson, *University of Bradford, UK*
Dr Michael D Waters, *Integrated Laboratory Systems, Inc, N Carolina, USA*
Dr Martin F Wilks, *University of Basel, Switzerland*
Dr Timothy C Marrs, *Edentox Associates, Kent, UK*

Titles in the Series:

How to obtain future titles on publication:
A standing order plan is available for this series. A standing order will bring
delivery of each new volume immediately on publication.

For further information please contact:
Book Sales Department, Royal Society of Chemistry, Thomas Graham House,
Science Park, Milton Road, Cambridge, CB4 0WF, UK
Telephone: +44 (0)1223 420066, Fax: +44 (0)1223 420247, Email: books@rsc.org
Visit our website at http://www.rsc.org/Shop/Books/

The Cellular Response to the Genotoxic Insult
The Question of Threshold for Genotoxic Carcinogens

Edited by

Helmut Greim
Institute of Toxicology and Environmental Hygiene,
Technical University of Munich, Germany
Email: helmut.greim@lrz.tu-muenchen.de

Richard J. Albertini
Department of Pathology, Genetic Toxicology Laboratory,
The University of Vermont, Burlington, VT, USA

RSC Publishing

Issues in Toxicology No. 13

ISBN: 978-1-84973-177-5
ISSN: 1757-7179

A catalogue record for this book is available from the British Library

Published by The Royal Society of Chemistry,
Thomas Graham House, Science Park, Milton Road,
Cambridge CB4 0WF, UK

Registered Charity Number 207890

For further information see our web site at www.rsc.org

Preface

So far there is general consensus that no threshold can be identified for genotoxic carcinogens. However, there is experimental evidence of no-observed-
effect-levels (NOELs) for carcinogenic and mutagenic effects in repeated dose
studies in animals and there are examples indicating that the shape of the
dose–responses of DNA adducts and mutations differ. It is also evident that
the dose–response for mutations will reach the background mutation
frequency. This implies that, at very low doses, the mutation rate induced by
a genotoxic carcinogen becomes indistinguishable from the background
mutation frequency. Moreover, the array of cellular defence mechanisms to a
genotoxic insult makes it unlikely that a single event overcomes these bar-
riers to cancer. Consequently, the notion that even low exposures to genotoxic agents present a cancer risk is to be questioned.

Besides metabolic inactivation there are two major cellular defence
mechanisms: DNA repair and apoptosis—both of which are triggered by
tumour suppressor genes. The mechanism by which the tumour suppressor
gene *p53* affects cell cycling and stimulates the DNA repair machinery after
activation by DNA damage had already been described by 1994. The expressed
protein p53 triggers the expression of two proteins p21 and Gadd45; p21
inhibits cyclin-dependent kinase activity, thus slowing down cell cycling and
DNA replication, while Gadd45 increases DNA repair. In addition p53 binds
to ERCC3, one of several excision repair molecules that together identify and
remove damaged segments from DNA.

Another function under the control of p53 is apoptosis. Apoptosis (programmed cell death) seems to represent the primary mechanism by which
mutant cells are continually removed from tissues. Intracellular sensors detect
DNA damage and signal imbalance by overexpressed oncogenes, survival
factor imbalance, or hypoxia. They activate death receptors or induce

Issues in Toxicology No. 13
The Cellular Response to the Genotoxic Insult: The Question of Threshold for
Genotoxic Carcinogens
Edited by Helmut Greim and Richard J. Albertini
© The Royal Society of Chemistry 2012
Published by the Royal Society of Chemistry, www.rsc.org

cytochrome c release, which activate caspases that trigger selective destruction of cellular structures. Up-regulation is controlled by tumour suppressor genes (*e.g. p53*). DNA damage activates *p53*, which up-regulates expression of CDK-inhibitor p21 and pro-apoptotic proteins (NOXA, PUMA). In normal cells Rb is activated and blocks deamidation of Bcl-x_L, inhibiting the NOXA and PUMA capabilities of Bax-activation, cytochrome c release and caspase activation. In tumour cells without Rb-activation, DNA damage activates Bcl deamidation and releases Bax block, leading to apoptosis. Cytochrome c release induces assembly of Apaf monomers to the apoptosome, which activates caspases. The inhibitor of apoptosis (IAP) family of proteins inhibit caspases. Smac co-released with cytochrome c releases this inhibition.

The increasing understanding of epigenetic regulation of gene expression by methylation of histones, which allow chromosomal regions to switch between on and off status, indicate further mechanisms by which cells can react to a genotoxic insult. Among others the methylation process itself is controlled by Argonaute proteins, which segregate into two clades, the Ago cade and the Piwi clade. Of these, Ago clade proteins complex with microRNAs and small interfering RNAs, which derive from double-strand RNA precursors. The microRNA–Ago complexes reduce the translation and stability of protein coding mRNAs, which results in a regulatory network that impacts about 30% of all genes.

Although cellular defence mechanisms are increasingly understood, the critical and rate limiting parameters and their dose–response to the insulting agent need to be evaluated. This book describes the different cellular defence mechanisms and their regulation. We have prepared a summarizing chapter (Chapter 1), which evaluates the plausibility of a dose-dependent threshold mechanism of genotoxic carcinogens and their rate-limiting parameters to allow determination of the onset of such counterbalancing reactions. Finally possible data gaps for further research are described.

Besides its scientific value, a better understanding of cellular defence is of regulatory importance since a scientifically defendable threshold concept for genotoxic carcinogens will allow the NOEL to be identified and from that the proposal of health-based exposure limits for genotoxic carcinogens.

Helmut Greim and
Richard J. Albertini

Contents

Issues in Toxicology No. 13
The Cellular Response to the Genotoxic Insult: The Question of Threshold for
Genotoxic Carcinogens
Edited by Helmut Greim and Richard J. Albertini
© The Royal Society of Chemistry 2012
Published by the Royal Society of Chemistry, www.rsc.org

**Chapter 4.2   Different Modes of Cell Death Induced by
                DNA Damage   239**
Olga Surova and Boris Zhivotovsky

**Chapter 4.3   Transcriptional Inhibition by DNA Damage as a Trigger
for Cell Death**   **266**
Mats Ljungman

5. Epigenetic Mechanisms

Introduction and Conclusion: The Rationale for Thresholds for Genotoxic Carcinogens

HELMUT GREIM*[a] AND RICHARD J. ALBERTINI[†b]

[a] Institute of Toxicology and Environmental Hygiene, Technical University of Munich, Hohenbachernstrasse 17-19, D-85350 Munich, Germany;
[b] Department of Pathology, Genetic Toxicology Laboratory, The University of Vermont, 665 Spear Street, Burlington, VT 05405, USA
*Email: helmut.greim@lrz.tum.de; [†]Ralbert315@aol.com

1 Introduction

The default assumption for risk assessment of DNA-reactive carcinogens is that exposure to even very small amounts may result in additional risk. While this low-dose linearity has often been supported by observations on the dose–response for DNA adduct formation, mutations and carcinogenesis of many chemicals, other studies have shown non-linearity in the low-dose range, suggesting the presence of thresholds. The biological plausibility of thresholds is supported by the existence of cellular defence mechanisms that protect against mutations so that at certain low doses the agent does not increase mutations, which are the initiating events for tumours resulting from DNA-reactive carcinogens, above their normal background frequencies.

In a published opinion, the Scientific Committee of the European Food and Safety Authority (EFSA)[1] has concluded that based on the current

Issues in Toxicology No. 13
The Cellular Response to the Genotoxic Insult: The Question of Threshold for Genotoxic Carcinogens
Edited by Helmut Greim and Richard J. Albertini
© The Royal Society of Chemistry 2012
Published by the Royal Society of Chemistry, www.rsc.org

understanding of cancer biology there are levels of exposure to substances, which are both DNA-reactive genotoxic and carcinogenic, below which the cancer incidence is not increased (biological thresholds in dose–response). These conclusions have been evaluated in a joint opinion by the three scientific committees of Directorate-General for Health & Consumers (DG SANCO) of the European Commission.[2] They concluded that a dose for a specific compound may exist below which the compound does not induce a genotoxic effect. This assumption is based on several cellular defence mechanisms such as:

- metabolic inactivation of the ultimate reactive compounds;
- shielding the genetic material by membranes or proteins against attack by the reactant;
- repair of DNA lesions;
- elimination of heavily damaged cells by apoptosis or necrosis.

In its *Guidelines for Carcinogenic Risk Assessment*,[3] the US Environmental Protection Agency (EPA) also indicates that direct, DNA-reactive carcinogens might act *via* a nonlinear mode of action (MOA), thereby supporting a nonlinear dose response extrapolation provided that the case for this can be demonstrated or expected. This document further proposes that narrative descriptions replace default assumptions for all carcinogens (including DNA-reactive carcinogens) to arrive at a more mechanistically based risk assessment.

However, without understanding the dose-dependent threshold mechanisms for DNA-reactive genotoxic carcinogens and their rate-limiting parameters for determination of the onset of such counterbalancing reactions, the biological expections required to make the case for nonlinearity remain hypothetic. It is the concept of this book to describe the knowledge on dose–response and mechanisms of mutagenic and carcinogenic events and the different components of cellular defence mechanisms to the genotoxic insult. This is specifically presented by the different authors. This introductory chapter summarizes this information and evaluates the plausibility of threshold effects for DNA-reactive genotoxic carcinogens.

2   Background

The concept of linearity in the dose–response of mutagenic effects derives from the basic studies on irradiated fruit flies by Stern and coworkers. In a recent re-evaluation, Calabrese[4] concluded that this generalization is insufficiently supported. Although Spencer and Stern (1948)[5] noted that it was not uncommon for control mutation rates to exceed those seen at 25 and 50 r due to background variation, Uphoff and Stern (1949)[6] in a summarizing evaluation of three critical studies concluded that the dose–response of irradiation is linear. Based on these results the first in the series of the so-called Biological Effects of

Ionizing Radiation (BEIR) reports of the US National Academy of Sciences[7] stated that 'Any radiation dose, however small, can induce some mutations'. That conclusion was repeated in the most recent BEIR report[8] where it is indicated that there is no convincing evidence to indicate deviation for the linear non-threshold model for ionizing radiation induced cancer. However, important to the argument for this conclusion is the biology of radiation induced mutations where a single radiation track traversing the nucleus of a cell has a finite probability of causing DNA damage.[8] For ionizing radiation, therefore, the current consensus is that there is no minimum amount of radiation dose which might be exceeded before any harmful mutations occur. According to BEIR 1, 'if we increase the radiation that reaches the reproductive glands by X percent, the number of mutations caused by radiation will also be increased by X percent'.[7] One year later these conclusions were generalized to radiation induced cancer by the US National Committee of Radiation Protection and Measurements (NCRP). In 1977 the US National Academy of Sciences Safe Drinking Water Committee extended this to all genotoxic chemical carcinogens. There is a critical difference, however, between ionizing radiations and DNA-reactive chemicals. Radiation tracks are a property of the former but not of the latter, where the DNA-reactive intermediates must in some instances first be formed and, in all instances, diffuse into the cellular location of the genetic material.

Since then numerous, mostly descriptive studies have been performed to support or question this general concept of a linear non-threshold low dose mutagenic and carcinogenic response for DNA-reactive carcinogens. In most cases *in vitro* or *in vivo* studies have been undertaken to determine the dose–response of genotoxic effects or the increase in tumour rates. Studies designed to evaluate a threshold effect showed that at low doses the effects did not differ from the control values and have been used to calculate a 'virtual safe dose', a 'practical threshold' or a 'no-adverse-effect-level' (NOAEL).[9,10]

Although such information adds to the assumption of a no effect level at low doses, they are challenged by the argument that the statistical power is insufficient to differentiate between the low dose effects and the controls. Even in a study using 40 800 trout in a ED_{001} tumour bioassay,[11] it was concluded that the use of over 30 000 animals did not provide *proof* that a threshold was reached. As a consequence other approaches have been applied to evaluate whether thresholds exist for the tumorigenicity of DNA-reactive carcinogens.

3 Studies in Rodents to Evaluate Low Dose Effects of Genotoxic Carcinogens

There have been repeated attempts to identify a threshold for the dose–response of long-term carcinogenicity studies in rodents. When forcing a linear extrapolation of the dose–response of the four carcinogens methyl eugenol (ME), nitrosodiethylamine (NDEA), ethylcarbamate (EC) and 2-acetylaminofluorene

(AF) through the zero point of origin, the extrapolation became incorrect and misleading.[12] In an additional evaluation of studies on the carcinogenicity of *N*-nitrosodiethylamine (NDEA) in rats, it was confirmed that DNA adducts and liver tumours fit an exponential curve better than a linear plot.[13] Moreover, glutathione-S-transferase placental form (GST-P) positive hepatocytes, which are considered to be initiated cells and adducts, were observed at doses below the threshold dose for the appearance of tumours.[14] This again stresses the conclusion that DNA adducts as well as initiated cells are indicators for a genotoxic event only, which needs promotion (*e.g.* by increased cell proliferation) to be transformed *via* mutations to a tumour.

To determine if there was also a threshold for initiated cells, Williams *et al.* (Chapter 1.2) used the model of preneoplasic lesions as a bioindicator of effects, because they precede the development of neoplasms, have phenotypic and genomic alterations indicative of neoplastic potential, and have higher rates of development into neoplasms than do normal target cells. In these studies the incidences of hepatocellular altered foci (HAF) together with other bioindicators of effect such as increased cell proliferation have been evaluated. These studies demonstrate that there is a dose–response at high doses, whereas at low doses of the DNA-reactive carcinogens, the HAF incidences are at or below the control values. Since HAFs are prerequisite to the eventual development of liver neoplasms it is concluded that the identified no-observed-effect-levels (NOELs) are no-observed-adverse-effect-levels (NOAELs) for liver tumour development. At these NOAELs, DNA adducts have still been observed, indicating that adducts are the more sensitive biomarkers and are indicators for exposure rather than for effects. It is further concluded that there is a level of DNA adduct formation of no biological relevance, which is estimated to be at about one adduct per 7000 genes or about three adducts per cell. This has been compared to the level of one lesion per 10^6 bases per cell so that this additional DNA modification is considered trivial. Moreover, most adducts would be in regions of DNA not coding for proteins and not all adducts are miscoding, which further supports the conclusion of a trivial level of DNA alterations.

Previously Williams *et al.* (2005) discussed the likelihood of a mutation at the permitted concentration of 20 parts per billion (ppb) aflatoxin in food.[15] Beyond considering species and individual differences in the capacity of metabolic activation and inactivation several aspects in DNA binding, DNA repair and mutations are discussed.

The DNA is embedded in nucleoproteins, which diminishes the chance of an electrophile to reach the DNA. The entire genome of a cell consists of approximately 3.2×10^9 base pairs so that an activated carcinogen has 3.2×10^9 potential targets. Since there are about 30 000 functional genes per cell with an average size of 3000 base pairs, there are 9×10^7 nucleotides for which binding may be of functional significance. Moreover, considering that an ultimate carcinogen needs to hit and mutate up to six genes to induce cancer[16,17] and assuming that the carcinogen is activated in the liver, cumulative lifetime exposure of 96 g is required corresponding to a daily dose of about 1.3 mg.

The daily intake of 30 ppb aflatoxin B_1 contaminated food results in an about 45-fold lower exposure. The lowest DNA adduct levels for aflatoxin B_1, dimethylnitrosamine or 2-amino-3,8-dimethylimidazo[4,5-*f*]quinxaline have been about one adduct per 10^8 nucleotides, respectively. This is far below the spontaneous DNA modifications of about one lesion per 10^6 bases.

Williams *et al.* (see Chapter 1.2) calculated a TIDI (toxicological insignificant intake level) of about one adduct in 10^9 nucleotides by applying the safety margin of 100 (10 for species to species extrapolation, 10 for individual variation) to the biological significant level of 14 adducts per 10^8 nucleotides. At this level hepatocellular foci have been increased in the absence of increased cell proliferation or induction of promotable hepatocellular neoplasia. This level of biological insignificance represents three adducts per cell or one adduct per 7000 genes. This is considered trivial compared with endogenous DNA lesions of one of 10^6 lesions per cell.

In evaluating the experimental and human data on vinyl chloride, Rozman *et al.*[18] reported that plotting the data on the logarithmic scale showed a linear dose–response, which allowed comparison of the doses used in the animal studies with the much lower human exposures. The logarithmic scale also showed a clear separation of the DNA adducts from the tumour data. The dose–response of the DNA adducts parallels the tumour dose–response data at lower doses and continues to even lower doses than the tumour data.

Similarly, Gocke and Müller (Chapter 1.3) provided evidence for different dose–responses of DNA and protein alkylations and mutations. After treatment of CD1 mice and Muta™ mice with ethylmethane sulfonate (EMS), NOELs for the induction of clastogenic effects (polychromated micronuclei) in CD1 mice were 80 mg kg^{-1} body weight per day and of around 25 mg kg^{-1} for the induction of mutations in bone marrow cells in the Muta mice, whereas the dose–response curves for the induction of ethyl adduct levels of terminal valine of globin in peripheral blood and of DNA in both test animals were linear to the lowest dose tested.

Swenberg *et al.* evaluated the relationships between macromolecular adducts and mutations induced by DNA-reactive genotoxic carcinogens.[19] The general conclusion is that DNA adducts as biomarkers of exposure extrapolate down to zero, whereas biomarkers of effect such as mutations can only be interpolated back to the background number of mutations. The likely explanation for this difference is that, at high exposures, the biology that results in mutagenesis is driven by DNA damage resulting from the chemical exposure. At very low exposures, the biology that results in mutagenesis is driven by endogenous DNA damage.

When studying cyproterone acetate (CPA) in the transgenic Big Blue™ rat, the dose–dependence of the DNA adduct levels showed a linear increase between doses of 25 and 75 mg kg^{-1}, whereas the mutation frequency was similar to the controls at 25 and 50 mg kg^{-1}.[20] The linear dose response started at 75 mg kg^{-1}. The authors assumed that to express the mutations, an additional effect of CPA operating at high doses only—probably the mitogenic activity—is required. In previous studies, Schulte-Hermann *et al.* showed an increase in the hepatic RNA and DNA synthesis, mitotic rate and liver growth after six consecutive oral doses of CPA between 40 and 100 mg kg^{-1}.[21] This indicates that as long there is no increase in cell proliferation, DNA adducts and possibly even mutations remain silent.

Although this is not the focus of the present dispute, this points to the dual role of chemicals tested at high doses as has been pointed out by Ames during his dispute about the relevance of high dose testing of chemicals.[22] When tested at the maximum tolerable dose (MTD) as requested by the guidelines for long-term exposure studies, the resulting cytotoxicity may induce cell proliferation and thus enhance silent or repaired DNA lesions that occur at low doses. This explains why about 50% of all tested natural and synthetic pesticides have been found carcinogenic in such studies. Ames *et al.* estimated that the daily intake of the US population of natural pesticides is about 1.5 g and by that is about 10 000 times higher than that of synthetic pesticides.[23] From this they concluded that, at the low doses of most human exposures, risk at least from synthetic pesticides is insignificant.

4   Threshold Mediated Mechanisms

Although the studies described above present experimental evidence for a threshold of mutagenic or carcinogenic effects, they are descriptive and do not sufficiently consider the cellular mechanisms to maintain the integrity of its genome. These are addressed subsequently.

4.1   Previous Conclusions

The dose–response and threshold-mediated mechanisms in mutagenesis and carcinogenesis were discussed during a joint symposium organized by the European Centre for Ecotoxicology and Toxicology of Chemicals (ECETOC) and the European Environmental Mutagen Society (EEMS).[24] There was general agreement that linearity of even DNA-reactive genotoxic effects cannot be applied universally and that a practical (pragmatic, biological) threshold for genotoxic mutagens or carcinogens can be justified for various genotoxic mechanisms.[25] These include aneuploidy,[26] indirect modes of action, extremely steep dose–effect relationship in combination with strong toxicity or specific toxicokinetic conditions, which lead to metabolic inactivation prior to a DNA attack.[27] Disruption of cell division and chromosome segregation, inhibition of DNA synthesis, overloading of oxidative defence mechanisms, metabolism and plasma binding capacity, disturbances of metal homeostasis, cytotoxicity and physiological perturbations are possible mechanisms that permit identification of a threshold genotoxic effect.[28]

The major conclusion of the ECETOC–EEMS workshop was that dose–response will be determined by the individual capacity of the host to metabolically activate and inactivate the genotoxic compound,[29] and by the capacity of DNA repair enzymes, which reduce the proportion of DNA lesions that are processed into DNA sequence changes.[30] Repair can be with fidelity or by introducing errors, in that base excision and nucleotide excision repair are error-free processes that act on simpler forms of DNA lesions. A special form of base excision repair removes mismatched DNA bases that occur as errors of DNA replication or from miscoding properties of damaged bases. Several types of recombinational processes such as homologous, illegitimate and site-specific recombination pathways repair severe DNA damage.[31]

4.2   Metabolic Inactivation of Genotoxic Compounds

The role of metabolic activation and inactivation, which determine the exposure of a cell to a genotoxic compound, is covered by the contributions of Bolt (Chapter 2.1) and Dekant and Mally (Chapter 2.2). In the first of these, Bolt considers the importance of detoxification mechanisms in the mutagenic consequences of DNA-reactive carcinogens that are also endogenously produced. Such compounds include ethylene oxide, formaldehyde, acetaldehyde, ethanol and isoprene as well as other agents as exemplified by vinyl acetate, which is metabolized to acetaldehyde. As these compounds are produced endogenously and are also formed from many external agents, intracellular levels are kept at physiological concentrations by homeostatic levels of detoxifying enzymes. However, when external exposures are high, physiological concentrations may be exceeded and adverse effects produced. Therefore, an additional mutational load resulting from exogenous sources can be made manifest only when physiological intracellular concentrations are exceeded.

Studies are beginning to explore the role of exogenously administered compounds that are also endogenously produced to determine their relative DNA reactivity. The use of highly sensitive liquid chromatography–mass spectrometry (LC-MS)/mass spectrometry–selected reaction monitoring (MS-SRM) methods to detect DNA adducts following exposure of rats to $[^{13}CD_2]$-formaldehyde was used to differentiation adducts originating from endogenous sources from those originating from exogenous exposures, *i.e.* the $[^{13}CD_2]$-adducts.[32] For formaldehyde, exogenous mono-adducts and guanine–guanine crosslinks were formed only at the site of contact and not at distant sites. The sensitivities of the method for detecting adducts approached amol amounts. These methods are now being used to make similar differentiations for other endogenous compounds. Data should soon become available that will clearly be able to quantify the additional DNA damage, if any, that may be attributable to exogenous sources of DNA-reactive chemicals that are also endogenously produced.

Both Chapters 2.1 and 2.2 consider oxidative stress, oxidative DNA damage and anti-oxidant defences. It is important to note that, although this book focuses on threshold mechanisms operative in carcinogenesis due to direct, DNA-reactive carcinogens, mutagenesis plays a role in most or all of cancer induction. The mechanisms that underlie a chemical's potential for inducing genotoxicity and producing mutations include direct DNA reactivity, where the earliest events are specific covalent binding of the chemical and/or its metabolite(s) to DNA, *i.e.* the production of agent specific DNA adducts, as well as a variety of indirect mechanisms. The biology of these indirect mechanisms leads to the reasonable expectation of their having a threshold relative to dose.

One important cause of indirect genotoxicity is the induction of oxidative stress and the production of oxidative DNA damage. Oxidative stress is induced in cells by reducing the content of defence molecules such as glutathione (GSH), by increasing the generation of reactive oxygen species (ROS), or by both mechanisms, to indirectly produce oxidative DNA damage. Reactive oxygen species (either arising normally but without adequate protection, or newly induced) are capable of reacting with the DNA by covalent binding to form DNA adducts that are also potentially mutagenic (see Chapters 2.1, 2.2 and 3.1). The roles of ROS in cells are complex and multiple, requiring homeostatic mechanism for survival. When these mechanisms are overwhelmed, oxidiative stress and downstream deleterious consequences can occur. DNA damage above background occurs, therefore, only after these homeostatic mechanisms are overwhelmed. A recent study of several mutational endpoints in human TK6 lymphoblastoid cells revealed a clear no-observed genotoxic effect level (NOGEL) following exposures to a range of concentrations of agents known to induce oxidative stress.[33]

Finally, as indicated in Chapters 2.1 and 2.2, indirect mutagenicity (especially that due to oxidative DNA damage) that is generated by many exogenous chemicals can be intimately associated with the direct, DNA reactive mutagenicity of these same agents. Metabolism is the critical link, especially phase II detoxification. For direct, DNA reactive carcinogens, mutagenicity may be made manifest by inhibition of detoxification due to depletion of critical molecules, such as glutathione. For the indirect mutagenicity resulting from these same chemicals, it is metabolic depletion of these same critical molecules that is important, except that the mechanism affected now is loss of protection from endogenously produced electrophiles.

An illustration of this has recently been made. Thompson *et al.* applied the US EPA Mode of Action framework noted above to determine a carcinogenic MOA for hexavalent chromium, an agent considered to have direct DNA-reactive mutagenicity, and concluded that in the sequence of key events leading to tumour induction, the production of oxidative stress preceded mutagenesis.[34]

4.3   Key Events in the Transformation of a Cell

Hanahan and Weinberg[35] originally described six characteristics essential to transform cells and for the development of a metastatic tumour. These include unlimited replicative potential, ability to develop blood vessels (angiogenesis), evasion of programmed cell death (apoptosis), self-sufficiency in growth signals, insensitivity to inhibitors of growth, and tissue invasion and metastasis. More recently, four additional hallmarks have been added, which comprise de-regulation of cellular energetics, genome instability and mutation, avoiding immune destruction, and tumour promoting inflammation.[36] Whereas these criteria characterize a malignant cell population, Preston and Williams[37] developed ten key events that lead to the transformation of a normal cell to a tumour with the Hanahan and Weinberg characteristics. Of these, reaction with DNA in target cells to produce DNA damage, misreplication on damaged DNA template or misrepair of DNA damage and mutations in critical genes in replicating target cell are specifically considered to evaluate the dose response of the critical events leading to mutations (see Chapter 1.1).

In brief the major conclusions are based on the assumption that the conversion of DNA damage by the majority of chemical carcinogens into mutations is by error of replications. Thus, the probability of inducing a chromosomal or gene mutation depends on the amount of induced DNA damage, the probability of this damage being unrepaired at the time of replication and its probability of being misreplicated. However, the background steady-state level of DNA damage is about 500 base alterations per cell. Assuming that there is no specific qualitative difference between the background and the chemical induced DNA damage, it is questionable whether a small additional amount of DNA damage would lead to an increase in mutations.

Preston (Chapter 1.1) differentiates between single base changes and chromosomal aberrations. Assuming that gene mutations for mutagenic chemicals are single base changes produced by replication errors on a damaged DNA template, the 'one hit' theory is applicable and by that the response will be linear with dose. On the other hand, chromosomal aberrations are assumed to require 'two hits' during replication. Since they occur independently the dose–response will be dose-squared. Thus depending on the relation between one hit and two hit events, the dose–response at low doses can vary between linear and non-linear—in the latter case resulting in very few if any genetic alterations. Sophisticated experimental tools are needed to differentiate between these two mechanisms in the evaluation of specific chemicals.

Duclos *et al.* (Chapter 3.1) explain that the predominant cause of DNA damage is the constant production of reactive oxygen species, which derive from mitochondria, cytochrome P450 dependent reactions, intracellular metabolism of xenobiotics and drugs, from inflammation, ionizing radiation and ultraviolet light (see above). The resulting H_2O_2 and $^{\bullet}NO$ readily diffuse through the cell and are detoxified by enzymes such as glutathione reductase and catalase. Since both H_2O_2 and ROS are used in cellular signalling pathways, a cellular steady-state level is maintained, which is 10 nM in case of H_2O_2. The most reactive ROS species is the $^{\bullet}OH$ radical, which is responsible for the majority of oxidative DNA damage resulting in DNA strand breaks or sites of base loss. Approximately 18 000 base losses or abasic sites per cell are formed by spontaneous depurination per day. Obviously the different oxidative DNA lesions are repaired to maintain the integrity of the genome and cause mutations. Moreover, during a four weeks' recovery segment after

2-acetylaminofluorene (AAF) treatment, the adduct values diminished by about 50% reflecting DNA damage repair (see Chapter 1.2).

5   The Cellular Response to the Genotoxic Impact

Maintenance of the cellular integrity is crucial for its physiological function, which is constantly threatened by DNA damage arising from numerous intrinsic and environmental sources. By transcribing the genetic information stored in the intact DNA, RNA polymerase II generates mRNA that instructs ribosomes to produce specific proteins. DNA damage will lead to interruption of mRNA synthesis with the potential production of unstable transcriptions and proteins. Alternatively DNA damage results in mutations with the possible consequence of cancer. To protect their integrity, eukaryotes including mammalian cells have developed different mechanisms including the so-called DNA damage response (DDR) to ensure protection of the genome. DDR identifies DNA lesions and, depending on the severity, leads to different responses. Mild DNA damage normally can be managed by DNA repair; more severe or irreparable DNA damage trigger the induction of cell death programmes such as apoptosis or necrosis.

The efficient mechanisms to maintain cellular integrity of tissue-specific adult stem cells (ASC) are detailed in Chapter 3.2. These cells are specifically present in highly regenerative organs and have evolved multiple systems to limit DNA damage. They eliminate toxic substances by pumping them across the membrane. When quiescent they use glycolysis for energy supply to avoid oxidative DNA damage. Depending on cell type, differentiation stage and age, the different repair systems ensure genome integrity. These include the array of DNA repair mechanisms (homologous recombination, non-homologous end joining, nucleotide excision repair, base excision repair and mismatch repair). As long as they remain in the S/G2 phase the high-fidelity homologous recombination (HR) is preferred for repair of double-strand breaks. When proliferating, the error-prone non-homologous end joining (NHEJ) repair mechanism becomes active, which is the predominant mechanism during all cell cycle phases of mammalian cells. The DNA damage response system co-ordinates repair and cell cycle progression and decides the cell fate by operating repair, cell cycle arrest, senescence or cell death.

5.1   DNA Repair

5.1.1   The BER System

Appropriate maintenance of the balance between ROS production and DNA repair is essential because ROS are both cell signalling and DNA damaging molecules.

As described in Chapter 3.1, most of the nuclear oxidative DNA lesions are repaired by the base excision repair (BER) pathway. This includes recognition of the lesion by an appropriate glycosylase, which cleaves the bond between the damaged base to the desoxyribose. At the resulting apurinic/apyrimidinic (AP) site, the DNA backbone is cleaved by a DNA AP endonuclease, which creates a nick in the AP site with the 3′OH and 5′deoxyribose phosphate termini. The latter blocking end is removed by polymerase ß, which then fills in the gap with the appropriate nucleotide. The remaining nick is sealed by a phosphodiester bond, which is catalysed by a DNA ligase III. Some DNA glycosilases are bifunctional and also cleave the DNA backbone. In this case the 3′ blocking end needs to be removed by APE1 exonuclease activity and polynucleotide kinase (PNK). More recently it has been shown that the formation of BER complexes at the lesion site is regulated by the protein XRCC1 (X-ray cross complementing factor 1).

In mammals, 11 glycosilases have been identified, five of which are specific for oxidized bases. Their specific functions and characteristics are described in Chapter 3.1. It is remarkable that the deficit of a single DNA glycosilase seems to be overcome by compensatory mechanisms involving redundant repair activities. This and other information indicate that, modified by additional proteins, mammalian BER enzymes interact with one another to increase repair efficiency. Double knock out mice turn to be tumour prone and by that establish the link between BER deficiency and tumorigenesis.

In mitochondria the DNA is associated with the inner membrane and closely located to the vicinity of the respiratory machinery, which constantly generates ROS. As a consequence the steady-state level of oxidative DNA damage is significantly higher than in nuclear DNA. Basically the same BER mechanisms remove the DNA damage induced by ROS.

At the transcriptional side, transcription blocking DNA lesions induce the following cascade of cellular defence (see Chapter 4.3). The blocked RNA polymerase II complexes trigger the activation of stress kinases, which induce stress responses *via* p53 leading to cell cycle arrest. RNA polymerase II complexes blocked at DNA lesions recruit BER enzymes to remove blocking DNA lesions *via* transcription coupled repair (TCR). Finally, the repaired DNA allows the transcription machinery to re-engage in mRNA synthesis. In the meantime the existing pool of mRNA may compensate for the loss of nascent RNA. Exhaustion of these 'short lived survival factors' leads to apoptosis.

As described in Chapter 3.3, the steady-state level of the BER system is tightly regulated and is linked to the amount of DNA lesions. When the levels of BER exceed the level of DNA lesions, the excessive BER enzymes are degraded. This is achieved by controlling the cytoplasmic pool of the three major BER enzymes XRCC1, Lig III and Pol β through targeted proteasomal degradation. Enzymes targeted for degradation are marked with a chain of ubiquitin, which serves as a target for additional ubiquitin chain attachment. The poly-ubiquitilated enzymes become recognized by the 26S proteasome and are degraded. Accumulation of Pol β is regulated by the ARF (Alternative Reading Frame) protein, which is a well-known tumour suppressor gene (p14 in humans) and is released in response to DNA damage. Its release inhibits ubiquitin ligases and by that increases Pol β availability and results in increased rate of DNA repair.

Moreover, the increasing understanding of epigenetic regulation of gene expression by methylation of histones, which allows chromosomal regions to switch between on and off status,[38] indicates further mechanism by which cells may react to a genotoxic insult. Among others the methylation process itself is controlled by Argonaute proteins,[39] which segregate into two clades, the Ago cade and the Piwi clade. Ago clade proteins complex with microRNAs and small interfering RNAs derived from double-strand RNA precursors. The microRNA–Ago complexes reduce the translation and stability of protein coding mRNAs, which results in a regulatory network that impacts on about 30% of all genes.

5.2   Post-translational Regulation

The three major functional processes of the BER system (recognition of the lesion and strand scission, gap tailoring, DNA synthesis and ligation) are co-ordinated *via* the XRCC1/ligase III and PARP1 scaffold proteins—essential in maintaining genomic integrity. Phosphorylation, acetylation, sumoylation, ubiquitylation and methylation of the various BER proteins can modify repair activities (for reviews see Almeida and Sobol,[40] Busso *et al.*[41]). Triggered by oxidative stress, these post-translational modifications activate and, in this way, may adjust the DNA repair capacity to the actual cellular environment. Interestingly, the oxoguanine DNA glycosilase (OGG1), which initiates repair of 8-oxoG, is phosphorylated by Cdk4, a serine–threonine kinase.[42] The cyclin dependent kinases inhibit progression of cell cycling through the G_1-S check-point to allow more time for DNA repair. Whereas Cdk2 activity is reduced upon H_2O_2 treatment of human diploid fibroblasts, Cdk4 activity does not change, so that this specific cyclin kinase is still available to phosphorylate and up-regulate OGG1 activity.

It also has been postulated that DNA repair in healthy control animals is sufficient to repair DNA adducts formed by acute exposures to DNA-reactive chemicals, whereas long-term exposure may reduce the repair capacity by modifying the expression of DNA repair proteins (*e.g.* Chan *et al.*[43]). This may result from impaired DNA repair proteins because of mutations caused by the slow accumulation of DNA adducts over a long period of time.[44] To determine the role of oxidative DNA damage in the toxicity of polychlorinated biphenyls, Jeong *et al.* showed that single exposure of rats to 2,3,7,8-tetrachlorodibenzo-*p*-dioxin (TCDD) or a mixture of polychlorinated biphenyls (PCBs) did not significantly increase the formation of DNA M_1dG adducts in liver and brain.[45] However, chronic exposure of single PCBs resulted in a dose dependent increase in the number of M_1dG adducts, which have been further increased by a mixture of the two PCBs (126 and 153) tested. Unfortunately neither the concentrations of the PCBs in the target tissues nor the NOEL of the effects have been determined, so that it remains unclear whether they are a consequence of the higher target doses or the result of altered DNA repair capacity.

Elimination rates differ for different DNA adducts. Recently Goggin *et al.*[46] determined the persistence and repair of several types of butadiene-specific DNA adducts, which results from the metabolic activation of butadiene to 1,2,3,4-di-epoxybutane (DEB). These lesions were detected in tissues of laboratory rodents exposed to butadiene (BD) by inhalation. The half-lives of the most abundant crosslinks, bis-N7G-butadiene, in mouse liver, kidney and lungs were 2.3–2.4 days, 4.6–5.7 days and 4.9 days, respectively. The *in vitro* half-lives of bis-N7G-butadiene were 3.5 days (*S,S* isomer) and 4.0 days (*meso* isomer) due to their spontaneous depurination. In contrast, tissue concentrations of the minor DEB adducts, N7G-N1A-BD and 1, N6-HMHP-dA, remained essentially unchanged during the course of the experiment, with an estimated half-life of 36–42 days. No differences were observed between DEB–DNA adduct levels in BD-treated wild-type mice and the corresponding animals deficient in methyl purine glycosylase or the Xpa gene. These results indicate that DEB-induced N7G-N1A-BD and 1,N6-HMHP-dA adducts persist *in vivo*, potentially contributing to mutations and cancer observed as a result of butadiene exposure. Although the study clearly demonstrates that specific DNA adducts may be persistent or repaired slowly, the butadiene exposures were high and the effect of lower doses was not measured.

5.3 The Role of the GJIC in Maintenance of Cellular Homeostasis

Gap junction intercellular communication (GJIC) plays a dominant role in the pre-transcriptional regulation of the cell. As described in Chapter 5.1, GJIC is a key player in the control of the cellular life cycle by providing an essential pathway for the intercellular exchange of small and hydrophilic molecules. It is assumed that numerous physiological processes are driven by substances that are exchanged *via* these channels contributing to the maintenance of tissue homeostasis. Gap junctions arise from the docking of two hemi-channels of adjacent cells, each of which are composed of six connexin units. Since DNA methylation maintains gene silencing, reduced connexion expression is associated with a high DNA methylation content of the connexion promotor in malignant cells of human and rodent cancer. Whereas DNA-reactive genotoxic carcinogens usually do not affect GJIC, non-genotoxic carcinogens, with several exceptions, decrease GJIC, which results in a disturbance of the equilibrium between cell growth and cell death. The extent to which GJIC-impairment is counterbalanced (*e.g.* by increased connexin activation), at least at low exposure concentrations to non-genotoxic carcinogens, remains to be elucidated.

5.4 Apoptosis

Apoptosis—programmed cell death—represents the primary mechanism by which, in cases of heavy DNA damage, the DNA damage response leads to elimination of mutant cells from the tissues. The mechanisms that trigger apoptosis are described in Chapter 4.1 and 4.3. In short, the extrinsic (receptor based) mechanism activates ATM (ataxia telangiectasia, mutated) and ATR (ATM and Rad3-related), which stabilizes p53. The role of the activated p53 is to induce cell cycle arrest, promote DNA repair or to trigger cells to undergo apoptosis or necrosis. Interaction of the activated p53 with the FasR/FADD receptor releases caspase-8 to activate the caspase-8 dependent apoptosis pathway. The intrinsic mitochondrial pathway is triggered by activation of p53, which activates Bax, and inactivation of Bcl-2. Both pathways increase the leakiness of mitochondria, causing release of cytochrome C from the mitochondria. Cytochrome C release induces assembly of Apaf monomers to the apoptosome resulting in the release of caspase-9, which activates caspase-3 and caspase-7. Both degrade the inhibitor of caspase-activated DNAse (ICAD) and the activated DNAse cleaves the DNA into nucleosomal fragments. There is interaction between the two pathways and more details about the activation of proteins that lead to apoptosis are understood. For example in normal cells Rb is activated and blocks de-amidation of bcl-x_L, inhibiting NOXA and PUMA capabilities of Bax-activation, cytochrome C release and caspase activation. DNA damage activates *p53* which up-regulates CDK-inhibitor p21 and the pro-apoptotic proteins NOXA and PUMA. In tumour cells without Rb-activation, DNA damage activates Bcl-de-amidation and releases the block of Bax leading to apoptosis.[47] IAP proteins inhibit caspases. Smac co-released with cytochrome C releases this inhibition.[48]

5.5   Necrosis

In response to more severe and irreparable DNA damage, the cellular response shifts towards additional cell death programmes including mitotic catastrophe, autophagy, which might be followed by apoptosis and necrosis—the latter is characterized by swelling of the cell, loss of membrane integrity and the release of cytoplasmic compartments (see Chapter 4.2). Upon sensing the damage a multiple network of signal transduction pathways trigger this cellular response through activation of a cascade of phosphorylation events, which initiate a number of cellular responses such as cell cycle transition, DNA replication, DNA repair and cell death (see Figure 4.2.1 in Chapter 4.2). 'Mitochondrial permeability transition' (MPT) may result from the following pathway. DNA damage causes hyper-activation of PARP-1, which reduces mitochondrial NAD and ATP levels. Cells with depleted ATP levels will undergo necrosis instead of apoptosis because the latter requires ATP (see Figure 4.1.3 in Chapter 4.1). Thus, the cellular energy status seems to decide whether severe DNA damage leads to apoptosis or necrosis.

6   Conclusions

Maintenance of cellular integrity, which is crucial for physiological function, is constantly threatened by DNA damage arising from numerous intrinsic and environmental sources. The background mutations may result in DNA replication errors or DNA damage arising from endogenously formed DNA-damaging chemicals such as ethylene oxide, endogenous formation of ROS, spontaneous depurination or errors of DNA polymerase. Superimposed on these are the mutations resulting from exogenous agents such as ionizing radiation, viruses or the different DNA-reactive chemical agents.

Eukaryotic cells have developed an array of defence mechanisms to maintain a balance between DNA damage and repair and to ensure cellular integrity. These include maintenance of the balance between physiological levels and detoxification of endogenously produced DNA-reactive chemicals. Phase II detoxifying enzymes play a critical role in keeping the endogenously produced chemicals at physiological levels. A variety of additional homeostatic mechanisms has evolved to closely regulate ROS production and repair oxidative DNA damage because ROS are both cell signalling and DNA damaging molecules.

Superimposed on the naturally occurring insults to cells are those that result from exposures to deleterious exogenous agents. Quantification of these effects and the extent to which they interact with background effects is difficult, though a quantitative estimate has been made by Williams *et al.*[49] Considering that the spontaneous DNA modifications of a mammalian cell is about one lesion per 10^6 bases, the lowest DNA adduct levels of three genotoxic carcinogens was determined to be one adduct per 10^8 nucleotides, which is far below the spontaneous DNA modifications and considered unlikely to be of biological significance.

The efficiency of the cellular defence mechanisms makes it unlikely that low exposure to genotoxic chemicals, which insignificantly affect the background rate of DNA damage, impairs cellular homeostasis. The increasing knowledge about the cellular defence renders sufficient evidence to assume a threshold for genotoxic carcinogens.

The information on the cellular reaction to DNA damage allows the following conclusions.

- Maintenance of the cellular integrity is crucial for its physiological function, which is constantly threatened by DNA damage arising from numerous intrinsic and environmental sources.
- Intracellular detoxification mechanisms protect from mutagenic consequences of endogenously produced DNA-reactive carcinogens to keep them at physiological concentrations by detoxifying enzymes, which also applies for exogenous genotoxins as long as they do not overwhelm detoxifying capacities.
- To protect their integrity, mammalian cells have developed mechanisms to ensure protection of the genome. Mild DNA damage is normally managed by DNA repair including impairment of cell proliferation to enhance repair efficiency. More severe or irreparable DNA damage trigger the induction of cell death programmes such as apoptosis or necrosis.
- The background steady-state level of DNA damage is about 500 base alterations per cell, which corresponds to about one lesion per 10^6 bases

and is much lower than the approximately 18 000 base losses or abasic sites per cell formed each day. This illustrates a remarkable efficiency of the DNA repair mechanism. Assuming that there is no specific qualitative difference between the background and the chemical induced DNA damage, a small additional amount of DNA damage should be equally efficiently repaired and would not lead to an increase in mutations.

All this supports the assumption of a threshold for the genotoxic impact of exogenous stressors. However, as indicated in Chapter 3.4, it is nearly impossible to provide an unambiguous answer to the question as to whether such a threshold can be generally assumed. For lesions that originate from ROS and simply add to the steady-state level of endogenous lesions, a threshold for deleterious effects would be lower than the existing background. Generally, repair is effected by a number of pathways with overlapping specificities and may become even more effective by activation of the complex regulatory system that includes cell cycle arrest and apoptosis. Thus, because of its central role in the regulation of the cellular defence programme the up-regulation of the GADD45 (Growth Arrest and DNA Damage) gene activity may be a promising tool to measure deviation from normality. Presently the high-throughput GreenScreen HC GADD45a-GFP genotoxicity assay is used to determine genotoxicity in mammalian cells.[50] Induction of apoptosis may be a meaningful measure for identification, which impact of a particular genotoxin causes significant damage (see Chapter 3.4). However, as stressed in Chapter 3.4, quantitative differences in the repair of lesions in different cells need to be considered. Due to the variety of specialized cells in different growth states and in different organs, which either rapidly proliferate while others are terminally differentiated and/or senescent, the sensitivities to particular genotoxins varies widely. Moreover, damage can occur in sequences of nucleotides that can form non-canonical DNA structures, within which the lesions may be refractory to repair; other damage may be hidden in tightly packed transcriptionally silent chromatin and are inaccessible to repair. In any case persistent DNA lesions lead to subsequent p53 activation and by that to cell cycle arrest, up-regulated repair and finally apoptosis, which reduce the probability of trans-lesion synthesis and mutagenesis.

In most cases the efficiency and adaptability of the cellular defence mechanisms and the dose dependency of their responses to the genotoxic challenge of a specific compound is not understood. This hampers identification of a true (or absolute) threshold for the genotoxic compound. Precise information as to the maximum level of an exogenous DNA-reactive genotoxic chemical that does not overwhelm cellular defence is urgently needed. This requires dose–response evaluation of the impact of these genotoxic agents on the regulation of the cellular defences.

Based on this evaluation and as discussed previously (Ames[22]), the relevance of high dose animal experiments when there is low exposure to humans is questionable. In such cases the high experimental exposure will overwhelm any cellular defence mechanism, a situation that most likely does not occur at the much lower human exposure. Any extrapolation from the dose response of such data to the lower human exposure is error prone. Unfortunately these high dose experiments are the basis for classification and labeling, which is hazard based and does not consider the difference between the effect levels in animal experiments and potential much lower human exposure.

References

1. European Food Safety Authority (EFSA), Opinion of the Scientific Committee on a request from EFSA related to a harmonised approach for risk assessment of substances which are both genotoxic and carcinogenic, *EFSA Journal*, 2005, **282**, 1–31.
2. European Commission, *Risk Assessment Methodologies and Approaches for Genotoxic and Carcinogenic Substances*, Joint opinion of the Scientific Committee on Health and Environmental Risks (SCHER), Scientific Committee on Health and Environmental Risks (SCCP) and Scientific Committee on Health and Environmental Risks (SCENIHT), Health & Consumer Protection Directorate-General, Brussels, 2009.
3. US Environmental Protection Agency, *Guidelines for Carcinogen Risk Assessment*, US Environmental Protection Agency, Washington DC, 2005.
4. E. J. Calabrese, Key studies used to support cancer risk assessment questioned, *Environ. Mol. Mutagen.*, 2011.
5. W. P. Spencer and C. Stern, Experiments to test the validity of the linear R-dose/mutation at low dosage, *Genetics*, 1948, **33**, 43–74.
6. D. E. Uphoff and C. Stern, The genetic effects of low intensity in irradiation, *Science*, 1949, **109**, 609–610.
7. US National Academy of Sciences, *Biological Effects of Atomic Radiation*, National Academy of Sciences, Washington DC, 1956.
8. Committee to Assess Health Risks from Exposure to Low Levels of Ionizing Radiation, *Health Risks from Exposure to Low Levels of Ionizing Radiation: BEIR VII Phase 2*, National Research Council, National Academies Press, Washington DC, 2006.
9. A. Maekawa, T. Ogiu, C. Matsuoka, H. Onodera, K. Furuta, Y. Kurokawa, M. Takahashi, T. Kokubo, H. Tanigawa, Y. Hayashi, M. Nakadate and A. Tanimura, Carcinogenicity of low doses of *N*-ethyl-*N*-nitrosourea in F344 rats: A dose–response study, *Gann*, 1984, **75**, 117–125.
10. A. Maekawa, H. Onodera, C. Matsuoka, T. Nagaoka, A. Todate, M. Shibutani, Y. Kodama and Y. Hayashi, Dose–response carcinogenicity in rats on low-dose levels of *N*-ethyl-*N*-nitrosourethane, *Jpn. J. Cancer Res.*, 1989, **80**, 632–636.
11. G. S. Bailey, A. P. Reddy, C. B. Pereira, V. Harttig, W. Baird, J. M. Spitzbergen, J. D. Hendrinks, G. A. Orner, D. E. Williams and J. A. Swenberg, Nonlinear cancer response of ultralow dose: a 40,800-animal ED_{001} tumour and biomarker study, *Chem. Res. Toxicol.*, 2009, **22**, 1264–1276.

12. W. J. Waddell, Comparisons of thresholds for carcinogenicity on linear logarithmic dosage scales, *Human Exp. Toxicol.*, 2005, **24**, 325–332.

13. W. J. Waddell, S. Fukushima and G. M. Williams, Concordance of thresholds for carcinogenicity of *N*-nitrosodiethylamine, *Arch. Toxicol.*, 2006, **80**, 305–309.

14. M. A. Moore, K. Nakagawa, K. Satoh, T. Ishikawa and K. Sato, Single GST-P positive cells—putative initiated hepatocytes, *Carcinogenesis*, 1987, **8**, 483–486.

15. G. M. Williams, M. J. Iatropoulos and A. M. Jeffrey, Thresholds for DNA-reactive (genotoxic) organic carcinogens, *J. Toxicol. Pathol.*, 2005, **18**, 69–77.

16. B. Vogelstein and K. W. Kinzler, The multistep nature of cancer, *Trends Genet.*, 1993, **9**, 138–141.

17. W. C. Hahn, C. M. Counter, A. S. Lundberg, R. L. Beijersbergen, M. W. Berooks and R. A. Weinberg, Creation of human tumour cells with defined genetic elements, *Nature*, 1999, **400**, 464–468.

18. K. K. Rozman, L. Kerecsen, M. K. Viluksela, D. Österle, E. Deml, M. Vilksela, B. U. Stahl, H. Greim and J. Doull, A toxicologist's view of cancer risk assessment, *Drug Metab. Rev.*, 1996, **28**, 29–52.

19. J. A. Swenberg, E. Fryar-Tita, Y. C. Jeong, G. Boysen, T. Starr, V. E. Walker and R. J. Albertini, Biomarkers in toxicology and risk assessment: Informing critical dose-response relationships, *Chem. Res. Toxicol.*, 2008, **21**, 253–265.

20. O. Krebs, B. Schäfer, T. Wolff, D. Oesterle, E. Deml and M. Sund, and J. Favor, The DNA damaging drug cyproterone acetate causes gene mutations and induces glutathione-S-transferase P in the liver of female Big Blue™ transgenic F344 rats, *Carcinogenesis*, 1998, **19**, 241–245.

21. R. Schulte-Hermann, V. Hoffmann, W. Parzefall, M. Kallenbach, A. Gerhard and J. Schuppler, Adaptive response of rat liver to the gestagen and antiandrogen cyproterone acetate and other inducers, II, Induction and growth, *Chem.-Biol. Interact.*, 1980, **31**, 287–300.

22. B. N. Ames, M. Profet and L. S. Gold, Nature's chemicals and synthetic chemicals: Comparative toxicology, *Proc. Natl. Acad. Sci. U. S. A.*, 1990, **87**, 7782–7786.

23. B. N. Ames, M. Profet and L. S. Gold, Dietary pesticides (99.99% all natural), *Proc. Natl. Acad. Sci. U. S. A.*, 1990, **87**, 7777–7781.

24. J. M. Parry and A. M. Sarrif (ed.), Dose–response and threshold-mediated mechanisms in mutagenesis. Proceedings of an international symposium, Salzburg, Austria, 7 September 1998, *Mutat. Res.*, 2000, **464**, 1–160.

25. J. A. Swenberg, A. Ham, H. Koc, E. Morinello, A. Ranasinghe, N. Tretyakova, P. B. Upton and K. Y. Wu, DNA-adducts: effects of low exposure of ethylene oxide, vinyl chloride and butadiene, *Mutat. Res.*, 2000, **464**, 77–86.

26. K. S. Bentley, D. Kirkland, M. Murphy and R. Marshall, Evaluation of thresholds for benomyl- and carbendazin-induced aneuploidy in cultured human lymphocytes using fluorescence in situ hybridization, *Mutat. Res.*, 2000, **464**, 41–51.
27. S. Madle, W. von der Hude, L. Broschinski and G. R. Jänig, Threshold effects in genetic toxicity: perspective of chemicals regulation in Germany, *Mutat. Res.*, 2000, **464**, 117–121.
28. L. Henderson, S. Albertini and M. Aardema, Thresholds in genotoxicity responses, *Mutat. Res.*, 2000, **464**, 123–128.
29. H. Autrup, Genetic polymorphisms in human xenobiotica metabolizing enzymes as susceptible factors in toxic response, *Mutat. Res.*, 2000, **464**, 65–76.
30. J. M. Parry, G. J. S. Jenkins, F. Haddad, R. Bourner and E. M. Parry, In vitro and in vivo extrapolations of genotoxin exposures: consideration of factors which influence dose-response thresholds, *Mutat. Res.*, 2000, **464**, 53–63.
31. E. Moustacchi, DNA damage and repair: consequences on dose-responses, *Mutat. Res.*, 2000, **464**, 35–40.
32. K. Lu, L. B. Collins, H. Ru, E. Bermudez and J. A. Swenberg, Distribution of DNA adducts caused by inhaled formaldehyde is consistent with induction of nasal carcinoma but not leukemia, *Toxicol. Sci.*, 2010, **16**, 441–451.
33. A. Platel, F. Nesslay, V. Gervais, N. Claude and D. Marzin, Study of oxidative damage in TK6 human lymphoblastoid cells by use of the thymidine kinse gene-mutation assay and the in vitro modified comet assay: Determination of no-observed-genotoxic-effect-levels, *Mutat. Res.*, 2011, **746**, 151–159.
34. C. M. Thompson, L. C. Haws, M. A. Harris, N. M. Gatto and D. M. Proctor, Application of the U.S. EPA Mode of Action framework for purposes of guiding future research: A case study involving the oral carcinogenicity of hexavalent chromium, *Toxicol. Sci.*, 2011, **119**, 20–40.
35. D. Hanahan and R. A. Weinberg, The hallmarks of cancer, *Cell*, 2000, **100**, 57–70.
36. D. Hanahan and R. A. Weinberg, The hallmarks of cancer: The next generation, *Cell*, 2011, **144**, 646–674.
37. R. J. Preston and G. M. Williams, DNA-reactive carcinogens: mode of action and human cancer hazard, *Crit. Rev. Toxicol.*, 2005, **35**, 673–683.
38. A. G. Rivenbark and B. D. Strahl, Unlocking cell fate, *Science*, 2007, **318**, 403–404.
39. A. A. Aravin, G. Hannon and J. Brennecke, The PiWi-piRNA pathway provides an adaptive defense in the transoson arms race, *Science*, 2007, **318**, 761–764.
40. K. H. Almeida and R. W. Sobol, A unified view of base excision repair: Lesion-dependent protein complexes regulated by post-translational modification, *DNA Repair*, 2007, **6**, 695–711.

41. C. S. Busso, M. W. Lake and T. Izumi, Posttranslational modification of mammalian AP endonuclease, *Cell. Mol. Life Sci.*, 2010, **67**, 3609–3620.
42. J. Hu, S. Z. Imam, K. Hashiguchi, N. C. Souza-Pinto and V. Bohr, Phosphorylation of human oxoguanine DNA glycosylase (α-OGG1) modulates its function, *Nucleic Acids Res.*, 2005, **33**, 3271–3282.
43. C. Y. Chan, P. M. Kim and L. M. Winn, TCCD affects DNA double strand break repair, *Toxicol. Sci.*, 2004, **81**, 849–855.
44. M. E. Wyde, V. A. Wong, A. H. Kim, G. W. Lucier and N. J. Walker, Induction of hepatic 8-oxo-deoxyguanosine adducts by 2,3,7,8-tetrachlorodibenzodioxin in Sprague-Dawley rats is female specific and estrogen-dependent, *Chem. Res. Toxicol.*, 2001, **14**, 849–855.
45. Y. C. Jeong, N. J. Walker, D. E. Burgin, G. Kissling, M. Gupta, L. Kupper, L. Birnbaum and J. A. Swenberg, Accumulation of M_1dG adducts after chronic exposure to PCBs, but not from acute exposure to poly-chlorinated aromatic hydrocarbons, *Free Radical Biol. Med.*, 2008, **45**, 585–591.
46. M. Gogin, S. Dewakar, V. E. Walker, J. Wickliffe, J. A. Swenberg and N. Tretyakova, Persistence and repair of bifunctional DNA adducts in tissues of laboratory animals exposed to 1,3-butadiene by inhalation, *Chem. Res. Toxicol.*, 2011, **24**, 809–817.
47. C. Li and C. B. Thompson, DNA damage, deamidation, and death, *Science*, 2002, **298**, 1346–1347.
48. D. W. Nicholson and N. A. Thornberry, Life and death decisions, *Science*, 2003, **299**, 214–215.
49. G. M. Williams, M. J. Iatropoulos and A. M. Jeffrey, Thresholds for DNA-reactive (genotoxic) organic carcinogens, *J. Toxicol. Pathol.*, 2005, **18**, 69–77.
50. P. W. Hastwell, L.-L. Chai, K. J. Roberts, T. W. Webster, J. S. Harvey, R. W. Rees and R. M. Walmsley, High-specifity and high-sensitive genotoxicity assessment in a human cell line: Validation of the GreenScreen HC GADD45a-GFP genotoxicity assay, *Mutat. Res.*, 2006, **607**, 106–175.

1. Threshold Effects Observed in Experimental Studies

CHAPTER 1.1

Mechanisms Responsible for the Chromosome and Gene Mutations Driving Carcinogenesis: Implications for Dose-Response Characteristics of Mutagenic Carcinogens

R. JULIAN PRESTON

National Health and Environmental Effects Research Laboratory, US Environmental Protection Agency, Research Triangle Park, NC 27711, USA
Email: preston.julian@epa.gov

1.1.1 Introduction

Over the past few years the development of sophisticated techniques for sequencing genomes and for establishing gene expression profiles, for example, has led to a much more detailed characterization of tumors at the molecular level.[1] What is apparent is that the carcinogenic process is complex and involves a range of interacting genotypic and phenotypic alterations to drive a normal cell to a transformed one and ultimately to a tumor. In addition, the role of

Issues in Toxicology No. 13
The Cellular Response to the Genotoxic Insult: The Question of Threshold for Genotoxic Carcinogens
Edited by Helmut Greim and Richard J. Albertini
© The Royal Society of Chemistry 2012
Published by the Royal Society of Chemistry, www.rsc.org

environmental agents in initiating or accelerating this process is becoming better understood. However, putting all of these types of information together to predict the likelihood of an exposure scenario inducing tumors above the background level is really a daunting task. This is necessary if it is going to be possible to characterize tumor dose-response curves in a qualitative and quantitative manner. The aim of this chapter is to consider the form of tumor dose-response curves for mutagenic chemicals and to determine whether or not there is the potential for non-linearity or perhaps a threshold response.

1.1.2   Genetic Alterations and Cancer

It is well-recognized that all cancers contain somatic mutations. In fact, it has been shown that a metastatic tumor contains in excess of 10 000 genetic alterations, both gene and chromosomal.[2] However, it has been determined that a subset of these alterations can be considered to be driver mutations because they confer a selective growth advantage and thus are integral to cancer development.[3] The non-driver genetic alterations are termed 'passengers' because they are a product of the cancer process that confers genomic instability on cells.

Recent studies have shed some light on the drivers for specific tumors and for cancers overall. Pleasance *et al.*[3] sequenced the genomes of a malignant melanoma and a lymphoblastoid cell line from one individual and thereby developed a catalogue of the somatic mutations from an individual cancer. Although the specific drivers could not be identified from a single cancer, the method could be applied to a large number of similar cancer types to identify the drivers. However, it was possible to identify mutations that were consistent with exposure to ultraviolet (UV) light—a known risk factor for melanoma.

Using a similar ultra high throughput sequencing approach, Greenman *et al.*[1] re-sequenced 210 cancer cell genomes and concluded, based on computational analysis, that there were driver mutations in about 120 genes, with the remainder (of the order of a thousand) being passengers. Based on this study, it needs to be noted that when developing models of cancer probability from environmental exposures leading to considerations of dose-response, the genetic 'target' should be considered to be no larger than this set of driver mutations.

Along similar lines, Stephens *et al.*[4] studied another major class of genetic alteration associated with cancer formation, namely somatic rearrangement. These investigators used a paired-end sequencing strategy to identify somatic rearrangements in breast cancer genomes. Such rearrangements were more frequent than anticipated and were largely intrachromosomal. This study exemplifies the role of chromosomal alterations in cancer formation. Bignell *et al.*[5] characterized homozygous deletions in cancer cells and concluded that many are in regions of inherent fragility, but that just a small subset overlies recessive cancer genes.

These two studies highlight the involvement of structural chromosome changes in the cancer process and demonstrate that they can also consist effectively of drivers and passengers. It remains uncertain whether or not genetic alterations in driver genes are produced by the same mechanism(s) as passenger alterations. For example, driver mutations as tumor inducers are more likely to be induced by the exposure, whereas passenger mutations are quite likely to be produced through changes in cellular housekeeping processes as a consequence of driver mutations. This could have an impact on dose-response characterization for tumors, especially for mutagenic chemicals. This issue of mechanisms of formation of genetic alterations will be returned to later in this chapter when discussions are introduced of how mechanism affects dose-response curve shape.

1.1.3 Carcinogenic Process

The carcinogenic process has for a long time been considered to be readily divided into three discrete phases: initiation, promotion and progression.[6] For mutagenic chemicals, it can be considered that initiation is driven by induced mutation and that promotion and progression are under the control of processes that are independent of the mutagenic exposure—perhaps other chronic exposures or endogenous processes. On the other hand, all three stages could be enhanced by chronic exposures to mutagenic chemicals (or mixtures) because the exposure extends over a time frame that is consistent with the extended time for a tumor to develop. The requirement for the latter is that exposure has to be extended over years (for humans), whereas in the case of initiation, exposure to the mutagenic agent could be acute or very short term. Such differences can have significant effects on dose-response considerations since the different exposure scenarios can have quite different effectiveness for inducing genetic alterations.

An alternate description of carcinogenesis is provided by Hanahan and Weinberg[7] in their 2000 paper, *Hallmarks of Cancer*. From considerations of some characteristics of a broad range of tumor types for different target tissues and in several different species, Hanahan and Weinberg described a set of six acquired characteristics essential to transform cells and for the development of a metastatic tumor. These characteristics are:

- unlimited replicative potential;
- ability to develop blood vessels (angiogenesis);
- evasion of programmed cell death (apoptosis);
- self-sufficiency in growth signals;
- insensitivity to inhibitors of growth;
- tissue invasion and metastasis.

Based on convincing evidence, Mantovani[8] proposed that cancer-related inflammation be considered a seventh acquired characteristic. In 2011 Hanahan and Weinberg re-evaluated these hallmarks in their review *Hallmarks of*

Cancer: The Next Generation.[9] Based on research carried out since their original publication, they added four hallmarks to make ten in total. These additions are:

- deregulating cellular energetic;
- genome instability and mutation;
- avoiding immune destruction;
- tumor-promoting inflammation (see Mantovani[8]).

In the context of the discussion in this chapter, it is proposed with a great deal of supporting evidence that each of these phenotypic characteristics is acquired, in part or in the whole, through the production of genetic alterations in proto-oncogenes or a tumor suppressor gene.[10] This view brings the concepts of drivers of carcinogenesis into the framework of the phenotypes representing the multi-stages in cancer development. A stage in the Hanahan and Weinberg concept is the acquiring of one of the characteristics. Such a process is likely to be driven by one or more genetic alterations of the type discussed in Section 1.1.2.

The objective of this chapter is to determine if an understanding of the mechanisms by which these genetic alterations can be formed in the presence of DNA damage can assist in defining the form of the tumor dose-response curve. There is a very wide gap between understanding the mechanism of formation of genetic alterations and the mechanism of formation of tumors, but using appropriate approaches based on key events and mode of action, this gap can be addressed. The next section begins this process.

1.1.4 Key Events for Mutagenic Mode of Action

The discussion concentrates on mutagenic carcinogens that are synonymous with DNA-reactive carcinogens. Preston and Williams[11] developed a set of key events for the formation of a tumor *via* a DNA reactive mode of action. The set of ten key events, presented below, begins with the interaction of a chemical with tumor target cells and progresses to malignant behavior of the target tissue. The critical key events for the present discussion are 2, 3 and 4. Key events 5–11 in this progression are along the lines of the production of additional mutations to enhance cell proliferation and the development of clonally expanded mutant cells to form preneoplastic lesions.

1. Exposure of target cells (*e.g.* stem cells) to ultimate DNA-reactive and mutagenic species—in some cases this requires metabolism.
2. Reaction with DNA in target cells to produce DNA damage.
3. Misreplication on damaged DNA template or misrepair of DNA damage.
4. Mutations in critical genes in replicating the target cell.
5. These mutations result in initiation of new DNA/cell replication.
6. New cell replication leads to clonal expansion of mutant cells.
7. DNA replication can lead to further mutations in critical genes.

8. Imbalanced and uncontrolled clonal growth of mutant cells may lead to preneoplastic lesions.
9. Progression of preneoplastic cells results in emergence of overt neoplasms, solid tumors (which require neoangiogenesis), or leukemia.
10. Additional mutations in critical genes as a result of uncontrolled cell division results in malignant behavior.

In this context, the mode of action for a toxicological effect is a biologically plausible sequence of key events that leads to an observed adverse outcome which is supported by robust experimental observations and mechanistic data. A key event is defined as an empirically observable, precursor step that is a necessary element of the mode of action or is a marker for such an element.[12]

Based on this definition, the key events noted above for DNA-reactive carcinogens and the role of genetic alterations in tumor development, it is reasonable to conclude that the induction of chromosomal and gene mutations represent key events in the carcinogenesis process for DNA-reactive (mutagenic) carcinogens. It is also important to note that information on key events, such as the form of their dose-response curve, can be used as a surrogate for the apical endpoint of cancer at exposure levels below those at which tumors can be assessed in laboratory animal or epidemiologic studies. It has to be borne in mind that there can be a considerable period of time between the production of a key event and the transformation of a cell and subsequent tumor development that can certainly influence the quantitative relationships between key events and tumors. Even then, for extrapolation to environmental exposure levels, it is necessary to utilize other information such as the mechanism of induction of a particular key event being used for extrapolation purposes. This moves the present discussion on to consider the mechanism of induction of chromosomal and gene mutations.

1.1.5 Mechanism of Induction of Chromosomal and Gene Mutations

In general terms, both chromosomal and gene mutations can be formed by errors of DNA repair or replication.[13] For ionizing radiation, again as a generalization, DNA repair errors are the main source of mutations in G1, G2 and unreplicated regions in S phase. This is because repair of ionizing radiation DNA damage is rapid and takes place before any damaged DNA can be replicated. Damage induced in DNA that is close in time to being replicated can be converted into mutations by errors of replication. For damage induced by the great majority of chemical carcinogens, conversion into mutations is by errors of replication, largely because DNA repair is relatively slow, taking many hours, and a cell has a high probability of replicating prior to repair of much of the induced DNA damage.[14] Thus, the probability of inducing a chromosomal or gene mutation is related to the amount of induced DNA damage, the probability of this damage being unrepaired at the time of replication and its probability of being misreplicated. The background error rate for DNA replication is low,[15] even though the steady-state level of DNA damage is about 500 base alterations per cell, resulting from hydrolysis, oxidation and nonenzymatic methylation of DNA.[16] If it is assumed that there is no specific qualitative difference between background and induced DNA damage, it can be argued that a small additional amount of DNA damage from exposure to a mutagenic carcinogen would be unlikely to lead to an increase in mutations—this would be an argument for a shallow slope to the dose-response curve for mutation, at least. Based on discussions of the key event nature of genetic alterations in the cancer process, it is reasonable to extend this threshold-like response to cancer.

It is possible using molecular combing techniques to assess DNA replication rates at different regions of the genome, even at the replicon level.[17] Thus, it should be possible to assess changes in rates of replication for specific cancer genes to determine if, for example, there are changes in response to exposure to mutagenic carcinogens. This might indicate changes in replication error rate or a propensity for the formation of chromosomal alterations. A faster initiation of replication or more rapid replication rate could lead to the balance between repair and replication tipping towards replication. In this case, replication would take place before repair was completed. Being able to make these types of molecular assessments is a real technical possibility.

These discussions illustrate how replication error rate and replication rate can influence the likelihood of inducing mutations and by extrapolation, tumors. They also illustrate that for environmental exposures, with the attendant low levels of induced DNA damage, this replication error rate is very unlikely to be increased significantly. Thus, the probability of inducing a mutation is very low and the subsequent tumor induction rate equally would be very low.

1.1.6   Mutation Induction and 'Hit Theory'

The notion of target or hit theory has its origins in radiation biology. In general terms, the production of chromosome alterations (this includes mutations formed by deletions) by ionizing radiation is the result of the interaction during DNA replication or repair of two DNA lesions (strand breaks, base damages or complex lesions) either intrachromosomally (deletions) or interchromosomally (interchanges).[13] These pairs of lesions can be produced by one ionization track ('one hit') or by two independent tracks ('two hit'). Thus, the dose-response for one-track alterations will be linear with dose, whereas that for two-track alterations will be proportional to (dose).[2] The overall response will be:

$$Y, \text{ Yield of alterations} = aD + bD^2 \qquad (1.1.1)$$

where D is dose and a and b are one-track and two-track coefficients, respectively. The resulting dose-response curve is a so-called linear quadratic curve, with the low dose region being linear.

It is possible to consider the genetic alterations induced by chemicals in a similar target theory manner. For mutagenic carcinogens, a 'hit' is DNA damage (*e.g.* DNA adduct) that upon replication can be converted into a gene mutation or a chromosomal alteration. However, because of the nature of chemical action, each hit is considered to be independently produced. For example, the two hits required for a chromosome mutation are essentially produced by 'two tracks' in radiation parlance. This argument leads to the following prediction. For mutagenic chemicals, gene mutations are single base changes produced by replication errors on a damaged DNA template. On the basis of hit theory, they will be produced by one hit and will be linear with dose. On the other hand, chromosomal aberrations (deletions and exchanges) require the interaction of two lesions during replication for their formation. Since these two hits are produced independently, the dose-response for chromosome aberrations will be a quadratic ($Y = aD^2$). The overall dose-response for genetic alterations (gene mutations and chromosomal aberrations) will be $Y = aD + bD^2$. Of particular note, the higher the proportion of gene mutations of the overall total of genetic alterations, the more the linear component prevails and if the proportion of chromosomal mutations increases, so too will the dose-squared component of the dose-response curve. Thus, the dose-response curve can vary between linear, low dose to quadratic, low dose depending on the specific chemical characteristics. In the latter case, there will be very few, if any, genetic alterations at low doses because of the form of a quadratic curve. Using the relationship between genetic alterations and cancer, discussed above, these considerations for mutations can be extended to those for tumors.

1.1.7   Next Steps

As discussed in Section 1.1.2, there is clear evidence that genetic alterations are involved in the carcinogenic process. There is also evidence that there is a limited set of cancer driver genes. It follows that learning more about the specific key events for tumors induced by specific environmental carcinogens will enhance our understanding of the form of dose-response curves and the possible rate-limiting steps that influence the low dose response. What is needed is to establish the dose-response for specific key events, especially at low dose levels that can be used to derive conclusions about tumor responses at these same low doses. It is likely that such information will be obtained for laboratory animal models that in turn will require extrapolation to predict human responses. It should be possible using genomic technologies (such as ultra high throughput sequencing[18]) to be able to enhance the extrapolation process because of the capability of comparing genomic responses in different species. The preceding sections highlight the expectation that dose-response curves for mutations induced by mutagenic carcinogens (when both gene and chromosome ones are produced) are likely to be non-linear with dose over a broad dose range, including low doses. A critical component of such considerations will be the development of sophisticated biologically based dose-response models that can provide the ability to predict responses at dose levels below those at which biological observations can be made or even are likely to be able to be made in the future. As an example of this, a recent paper by Calabrese and Shibata[19] describes an algebraic cancer equation that demonstrates how age-related increases in cancer frequencies can result from relatively normal cell division and mutation rates. The authors use inputs of stem cells, mutation and division rates, and the number of driver mutations to predict outcomes of cancer epidemiology. This represents a good fit with the information presented in this condensed review of the topic of mutations, mechanisms and cancer.

1.1.8   Conclusions

The concept of an association of specific genetic alterations with tumor development has been around for some time. However, with sophisticated next generation sequencing techniques, it has been possible to analyse tumors at the whole genome level. This has led to the concept of drivers and passengers, with the drivers being identified for several tumor types. Such drivers are consistent with the concept of tumor development being the result of a set of ten acquired characteristics that in essence provide a growth advantage to the mutated cells. The necessary genetic alterations for cell transformation and tumor development (*i.e.* gene and chromosomal mutations) induced by mutagenic carcinogens are largely the result of errors of DNA replication on a damaged template. It is reasonable using classical target theory to demonstrate that the dose-response curve for gene mutations is linear, whereas that for chromosome mutations (deletions and exchanges) is quadratic. The dose-response curve for genetic alterations (when both gene and chromosome mutations are produced) is non-linear, with its specific form being determined by the relative contributions of the two types of genetic alteration. By following the arguments presented, the dose-response curve overall is also non-linear for tumors although it is effectively linear as zero dose is approached. Such non-linearity is unlikely to lead to a threshold at very low doses, although other mechanistic considerations (*e.g.* failure of chemical to reach target tissue, error-free DNA repair at very low doses) would allow for such low dose threshold responses for mutagenic carcinogens.

Acknowledgements

This manuscript has been reviewed by the National Health and Environmental Effects Research Laboratory, US Environmental Protection Agency and approved for publication. However, it does not necessarily reflect Agency policy. The author thanks Dr Andrew Kligerman and Dr Rory Conolly for their helpful comments during the review of this chapter.

References

1. C. Greenman, P. Stephens, R. Smith, G. L. Dalgliesh, C. Hunter, G. Bignell, H. Davies, J. Teague, A. Butler, C. Stevens, S. Edkins, S. O'Meara, I. Vastrik, E. E. Schmidt, T. Avis, S. Barthorpe, G. Bhamra, G. Buck, B. Choudhury, J. Clements, J. Cole, E. Dicks, S. Forbes, K. Gray, K. Halliday, R. Harrison, K. Hills, J. Hinton, A. Jenkinson, D. Jones, A. Menzies, T. Mironenko, J. Perry, K. Raine, D. Richardson, R. Shepherd, A. Small, C. Tofts, J. Varian, T. Webb, S. West, S. Widaa, A. Yates, D. P. Cahill, D. N. Louis, P. Goldstraw, A. G. Nicholson, F. Brasseur, L. Loojenga, B. L. Weber, Y. E. Chiew, A. DeFazio, M. F. Greaves, A. R. Green, P. Campbell, E. Birney, D. F. Easton, G. Chenevix-Trench, M. H. Tan, S. K. Khoo, B. T. The, S. T. Yuen, S. Y. Leung, R. Wooster, P. A. Futreal and M. R. Stratton, Patterns of somatic mutation in human cancer genomes, *Nature*, 2007, **446**, 153–158.
2. D. L. Stoler, N. Chen, M. Basik, M. S. Kahlenberg, M. A. Rodriguez-Bigas, N. J. Petrelli and G. R. Anderson, The onset and extent of genomic instability in sporadic colorectal tumor progression, *Proc. Natl. Acad. Sci., U. S. A.*, 1999, **96**, 15121–15126.
3. E. D. Pleasance, R. K. Cheetham, P. J. Stephens, D. J. McBride, S. J. Humphray, C. D. Greenman, I. Varela, M.-L. Lin, G. R. Ordonez, G. R. Bignell, K. Ye, J. Alipaz, M. J. Bauer, D. Beare, A. Butler, R. J. Carter, L. Chen, A. J. Cox, S. Edkins, P. I. Kokko-Gonzales, N. A. Gormley, R. J. Grocock, C. D. Haudenschild, M. M. Hims, T. James, M. Jia, Z. Kingsbury, C. Leroy, J. Marshall, A. Menzies, L. J. Mudie, Z. Ning, T. Royce, O. B. Schulz-Trieglaff, A. Spiridou, L. A. Stebbings, L. SzajKowski, J. Teague, D. Williamson, L. Chin, M. T. Ross, P. J. Campbell, D. R. Bentley, P. A. Futreal and M. R. Stratton, A comprehensive catalogue of somatic mutations from a human cancer genome, *Nature*, 2010, **463**, 191–197.
4. P. J. Stephens, D. J. McBride, M. L. Lin, I. Varela, E. D. Pleasance, J. T. Simpson, L. A. Stebbings, C. Leroy, S. Edkins, L. J. Mudie, C. D. Greenman, M. Jia, C. Latimer, J. W. Teague, K. W. Lau, J. Burton, M. A. Quail, H. Swardlow, C. Churcher, R. Natrajan, A. M. Sieuwerts, J. W. Martens, D. P. Silver, A. Langerød, H. E. Russnes, J. A. Foekens, J. S. Reis-Filho, L. van't Veer, A. L. Richardson, A. L. Børresen-Dale, P. J. Campbell, P. A. Futreal and M. R. Stratton, Complex landscapes of somatic rearrangements in human breast cancer genomes, *Nature*, 2009, **462**, 1005–1010.
5. G. R. Bignell, C. D. Greenman, H. Davies, A. P. Butler, S. Edkins, J. M. Andrews, G. Buck, L. Chen, D. Beare, C. Latimer, S. Widaa, J. Hinton, C. Fahey, B. Fu, S. Swamy, G. L. Dalgliesh, B. T. The, P. Deloukas, F. Yang, P. J. Campbell, P. A. Futreal and M. R. Stratton, Signatures of mutation and selection in the cancer genome, *Nature*, 2010, **463**, 893–898.
6. J. E. Klaunig and L. M. Kamendulis, Chemical carcinogenesis, in *Casarett and Doull's Toxicology: The Basic Science of Poisons*, ed. C. D. Klaassen, McGraw-Hill, New York, 2008, pp. 329–379.

7. D. Hanahan and R. A. Weinberg, The hallmarks of cancer, *Cell*, 2000, **100**, 57–70.
8. A. Mantovani, Cancer: Inflaming metastasis, *Nature*, 2009, **457**, 36–37.
9. D. Hanahan and R. A. Weinberg, Hallmarks of cancer: The next generation, *Cell*, 2011, **144**, 646–674.
10. N. Beerenwinkel, T. Antal, D. Dingli, A. Traulsen, K. W. Kinzler, V. E. Velculescu, B. Vogelstein and M. A. Nowak, Genetic progression and the waiting time to cancer, *PLoS Comput. Biol.*, 2007, **3**, e225.
11. R. J. Preston and G. M. Williams, DNA-reactive carcinogens: Mode of action and human cancer hazard, *Crit. Rev. Toxicol.*, 2005, **35**, 673–683.
12. A. R. Boobis, G. P. Daston, R. J. Preston and S. S. Olin, Application of key events analysis to chemical carcinogens and noncarcinogens, *Crit. Rev. Food Sci. and Nutr.*, 2009, **49**, 690–707.
13. R. J. Preston, A consideration of the mechanisms of induction of mutations in mammalian cells by low doses and dose rates of ionizing radiation, in *Advances in Radiation Biology*, ed. O. F. Nygaard, W. K. Sinclair and J. T. Lett, Academic Press, New York, 1992, Vol. 16, pp. 125–135.
14. R. J. Preston and G. R. Hoffmann, Genetic toxicology, in *Casarett and Doull's Toxicology: The Basic Science of Poisons,* ed. C. D. Klaassen, McGraw-Hill, New York, 2008, pp. 381–413.
15. M. Budzowska and R. Kanaar, Mechanism of dealing with DNA damage-induced replication problems, *Cell Biochem. Biophys.*, 2009, **53**, 17–31.
16. T. Lindahl, Instability and decay of the primary structure of DNA, *Nature*, 1993, **362**, 709–715.
17. J. Herrick and A. Bensimon, Introduction to molecular combing: genomics, DNA replication, and cancer, in *Methods in Molecular Biology, DNA Replication*, ed. S. Vengrova and J. Z. Dalgaard, Humana Press, New York, 2009, Vol. 521, pp. 71–101.
18. P. J. Campbell, E. D. Pleasance, P. J. Stephens, E. Dicks, R. Rance, I. Goodhead, G. A. Follows, A. R. Green, P. A. Futreal and M. R. Stratton, Subclonal phylogenetic structures in cancer revealed by ultra-deep sequencing, *Proc. Natl. Acad. Sci., U. S. A.*, 2008, **105**, 13081–13086.
19. P. Calabrese and D. Shibata, A simple algebraic cancer equation: calculating how cancers may arise with normal mutation rates, *BMC Cancer*, 2010, **10**, 3–14.

CHAPTER 1.2

Dose-Effect Relationships for DNA-reactive Liver Carcinogens

G. M. WILLIAMS,* M. J. IATROPOULOS AND A. M. JEFFREY

Chemical Safety Laboratory, Department of Pathology, New York Medical College, Valhalla, NY, USA
*Email: gary_williams@nymc.edu

1.2.1 Introduction

The concepts of thresholds or virtually safe doses in chemical carcinogenesis have long been under consideration.[1–5] Central to current application of this concept is the mode of action (MoA) of carcinogens.[6–12] The MoA of one type of carcinogen entails epigenetic or non DNA-reactive mechanisms, which do not involve direct reaction of the chemical with target cell DNA, but rather produce other cellular effects, such as indirect genotoxicity, changes in gene expression or enhanced cell proliferation, which underlie the carcinogenicity in target tissues.[11,13,14] Epigenetic carcinogens are generally accepted to have cancer thresholds[11,12,15,16] at exposures below which they do not elicit the cellular effects that lead to carcinogenicity. In contrast to epigenetic carcinogens, DNA-reactive or genotoxic carcinogens operate either by bioactivation-independent, bioactivation-dependent, or in some cases photo-activated, direct reactions with DNA in target tissues to form chemical-specific adducts or other DNA lesions which in turn produce mutations in critical genes.[14,17–20]

It has been postulated that a DNA-reactive MoA in mutagenesis and carcinogenesis could involve a single direct biochemical reaction, specifically, a

Issues in Toxicology No. 13
The Cellular Response to the Genotoxic Insult: The Question of Threshold for Genotoxic Carcinogens
Edited by Helmut Greim and Richard J. Albertini
© The Royal Society of Chemistry 2012
Published by the Royal Society of Chemistry, www.rsc.org

single hit at a critical target in the DNA of a single target cell[21] and thus no threshold would exist. Nevertheless, biological processes exist which ameliorate the genotoxicity of DNA-reactive carcinogens, including limited uptake, detoxication, DNA repair and apoptosis.[22–25] Accordingly, thresholds also have been postulated for DNA-reactive carcinogens.[4,12,23,24] Currently, however, this type of carcinogen is not generally accepted as having thresholds for carcinogenicity, in part because of the nature of its MoA.[11,21] Indeed, most of the carcinogenicity studies of DNA-reactive agents that used relative low cumulative doses (CDs) have been interpreted as not showing a threshold[26–31], although some investigators have calculated 'virtually safe doses', 'practical thresholds' or 'no-observed-adverse-effect-levels' (NOAEL).[27,29]

Carcinogenicity studies, however, are unlikely to provide evidence of thresholds because they typically involve the measurement of neoplasms in chronic (two-year) rodent bioassays that do not incorporate the large number of animals required to establish the absence of a tumor increase over background at low doses. Even in the large study by Bailey *et al.*[32] using 40 800 trout encompassing a dose range of 0–225 parts per million (ppm) dibenzo[a]pyrene, the authors concluded that 'although the data were consistent with a threshold interpretation, even the use of over 30 000 animals did not provide *proof* that a threshold was reached, or would exist, in either target organ for this carcinogen'. Thus, other approaches must be sought to determine whether thresholds exist for the tumorigenicity of DNA-reactive carcinogens. One approach involves measurements of bioindicators of effect, such as preneoplastic lesions, as discussed below.

In this review of dose-effect relationships for DNA-reactive liver carcinogens, we define a threshold as a CD below which are no-effect levels for bioindicators of carcinogenic effect and subsequently promotable neoplasms. The experimental design, modulating the factors of time and size of target tissue, can, of course, influence the effects of identical CD. Non-linearities may be obvious in a dose-effect curve, but this alone cannot be interpreted as evidence of a threshold since it may only indicate a change in the slope of the curve.

The observation of NOAEL allows for the possibility of practical thresholds.[5] In most dose-effect studies, the dose range to detect thresholds or non-linearities is insufficient. For example, in the standard US National Toxicology Program (NTP) bioassay, only two or three doses are normally tested. In wider range dose-effect studies, the low CDs are often difficult to depict graphically on a linear scale, particularly with regard to extrapolation to a US Environmental Protection Agency (EPA) acceptable risk of one tumor in 10^6 individuals (Figure 1.2.1A). One recommendation that has been advanced[33] is to plot the data with log dose and linear effect scales (Figure 1.2.1B). Although the data remain the same, *i.e.* x (dose) $= y$ (% animals with tumors), in such a plot, it now appears that the 'actual' data (*i.e.* that which could readily be achieved in a bioassay with the effect being between 10 and 100% tumor incidences as represented with filled circles in Figure 1.2.1B) could be extrapolated to an intersect for zero tumor incidence at a dose of about 1e+1, which is clearly not the case for linear data. If both the dose and effect are plotted on log scales then, provided there is no background to subtract, a straight line is achieved

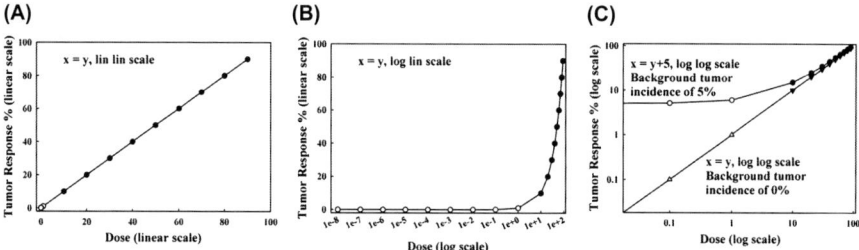

Figure 1.2.1 Graphical representations of hypothetical tumor response data. **A.** The response is directly proportional to dose. • Doses at which a measurable tumor incidence is likely to occur with 50 animals (10% response). ○ Doses below which a measurable tumor incidence is unlikely to occur with 50 animals (<10% response). If additional animals were incorporated into the study such that reliable data were collected at lower doses, the results would be difficult to see in the bottom left corner of the graph. **B.** A log linear (lin) plot makes the low doses (○) more readily visible but the very low, but albeit real, tumor response is still left compressed and invisible. However, at the higher doses (•), it appears the response may be reasonably extrapolated to the *x* axis and provide information regarding a dose at which the tumor response would be zero: this is not the case. **C.** A log/log plot avoids the problem of the log linear plot shown in B, provided there is no background response (▼△). However, in the case of a background response, in this case 5% tumor incidence, an apparent threshold below which there is no response appears at a dose of about 1. Again, this is not the case but rather that the predicted response (△) is much smaller than the background 5% value. Subtraction of the background value prior to the log transformation of the data would satisfactorily correct this problem except that the variation in the background value is often much greater than the predicted response and negative values result.

and the very low dose values (open triangles) can be seen (Figure 1.2.1C). More often, a background tumor incidence is present, *e.g.* 5% in the theoretical example (in controls) shown in Figure 1.2.1C. In this case, a log/log plot shows what has been interpreted as no effect doses while, in reality, the predicted magnitude of the linear effect data are smaller than the background and, therefore, simply become immeasurably small against that background. Thus, when these types of plots are used, care is needed to ensure that a misinterpretation of the results with respect to possible thresholds is not made.

1.2.2 Dose-effect Studies of Initiation of Liver Carcinogenesis

Preneoplastic lesions, as the terminology implies, precede the development of neoplasms, have phenotypic and genomic alterations indicative of neoplastic potential, and have higher rates of development into neoplasms than do normal target cells.[34–36] Accordingly, induction of preneoplastic lesions can be used as a bioindicator to monitor initiation of carcinogenesis. Preneoplasia has proven useful for this purpose because it occurs at lower CD and with shorter durations of dosing than that which is needed for tumor induction.[36–38]

In liver carcinogenesis, hepatocellular altered foci (HAF) are accepted preneoplastic lesions.[35–38] Dose-effect studies aimed at delineating NOAEL have been conducted with liver carcinogens using preneoplastic lesions, together with other bioindicators of effect, such as increased cell proliferation, as endpoints for measuring tumor initiation. The concept underlying this approach is that a NOAEL for induction of preneoplasia will necessarily be a NOAEL for neoplasia. This concept is supported by the demonstration in a series of studies by this laboratory,[24,39–43] that 8–16 weeks of carcinogen administration followed by tumor promotion with phenobarbital yielded neoplasms only with doses of carcinogen that produced preneoplastic effects. The sensitivity of an initiation/promotion protocol is demonstrated by the fact that at lower CD it yields higher tumor incidences than with chronic administration of carcinogen alone[43] (Table 1.2.1), reflecting the acceleration of tumor development by the promoter.

Table 1.2.1 Carcinogenicity in a rat initiation/promotion (I/P) protocol using phenobarbital[a] as promoter compared to that resulting from chronic administration of carcinogen alone.

Protocol and dosage	Cumulative dose $(mg\ kg^{-1})$	Incidence of liver neoplasms (%)
Diethylnitrosamine (DEN)		
I/P protocol[b]		
Williams et al., 1996[39]: 0 mg kg^{-1}	0	0
0 mg kg^{-1}/P[a]	0	0
5 mg kg^{-1b}	50	0
5 mg kg^{-1b}/P[a]	50	3 (one adenoma only)
10 mg kg^{-1b}	100	9 (adenomas)
10 mg kg^{-1b}/P[a]	100	32 (adenomas and carcinomas)
20 mg kg^{-1b}	200	80 (adenomas and carcinomas)
20 mg kg^{-1b}/P[a]	200	100 (only carcinomas)
Continuous administration protocol		
Peto et al., 1991[30,31]: 6.3 ppm[c]	287	45
16.9 ppm[c]	771	78
2-Acetylaminofluorene (AAF)		
I/P protocol[b]		
Williams et al., 1998[40]: 0 mg kg^{-1}/P[a]	0	0
1.1 mg kg^{-1b}/P[a]	94	3 (1 adenoma only)
3.4 mg kg^{-1b}/P[a]	288	100 (adenomas and carcinomas)
Continuous administration protocol		
Williams et al., 1991[88]: 50 ppm in diet[d]	1772	100

[a]Phenobarbital at 500 ppm.
[b]Intragastric instillation, for DEN once per week for ten weeks, and for AAF seven days per week for 12 weeks.
[c]In drinking water for 2.5 years. Approximately equivalent to 0.5 and 1.4 mg day^{-1} kg^{-1} bw.
[d]In diet for 76 weeks. Approximately equivalent to 5 mg day^{-1} kg^{-1} bw.

Hino and Kitagawa[44] conducted a dose-effect study on 3'-methyl-4-(dimethylamino)azobenzene (MDAB) using adenosine triphosphatase (ATPase) deficiency as a marker for HAF. MDAB was fed to groups of 5–9 male Donryu rats for 24 weeks at concentrations of 0, 1, 5, 10, 20, 60, 100 or 300 ppm. The incidences of HAF were dose-related with the highest dose producing a 23-fold increase over the control. With doses of 1–20 ppm, the incidences of HAF were below the control background, which the authors suggested could be due to inhibition of background carcinogenesis by MDAB, an effect described as hormesis which has been reviewed by Calabrese.[45] The data were interpreted to show a 'practical threshold' between 20 and 60 ppm. In general, hormetic (low dose) effects can either (depending on the target tissue) result from inhibitory signals that trigger stimulatory overcompensation hormesis, or stimulatory signals that trigger inhibitory overcompensation (hormoligosis).[46,47]

Root *et al.*[48] conducted a dose-effect study on aflatoxin B$_1$ (AFB) using γ-glutamyltransferase (GGTase) as a marker for HAF. AFB was administered to groups of ten male F344 rats by gavage for ten days at daily doses of 50, 100, 150, 200, 250 or 350 μg kg^{-1} body weight (bw), after which rats were maintained for 12 weeks for development of HAF. AFB-guanine adduct levels were directly proportional to dose after the first dose, but after the tenth dose were much lower in the top three dose groups than after a single dose. The dose-effect relationship for HAF was sublinear, and achieved over a 1000-fold increase in HAF at the highest dose. A threshold was claimed at a dose of about 150 μg kg^{-1} bw day^{-1}. From Figure 5 of the paper, however, some values at 100 μg kg^{-1} bw day^{-1} are greater than those at 50 μg kg^{-1} bw day^{-1}, and thus a threshold may not have been identified, but rather a NOAEL. There is no mention of control rats and thus the incidence of background HAF in this study is unknown.

Enzmann *et al.*[49] conducted a dose-effect study on *N*-nitrosomorpholine (NNM) using glucose-6-phosphate dehydrogenase (G6PDH), glycogen phosphorylase (PHO) and glycogen content measured by periodic acid–Schiff stain (PAS) as markers for HAF. NNM was administered to groups of 9–23 male Sprague–Dawley rats in the drinking water at concentrations of 0, 0.006, 0.06, 0.60, 6.0 or 60.0 mg L^{-1} for 6 or 12 weeks. The dose-effect curves were non-linear with a slight positive slope at the low doses and a markedly increased slope at higher doses. Quantitation of hepatocytic replication by proliferating cell nuclear antigen (PCNA) at up to 6.0 mg L^{-1} revealed a dose-dependent increase at 12 weeks, but not six weeks.

The apparent non-linear shape of the dose–response curves of the HAF was interpreted to suggest that some mechanisms contributed to carcinogenesis over the whole dose range, whereas other mechanisms enhanced carcinogenesis only at higher doses. Of the three HAF markers used, the G6PDH$^+$ and PAS$^+$ HAF were increased at ≥6 mg L^{-1} (at both six and 12 weeks interval). In contrast, PHO$^+$ HAF were not increased. Thus, G6PDH$^+$ and PAS$^+$ HAF incidence curves were suggested to indicate non-linearity, but this, at least in part, was a consequence of log/log plotting of the data (Figure 1 in the paper), with the background value not having been first subtracted. Based on a linear decrease in the HAF incidence from of 4.4 per cm^2 of liver tissue at 6 mg L^{-1} to 0.44 (0.22 if the background is first subtracted) at 0.6 mg L^{-1}, lower doses would not have been expected to produce a measurable increase above the observed background value of 2.19 HAF per cm^2.

Fukushima[50] conducted a dose-effect study on 2-amino-3,8-dimethylimidazo[4,5-*f*]quinoxaline ($MeIQ_x$) using placental-type glutathione *S*-transferase (GST-P) as the marker for HAF. $MeIQ_x$ was administered to groups of 50–150 male F344 rats for 16 weeks at doses of 0, 0.001, 0.01, 0.1, 1.0, 10.0 or 100 ppm in the diet. The numbers of GST-P$^+$ HAF per cm^2 of the rat livers of the four groups receiving up to 1 ppm of the carcinogen did not differ from the control value and hence were NOAEL. In contrast a measurable increase was observed with 10 ppm and a substantial elevation with 100 ppm $MeIQ_x$ (both significant at $p < 0.001$). The apparent non-linearity in the induction of HAF was again, at least in part, a consequence of a log/log plot of the data (Figure 1 in the paper) in which there was a background value that was not first subtracted. Transplacental and breast milk-mediated exposures resulting from feeding diets containing up to 10 ppm $MeIQ_x$ did not increase GST-P$^+$ HAF, whereas caloric restriction for 15 weeks reduced GST-P$^+$ HAF by 20%.

Fukushima *et al.*[51] conducted a longer duration dose-effect study on $MeIQ_x$ using GST-P as a marker for HAF. $MeIQ_x$ was administered in the diet to groups of 30–50 male F344 rats for 32 weeks at concentrations of 0, 0.001, 0.01, 0.1, 1.0, 10 or 100 ppm in the diet. The lowest dose was estimated to provide intake equivalent to the daily intake of this carcinogen in humans.[52] The numbers of GST-P$^+$ HAF per cm^2 of the rat livers of the four groups receiving up to 1 ppm of the carcinogen did not differ from the control value and hence were NOAEL. In contrast an increase was observed with 10 ppm and a substantial elevation ($p < 0.01$) with 100 ppm $MeIQ_x$. Thus, $MeIQ_x$ at doses below 10 ppm did not induce measurable increase in GST-P$^+$ HAF although they did form DNA adducts at 0.1 ppm and above. Moreover, as discussed above, the log/log plots (Figure 1 in the paper) give deceptive-looking responses when control values are not subtracted. The effect at 10 ppm for total $MeIQ_x$-induced HAF (Table II in the paper), however, is significantly lower than would be projected from the 100 ppm values with a linear model. In a follow-up two-year study with $MeIQ_x$ in male F344 rats using 0.001, 1 or 100 ppm in the diet, increased frequency of hepatocellular carcinomas, adenomas and GST-P$^+$ HAF and increased levels of adducts at 100 ppm were reported.[53] With 0.001 and 1 ppm, no significant induction of hepatocellular preneoplastic or neoplastic lesions were evident, consistent with no significant increase in DNA adducts at 1 ppm. The authors state that 1 ppm may have been a NOAEL for $MeIQ_x$ carcinogenicity, although 10 ppm, which was not tested, would have been predicted by a linear model to have shown less than one tumor, *i.e.* also a NOAEL.

Fukushima *et al.*[51] also conducted a dose-effect study of *N*-nitrosodiethylamine, also known as diethylnitrosamine (DEN), in which the carcinogen was administered in the drinking water to groups of 151–326 male F344 rats for 16 weeks at concentrations of 0, 0.0001, 0.001, 0.01, 0.1, 1 or 10 ppm. Numbers of GST-P$^+$ HAF in the liver in groups receiving DEN at 0.0001–0.01 ppm were not different from that of the control, while the groups administered 0.1 or 1 ppm DEN showed significant increases in HAF. Extracting the data from Figure 4 in the paper, however, the effects at 0.1 and 1 ppm were approximately 0.48 and 5.8 GST-P$^+$ foci/cm^2, respectively. The projected value, based on a linear response, at 0.01 ppm would have been 0.05, while the observed value was 0.12, which is similar to the background value. Thus, the expected value was too small to add significantly to the background and thus provided a NOAEL, as the authors stated, but not evidence for a threshold.

Laib *et al.*[54] conducted two studies with vinyl chloride (VC) in which groups of 7–21 newborn Wistar and Sprague–Dawley rats were exposed to VC at 0, 2.5, 5, 10, 20, 40, 70, 80, 150, 500 or 2000 ppm by inhalation. In one, the exposure was for 10 weeks (eight hours per day, five days per week), followed by one week of recovery and termination, and in the second, the exposure was for three weeks (eight hours per day, five days per week) followed by ten weeks of recovery and termination. Hepatocellular ATPase-deficient HAF were evaluated and yielded a linear relationship between the dose of VC and the percentage of induced HAF. No threshold for the induction of HAF by VC was observed.[54]

Williams and coworkers provided further evidence of NOAEL for DNA-reactive carcinogens in male F344 rat liver carcinogenesis in a series of studies using the hepatocarcinogens 2-acetylaminofluorene (AAF)[23,40,42,43] and DEN.[23,39,41,43,55] In these investigations, the effects of AAF and DEN were quantified by measuring DNA adducts, hepatocellular cytotoxicity, cell proliferation (quantified as the proliferating cell nuclear antigen positive replicating fraction) and formation of GST-P$^+$ HAF in the initiation phase of carcinogenesis, with the use of phenobarbital (PB) promotion to elicit manifestation of initiation of liver carcinogenesis by the formation of neoplasms after 24 weeks.

With AAF,[24,40,42,43] the CD of 28 mg kg^{-1} bw, delivered over 12 weeks, was a NOAEL for both hepatocellular proliferation and GST-P$^+$ HAF (Table 1.2.2), as were three lower CDs. At 28 mg kg^{-1}, no promotable (with 24 weeks of PB) hepatocellular neoplasia was produced (Table 1.2.2), although about six DNA adducts in 10^8 normal nucleotides (nts) were formed as measured by ^{32}P-post-labeling. A NOAEL for DNA adduct formation was not found owing to the presence of a background in C rats, possibly due to cross-contamination, as has been observed by others. The adduct level of biological significance (as reflected by induction of promotable neoplasia) was found to be approximately three adducts in 10^7 nts. At 12 weeks, the adducts found by ^{32}P-postlabeling were *N*-(deoxyguanosine-8-yl)-2-aminofluorene (dG-C-8-AF) and *N*-(deoxyguanosine-*N^2*-yl)-2-acetylaminofluorene (dG-*N^2*-AAF)—the latter an accepted miscoding lesion. During a four-week recovery segment, after AAF at CDs from 9.41 to 288.2 mg kg^{-1}, the adduct values diminished by ≥50% reflecting DNA damage repair.

With DEN,[24,39,41,44] the CD of 25.5 mg kg^{-1} bw (the lowest dose tested), delivered over ten weeks, was a NOAEL for cell proliferation. HAF were increased, although no promotable hepatocellular neoplasia was produced (Table 1.2.3). At this biological effect level, about 14 DNA adducts in 10^8 nts were formed, and at 51.1 mg kg^{-1} bw, which yielded promotable neoplasia, two adducts per 10^6 nts were formed. The adducts quantified by high-performance

Table 1.2.2 Liver DNA adducts, hepatocellular percent replicating fraction (RF) and GST-P$^+$ HAF per cm^2 of liver tissue of rats exposed to 2-acetylaminofluorene (AAF).

Cumulative doses (CD) at 12 weeks (mg kg^{-1})	DNA adducts in 10^8 nts	RF	HAF per cm^2	CD at 16 weeks (mg kg^{-1})	DNA adducts in 10^8 nts	RF	HAF per cm^2	Percentage of phenobarbital (PB) promotable (for 24 weeks) neoplasms	Reference
Control	0	0.2–3.1	0.9	0.2	0	0.2–1.2	1.4	1.1	0	23, 40, 42
CD1	0.094[a]	0.6	1.5	0.8	0.125	0.6	1.6	1.5	ND	42
CD2	0.94[a]	6.0	1.8	1.2	1.25	1.6	1.8	1.3	ND	42
CD3	9.41[b]	21.4	ND	ND	9.41[b]	11.4	ND	ND	0	23, 40
CD4	28.0[c]	5.6	1.7	0.6	ND	ND	ND	ND	0	40, 42
CD5	94.1[b,c]	29.6	1.9	1.1	94.1[b]	14.4	ND	ND	3	40
CD6	288.2[b,c]	31.7	8.2	19.4	288.2[b]	12.6	ND	ND	100	40

[a]AAF was administered by intragastric instillation in 0.5% carboxymethylcellulose either three days per week for up to 12 weeks followed by [b]four weeks of recovery or [c]seven days per week for up to 12 weeks followed by 24 weeks of 500 ppm phenobarbital (PB) in the diet for 24 weeks. GST-P$^+$ = glutathione S-transferase placental-type. HAF = hepatocellular altered foci. ND = not done. RF = replicating fraction measured by proliferating cell nuclear antigen (PCNA).

Table 1.2.3 Liver DNA adducts, hepatocellular percent replicating fraction (RF) and GST-P$^+$ HAF per cm^2 of liver tissue of rats exposed to diethylnitrosamine (DEN).[a]

Cumulative doses (CD) at ten weeks (mg kg^{-1})		DNA adducts in 10^8 nts	RF	HAF per cm^2	Percentage of phenobarbital (PB) promotable (for 24 weeks) neoplasms	Reference
Control	0	3	4.6	0.1	0	23, 39, 41
CD1	25.5	14	4.7	3.7	0	23, 41
CD2	51.1	203	5.8	7.3	3	23, 39, 41
CD3	102.2	287	5.8	12.8	32	23, 39, 41
CD4	204.4	434	7.0	36.5	100	23, 30, 41
CD5	306.6	5040	11.6	69.4	100	23, 39

[a]DEN was administered by intragastric instillation in 0.5% carboxymethylcellulose once a week for ten weeks followed by 24 weeks of 500 ppm phenobarbital (PB) in the diet. GST-P$^+$ = glutathione *S*-transferase placental-type. HAF = hepatocellular altered foci. RF = measured by proliferating cell nuclear antigen (PCNA).

liquid chromatography (HPLC) analysis with fluorescence detection were 7-ethylguanine and O^6-ethylguanine—the latter being an accepted miscoding lesion. A NOAEL for DNA adduct formation was not found.

In these studies, DNA adducts were formed at CD that were below those that elicited measureable increases in either hepatocellular proliferation or HAF (Tables 1.2.2 and 1.2.3), as has been noted in other studies.[48,50,51] Thus, a NOAEL for adducts was not achieved. The adduct levels at NOAEL for other effects were at or above one in 10^9 nts.[43]

1.2.3 Conclusions

Most of the experiments reviewed in this chapter which monitored preneoplasia in the rat liver demonstrated NOAEL for critical effects of the DNA-reactive carcinogens studied. Importantly NOAEL were found for induction of HAF, which are prerequisite to the eventual development of liver neoplasms.[37,56] Quantification of HAF provides more robust data than does measurement of tumors because, with hepatocarcinogen dosing, the numbers of HAF per liver greatly exceed the numbers of tumors.[57] We conclude that NOAEL for HAF can be considered NOAELs for liver tumor development. Moreover, NOAELs for induced cell proliferation, a response to hepatocellular injury and an enhancing factor in hepatocarcinogenesis,[58–60] were demonstrated.

These NOAELs for HAF and increased proliferation were found at CD that still produced DNA adducts, some of which are potentially miscoding (*i.e.* dG-N^2-AAF and O^6-ethylguanine), while others are not involved in base pairing. This indicates that adducts are a more sensitive biomarker of exposure and that there is a level of DNA adduct formation which appears not to be of biological significance. We conclude this to be at about one in 10^9 nts, which represents only about three adducts per cell, or about one adduct per 7000 genes.[43] Since only 1–2% of the genome is functionally active,[61] most adducts would be in regions of DNA not coding for proteins and not all adducts are miscoding.

Moreover, this level of DNA modification is small compared with the level of endogenous DNA lesions per cell, estimated to be in the order of 10^4 to 10^6 per cell,[62–64] much of which is oxidative damage.

Cancer is considered to be a multi-hit and multi-step process involving changes in the structure or function of oncogenes and tumor suppression genes, leading to initiation of tumor development, which can be enhanced by promotion. Such changes require, in general, substantial and sustained exposures, although there are examples where a single large dose can be tumorigenic, especially for carcinogens that are not effectively detoxicated.[65–67] The possibility of multiple effective hits in critical genes at extremely low levels of DNA adduct formation is highly unlikely, although in the whole body 6×10^{13} cells (Table 1.2.4) are at risk for such rare events. In spite of this, most cells are not actively replicating and hence are not susceptible to mutagenesis.

Thus, the totality of the biological evidence supports the possibility of NOAEL and practical thresholds for DNA-reactive carcinogens, similar to epigenetic carcinogens, but likely at lower CD, because of their MoA. The possibility of a hormetic effect at low doses of hepatocarcinogen, as reported in the study by Hino and Kitagawa,[44] further supports the possibility of thresholds.

1.2.4 Human Relevance

Translation of the implications from this review to human relevance involves several considerations. In order to extrapolate meaningfully from rats to humans, differences in adult liver anatomy and physiology should be considered. For example, the rat liver accounts for 4.8% of body weight, whereas the human liver is only 2.8%.[60] Moreover, the absence of a gall bladder in the rat influences the intrahepatobiliary adaptive homeostasis by enlargement of the periportal biliary plexus.[60] In addition, there are known to be important differences in xenobiotic metabolism between rat and human livers. For example, tamoxifen, which is a potent rat hepatocarcinogen,[68] has not been associated with liver cancer in humans,[69] apparently owing to insufficient hepatic activation in humans by sulfotransferase.[70]

The epidemiology of chemical-induced cancer in humans establishes the need for sustained exposure, except for *in utero* exposure to diethylstilbesterol (DES).[71] All chemicals judged by the International Agency for Research on Cancer (IARC) to be human carcinogens based on sufficient epidemiological evidence are characterized by repeat exposure to toxicologically significant (>1.3 mg day^{-1}) levels (see below),[72] apart from the aforementioned DES. Thus, for the many DNA-reactive carcinogens for which there is human exposure, there are apparently conditions of intake that pose no significant (or measurable) cancer risk, although this 'virtually safe dose' can vary over several orders of magnitude, depending on how it is calculated.[73]

To estimate intakes that do not convey a significant cancer risk, a procedure for calculating a 'toxicologically insignificant daily intake' (TIDI)[12] is given in Table 1.2.5. For a DNA-reactive carcinogen, the molecular effect for which a

Table 1.2.4 Estimated minimum dose of a chemical carcinogen needed to produce a neoplasm in humans.

Line	Value	Calculation[a]	Value identification	Comments
1	6×10^{13}		Number of cells in the body	
2	3.2×10^9		Bases in the human genome	
3	0.025		Liver weight as fraction of body weight	For a 70 kg person (average of male and female)
4	6×10^{23}		Avogadro's number	
5	0.006		Stem (replicating) cells as fraction of liver cell population	Less than 1%
6	2.5	$\underline{1} \times \underline{3} \times 10^{12}/\underline{4}$	pmole carcinogen to distribute one molecule per liver cell, based on Avogadro's number	Assumes all activation occurs on first liver passage
7	0.75	$\underline{6} \times 300/1000$	ng for 1 molecule/cell for a carcinogen of 300 MW	Assume average MW for carcinogens of 300
8	3.75	$\underline{7} \times 100/20$	ng if 20% is bioactivated to a DNA-reactive metabolite	For some carcinogens *all* of the metabolism may be through a reactive intermediate while for others <1% may be bioactivated
9	7.5	$\underline{8} \times 100/50$	ng if 50% of bioactivated carcinogen binds to DNA	Only a very small % will actually bind to DNA, usually <10%, assume 50%
10	4		Genes needing to be mutated to induce a human neoplasm	Neoplasia is a multistep process
11	1–3		Sites at which a mutation can activate an oncogene	*ras*, for example, has 3 possible sites
12	Many		Sites at which a mutation can inactivate a tumor suppressor gene	Not all mutations will inactivate: some codon changes will not change amino acid sequence, and some amino acid changes will not affect function
13	96	$\underline{9} \times \underline{2} \times \underline{10} \times 10^9$	g of carcinogen required over a lifetime to hit four critical bases per cell in liver	Assume only one site per gene capable of activating an oncogene or inactivation of a tumor suppressor gene
14	2.8	$\underline{1} \times \underline{3} \times \underline{5}/2$	More stem (labile) cells in the liver than bases in the genome	Multiple cells will have equivalent damage

Table 1.2.4 (*Continued*)

Line	Value	Calculation[a]	Value identification	Comments
15	34	13/14	g to hit the required four bases in one stem (labile) cell in the liver over a lifetime (70 years) exposure	Underestimated based on DNA repair; over estimated based on clonal expansion
16	1.3	$15 \times 10^6/(70 \times 365.25)$	mg per day	~ 30 μg per day limit set for AFB_1, a potent human liver carcinogen and provides a safety margin of ~ 45
17	3.2	$2/10^9$	1 DNA adduct in 10^9 bases represents only about 3 adducts per cell in the body affecting only 3 of the $< 30,000$ genes estimated to be present per human cell.	
18	$< 30,000$		Endogenous DNA lesions arising and repaired in a diploid mammalian cell in 24 hours	
19	1	$2 \times 10^5/18$	Spontaneous DNA modifications formed and repaired in 24 hours per 10^5 normal nucleotides	

[a]Underline indicates reference to the line number, plain text indicates numerical value; modified from ref. 43.

Table 1.2.5   Toxicologically insignificant daily intake (TIDI) for carcinogens.[a]

- No-adverse-effect-level (NOAEL) for molecular/cellular effect that is the basis for carcinogenicity.
- Safety margin (SM)
    - Multiple of uncertainty factors (UF)[b]
        - 10 for species to species extrapolation
        - 10 for individual variation
- TIDI is NOAEL divided by SM

[a]Based on ref. 12.
[b]UF for short-term to long-term intake not needed if molecular/cellular effect has reached steady state.

NOAEL would be used to calculate the TIDI is DNA binding of less than one in 10^9 nts,[12] which was concluded above to be biologically insignificant. Importantly, this assessment applies only to exposures to a single carcinogen. It is well-established experimentally that combinations of DNA-reactive carcinogens given at effective levels can produce additive or synergistic effects.[74,75] Additionally, Schmähl[74] has described the phenomenon of 'syn-carcinogenesis' resulting from carcinogens given in sequence. For example, sequential administration of AAF and DEN, in either order, produced a syn-carcinogenic effect in rat liver carcinogenicity.[76,77] These types of interactions indicate that overall cancer risk assessment in humans is more complex than simple modeling of dose-effect data from experimental animals administered single compounds. Whether such interactions would occur at TIDIs is unknown, but has been reported for teratogenicity.[78]

Non-carcinogenic compounds can also enhance the effects of carcinogens. As an example of co-carcinogenicity, benzo[e]pyrene (B[e]P), pyrene and fluoranthene increase the carcinogenicity of benzo[a]pyrene (B[a]P) on mouse skin while not being carcinogenic themselves.[79] The B[a]P/B[e]P result was confirmed by DiGiovanni *et al.*,[80] who also noted that in contrast B[e]P inhibited the tumor initiating activities of 7,12-dimethylbenz[a]anthracene (DMBA) and other polycyclic aromatic hydrocarbons.[81] In the few studies undertaken, these results correlate well with the levels of DNA adducts formed.[81,82] In addition, experimentally, tumor promotion by a compound administered after a carcinogen is well recognized,[75,79] and this can enhance the carcinogenicity of initiating agents (Table 1.2.1).

Human exposures to trace levels of organic DNA-reactive (genotoxic) carcinogens occur through a variety of sources. Several types of DNA-reactive carcinogens are food-borne,[83] including mycotoxins[52,84] and the heterocyclic amines formed during cooking of food.[52] Drinking water can contain minute amounts of reactive chlorination byproducts.[85] Environmental exposures also occur through other natural and industrial sources.[86] Most such exposures are less than lifetime.[71,87] We have calculated that lifetime intake of about 1.3 mg day^{-1} of a DNA-reactive carcinogen is unlikely to pose a human cancer risk (Table 1.2.4, line 16), although these calculations involve several assumptions.[43] For a specific carcinogen, the calculation of a TIDI based on adducts would likely yield a much lower value for a safe intake. The research reviewed here supports the conclusion that very low levels of intake of individual DNA-reactive carcinogens are unlikely to pose significant (>1 in 10^6) cancer risk, even assuming lifetime exposure, and thus may represent practical thresholds. However, concurrent intakes of several carcinogens may occur, for example with cigarette smoking,[79] and this must be considered in overall risk assessment.

References

1. N. Mantel, The concept of threshold in carcinogenesis, *Clin. Pharm. Ther.*, 1963, **4**, 104–109.
2. C. C. Brown, Mathematical aspects of dose-response studies in carcinogenesis - the concept of thresholds, *Oncology*, 1976, **33**, 62–65.
3. F. W. Carlborg, The threshold and the virtually safe dose, *Food Chem. Toxicol.*, 1982, **20**, 219–221.
4. W. K. Lutz and A. Kopp-Schneider, Threshold dose response for tumor induction by genotoxic carcinogens modeled *via* cell-cycle delay, *Toxicol. Sci.*, 1999, **49**, 110–115.
5. W. Slob, Thresholds in toxicology and risk assessment, *Int. J. Toxicol.*, 1999, **18**, 259–268.

6. J. H. Weisburger and G. M. Williams, Types and amounts of carcinogens as potential human cancer hazards, *Cell Biol. Toxicol.*, 1989, **5**, 377–391.
7. G. M. Williams, Mechanistic considerations in cancer risk assessment, *Inhalation Toxicol.*, 1999, **11**, 549–554.
8. G. M. Williams, Mechanisms of chemical carcinogenesis and application to human cancer risk assessment, *Toxicology*, 2001, **166**, 3–10.
9. C. Sonich-Mullin, R. Fielder, J. Wiltse, K. Baetke, J. Dempsey, P. Fenner-Crisp, D. Grant, M Hartley, A. Knaap, D. Kroese, I. Mangelsdorf, E. Meek, J. M. Rice and M. Younes, IPCS conceptual framework for evaluating a mode of action for chemical carcininogenesis, *Regul. Toxicol. Pharmacol.*, 2001, **34**, 146–152.
10. M. E. Meek, J. R. Bucher, S. M. Cohen, V. DeMarco, R. N. Hill, L. D. Lehman–McKeeman, D. G. Longfellow, T. Pastoor, J. Seed and D. E. Patton, A framework for human relevance analysis of information on carcinogenic modes of action, *Crit. Rev. Toxicol.*, 2003, **33**, 591–654.
11. S. Barlow, A. G. Renwick, J. Kleiner, J. W. Bridges, L. Busk, E. Dybing, L. Edler, G. Eisenbrand, J. Fink-Gemmels, A. Knaap, R. Kroes, D. Liem, D. J. G. Müller, S. Page, V. Rolland, J. Schlatter, A. Tritscher, W. Tueting, and G. Würzen, Introduction, in *Risk assessment of Substances that are both Genotoxic and Carcinogenic*, EFSA/WHO International Conference with support of ILSI Europe, Brussels, 2005, pp. 1–286.
12. G. M. Williams, Application of mode-of-action considerations in human cancer risk assessment, *Toxicol. Lett.*, 2008, **180**, 75–80.
13. J. H. Weisburger and G. M. Williams, Classification of carcinogens as genotoxic and epigenetic as basis for improved toxicologic bioassay methods, in *Environmental Mutagens and Carcinogens*, ed. T. Sugimura, S. Kondo and H. Takebe, University of Tokyo Press, Tokyo/Alan R. Liss, New York, 1982, pp. 283–294.
14. G. M. Williams, DNA reactive and epigenetic carcinogens, *Exp. Toxicol. Pathol.*, 1992, **44**, 457–464.
15. G. M. Williams, E. Karbe, P. Fenner-Crisp, M. J. Iatropoulos and J. H. Weisburger, Risk assessment of carcinogens in food with special consideration of non-genotoxic carcinogens: Scientific arguments for use of risk assessment and for changing the Delaney Clause specifically, *Exp. Toxicol. Pathol.*, 1996, **48**, 209–215.
16. E. Dybing, J. Doe, J. Groten, J. Kleiner, J. O'Brien, A. G. Renwick, J. Schlatter, P. Steinberg, A. Tritscher, R. Walker and M. Younes, Hazard characterisation of chemicals in food and diet: dose response, mechanism and extrapolation issues, *Food Chem. Toxicol.*, 2002, **40**, 237–282.
17. E. R. Nestmann, D. W. Bryant, C. J. Carr, T. T. Fennell, J. E. Gallagher, N. J. Gorelick, J. A. Swenberg and G. M. Williams, Toxicological significance of DNA adducts: Summary of discussion with an expert panel, *Regul. Toxicol. Pharmacol.*, 1996, **24**, 9–18.
18. R. J. Preston and G. M. Williams, DNA-reactive carcinogens: mode of action and human cancer hazard, *Crit. Rev. Toxicol.*, 2005, **35**, 673–683.
19. A. M. Jarabek, L. H. Pottenger, L. S. Andrews, D. Casciano, M. R. Embry, J. H. Kim, R. J. Preston, M. V. Reddy, R. Schoeny, D. Shuker, J.

Skare, J. Swenberg, G. M. Williams and E. Zeiger, Creating context for the use of DNA adduct data in cancer risk assessment: I. Data organization, *Crit. Rev. Toxicol.*, 2009, **39**, 659–678.

20. L. H. Pottenger, N. Carmichael, M. Banton, P. Boogaard, J. Kim, D. Kirkland, R. D. Phillips, J. van Benthem, G. M. Williams and A. Castrovinci, ECETOC workshop on the biological significance of DNA adducts: Follow-up from an expert panel meeting, *Mutat. Res.*, 2009, **678**, 152–157.

21. M. Kirsch-Volders, M. Aardema and A. Elhajouji, Concepts of threshold in mutagenesis and carcinogenesis, *Mutat. Res.*, 2000, **464**, 3–11.

22. J. A. Swenberg, D. K. La, N. A. Scheller and K. Y. Wu, Dose–response relationships for carcinogens, *Toxicol. Lett.*, 1995, **82–83**, 751–756.

23. F. Oesch, M. E. Herrero, J. G. Hengstler, M. Lohmann and M. Arand, Metabolic detoxification: Implications for thresholds, *Toxicol. Pathol.*, 2000, **28**, 382–387.

24. G. M. Williams, M. J. Iatropoulos and A. M. Jeffrey, Mechanistic basis for nonlinearities and thresholds in rat liver carcinogenesis by the DNA-reactive carcinogens 2-acetylaminofluorene and diethylnitrosamine, *Toxicol. Pathol.*, 2000, **28**, 388–395.

25. J. H. Hengstler, M. S. Bogdanffy, H. M. Bolt and F. Oesch, Challenging dogma: Thresholds for genotoxic carcinogens? The case of vinyl acetate, *Annu. Rev. Pharmacol. Toxicol.*, 2003, **43**, 485–520.

26. J. H. Farmer, R. L. Kodell, D. L. Greenman and G. W. Shaw, Dose and time responses models for the incidence of bladder and liver neoplasms in mice fed 2-acetylamino-fluorene continuously, *J. Environ. Pathol. Toxicol.*, 1980, **3**, 55–68.

27. A. Maekawa, T. Ogiu, C. Matsuoka, H. Onodera, K. Furuta, Y. Kurokawa, M. Takahashi, T. Kokubo, H. Tanigawa and Y. Hayashi, Carcinogenicity of low doses of *N*-ethyl-*N*-nitrosourea in F344 rats: A dose–response study, *GANN*, 1984, **75**, 117–125.

28. W. Lijinsky, R. M. Kovatch, C. W. Riggs and P. T. Walters, Dose–response study with *N*-nitrosomorpholine in drinking water of F-344 rats, *Cancer Res.*, 1988, **48**, 2089–2095.

29. A. Maekawa, H. Onodera, Y. Matsushima, T. Nagaoka, A. Todate, M. Shibutani, Y. Kodama and Y. Hayashi, Dose-response carcinogenicity in rats on low-dose levels of *N*-ethyl-*N*-nitrosourethane, *Jpn. J. Cancer Res.*, 1989, **80**, 632–636.

30. R. Peto, R. Gray, P. Brantom and P. Grasso, Effects on 4080 rats of chronic ingestion of *N*-nitrosodiethylamine or *N*-nitrosodimethylamine: A detailed dose–response study, *Cancer Res.*, 1991, **51**, 6415–6451.

31. R. Peto, R. Gray, P. Brantom and P. Grasso, Dose and time relationships for tumor induction in the liver and esophagus of 4080 inbred rats by chronic ingestion of *N*-nitrosodiethylamine or *N*-nitrosodimethylamine, *Cancer Res.*, 1991, **51**, 6452–6469.

32. G. S. Bailey, A. P. Reddy, C. B. Pereira, V. Harttig, W. Baird, J. M. Spitzbergen, J. D. Hendrinks, G. A. Orner, D. E. Williams and J. A. Swenberg, Nonlinear cancer response of ultralow dose: A 40,800-animal ED_{001} tumor and biomarker study, *Chem. Res. Toxicol.*, 2009, **22**, 1264–1276.

33. W. J. Waddell, Thresholds in chemical carcinogenesis: what are animal experiments telling us? *Toxicol. Pathol.*, 2003, **31**, 260–262.
34. G. M. Williams, Functional markers and growth behavior of preneoplastic hepatocytes, *Cancer Res.*, 1976, **36**, 2540–2543.
35. P. Bannasch, Preneoplastic lesions as end points in carcinogenicity testing. I. Hepatic preneoplasia, *Carcinogenesis*, 1986, **7**, 689–695.
36. G. M. Williams, Chemically induced rodent preneoplastic lesions as indicators of carcinogenic activity, in *The Use of Short- and Medium-term Tests for Carcinogens and Data on Genetic Effects in Carcinogenic Hazard Evaluation*, ed. D. B. McGregor, J. M. Rice and S. Venitt, IARC Scientific Publications No. 146, International Agency for Research on Cancer, Lyon, 1999, pp. 185–202.
37. G. M. Williams, The significance of chemically-induced hepatocellular altered foci in rat liver and application to carcinogen detection, *Toxicol. Pathol.*, 1989, **17**, 663–674.
38. H. Tsuda, S. Fukushima, H. Wanibuchi, K. Morimura, D. Nakae, K. Imaida, M. Tatematsu, M. Hirose, K. Wakabayashi and M. A. Moore, Value of GST-P positive preneoplastic hepatic foci in dose-response studies of hepatocarcinogenesis: evidence for practical thresholds with both genotoxic and nongenotoxic carcinogens. A review of recent work, *Toxicol. Pathol.*, 2003, **31**, 80–86.
39. G. M. Williams, M. J. Iatropoulos, C. X. Wang, N. Ali, A. Rivenson, L. A. Peterson, C. Schulz and R. Gebhardt, Diethylnitrosamine exposure-responses for DNA damage, centrilobular cytotoxicity, cell proliferation and carcinogenesis in rat liver exhibit some non-linearities, *Carcinogenesis*, 1996, **17**, 2253–2258.
40. G. M. Williams, M. J. Iatropoulos, C. X. Wang, A. M. Jeffrey, S. Thompson, B. Pittman, M. Palasch and R. Gebhardt, Non-linearities in 2-acetylaminofluorene exposure-responses for genotoxic and epigenetic effects leading to initiation of carcinogenesis in rat liver, *Toxicol. Sci.*, 1998, **45**, 152–161.
41. G. M. Williams, M. J. Iatropoulos, A. M. Jeffrey, F. Q. Luo, C. X. Wang, S. Thompson and B. Pittman, Diethylnitrosamine exposure-responses for DNA ethylation, hepatocellular proliferation and initiation of carcinogenesis in rat liver display non-linearities and thresholds, *Arch. Toxicol.*, 1999, **73**, 394–402.
42. G. M. Williams, M. J. Iatropoulos and A. M. Jeffrey, Thresholds for the effects of 2-acetylaminofluorene in rat liver, *Toxicol. Pathol.*, 2004, **32**(Suppl. 2), 85–91.
43. G. M. Williams, M. J. Iatropoulos and A. M. Jeffrey, Thresholds for DNA-reactive (genotoxic) organic carcinogens, *J. Toxicol. Pathol.*, 2005, **18**, 69–77.
44. O. Hino and T. Kitagawa, Existence of a practical threshold dose for the hepatocarcinogen 3'-methyl-4-(dimethylamino) azobenzene in rat liver, *GANN*, 1981, **72**, 637–638.
45. E. J. Calabrese, Hormesis is central to toxicology, pharmacology and risk assessment, *Hum. Exper. Toxicol.*, 2010, **29**, 249–261.
46. K. K. Rozman and J. Doull, Scientific foundations of hormesis. Part 2. Maturation, strengths, limitations, and possible applications in

toxicology, pharmacology, and epidemiology, *Crit. Rev. Toxicol.*, 2003, **33**, 451–462.

47. K. K. Rozman, Hormesis and risk assessment, *Human Exp. Toxicol.*, 2005, **24**, 255–257.

48. M. Root, T. Lange and T. C. Campbell, Dissimilarity in aflatoxin dose–response relationships between DNA adduct formation and development of preneoplastic foci in rat liver, *Chem. Biol. Interact.*, 1997, **106**, 213–227.

49. H. Enzmann, H. Zerban, A. Kopp-Schneider, E. Loser and P. Bannach, Effects of low doses of *N*-nitrosomorpholine on the development of early stages of hepatocarcinogenesis, *Carcinogenesis*, 1995, **16**, 1513–1518.

50. S. Fukushima, Low-dose carcinogenicity of a heterocyclic amine, 2-amino-3,8-dimethylimidazo[4,5-*f*]quinoxaline: Relevance to risk assessment, *Cancer Lett.*, 1999, **143**, 157–159.

51. S. Fukushima, H. Wanibuchi, K. Morimura, M. Wei, D. Nakae, Y. Konishi, H. Tsuda, N. Uehara, K. Imaida, T. Shirai, M. Tatematsu, T. Tsukamoto, M. Hirose, F. Furukawa, K. Wakabayashi and Y. Totsuka, Lack of a dose-response relationship for carcinogenicity in the rat liver with low doses of 2-amino-3,8-dimethylimidazo[4,5-*f*]quinoxaline or *N*-nitrosodiethylamine, *Jpn. J. Cancer Res.*, 2002, **93**, 1076–1082.

52. International Agency for Research on Cancer (IARC), *Some Naturally Occurring Substances: Food Items and Constituents, Heterocyclic Aromatic Amines and Mycotoxins*, IARC Monographs on the Evaluation of Carcinogenic Risks to Humans, Vol. 56, IARC, Lyon, 1993.

53. T. Murai, S. Mori, J. S. Kang, K. Morimura, H. Wanibuchi, Y. Totsuka and S. Fukushima, Evidence of a threshold-effect for 2-amino-3,8-dimethylimidazo[4,5-*f*]quinoxaline liver carcinogenicity in F344/DuCrj rats, *Toxicol Pathol.*, 2008, **36**, 472–477.

54. R. J. Laib, T. Pellio, V. M. Wünschel, N. Zimmermann and H. M. Bolt, The rat liver foci bioassay: II. Investigations on the dose-dependent induction of ATPase-deficient foci by vinyl chloride at very low doses, *Carcinogenesis*, 1985, **6**, 69–72.

55. G. M. Williams, R. Gebhardt, H. Sirma and F. Stenbäck, Non-linearity of neoplastic conversion induced in rat liver by low exposures to diethylnitrosamine, *Carcinogenesis*, 1993, **14**, 2149–2156.

56. G. M. Williams, Chemicals with carcinogenic activity in rodent liver, in *Comprehensive Toxicology*, Academic Press, Oxford, 2010, Vol. 9, pp. 221–250.

57. G. M. Williams and K. Watanabe, Quantitative kinetics of development of N-2-fluroenylacetanide-induced altered (hyperplastic) hepatocellular foci resistant to iron accumulation and of their reversion or persistence following removal of carcinogen, *J. Natl. Cancer Inst.*, 1978, **61**, 113–121.

58. D. J. Loury, T. L. Goldsworthy and B. E. Butterworth, The value of measuring cell replication as a predictive index of tissue-specific tumorigenic potential, in *Nongenotoxic Mechanisms in Carcinogenesis*, ed. B. E. Butterworth and T. J. Slaga, The Banbury Report, Cold Spring Harbor Laboratory, NY, 1987, pp. 119–136.

59. S. M. Cohen and L. B. Ellwein, Genetic errors, cell proliferation, and carcinogenesis, *Cancer Res.*, 1991, **51**, 6493–6505.

60. G. M. Williams and M. J. Iatropoulos, Alteration of liver cell function and proliferation: differentiation between adaptation and toxicity, *Toxicol. Pathol.*, 2002, **30**, 41–53.
61. C. Biémont, A brief history of the status of transposable elements: From junk DNA to major players in evolution, *Genetics*, 2010, **186**, 1085–1093.
62. T. Lindahl, Instability and decay of the primary structure of DNA, *Nature*, 1993, **362**, 709–715.
63. L. J. Marnett and P. C. Burcham, Endogenous DNA adducts: Potential and paradox, *Chem. Res. Toxicol.*, 1993, **6**, 771–785.
64. A. C. Povey, DNA adducts: endogenous and induced, *Toxicol. Pathol.*, 2000, **28**, 405–414.
65. H. Druckrey, D. Steinhoff, R. Preussmann and S. Ivankovic, Induction of cancer by a single dose of methylnitrosourea and various dialkylnitrosamines in rats, *Z. Krebsforsch.*, 1964, **66**, 1–10.
66. H. C. Pitot, L. Barsness, T. Goldworthy and T. Kitagawa, Biochemical characterization of stages of hepatocarcinogenesis after a single dose of diethylnitrosamine, *Nature*, 1978, **271**, 456–458.
67. T. Tanaka, H. Mori, N. Hirota, K. Furuya and G. M. Williams, Effect of DNA synthesis on induction of preneoplastic and neoplastic lesions in rat liver by a single dose of methoxyazoxymethanol acetate, *Chem.-Biol. Interact.*, 1986, **58**, 13–27.
68. G. M. Williams, M. J. Iatropoulos, M. V. Djordjevic and O. P. Kaltenberg, The triphenylethylene drug, tamoxifen is a strong liver carcinogen in the rat, *Carcinogenesis*, 1993, **14**, 315–317.
69. International Agency for Research on Cancer (IARC), *Evaluation of Carcinogenic Risks to Humans: Some Pharmaceutical Drugs, Tamoxifen*, IARC Monographs on the Evaluation of Carcinogenic Risks to Humans, Vol. 66, IARC, Lyon, 1996, pp. 367–388.
70. H. Glatt, W. Davis, W. Meinl, H. Hermersdörfer, S. Venitt and D. H. Phillips, Rat, but not human, sulfotransferase activates a tamoxifen metabolite to produce DNA adducts and gene mutations in bacteria and mammalian cells in culture, *Carcinogenesis*, 1998, **19**, 1709–1713.
71. G. M. Williams, B. Reiss and J. H. Weisburger, A comparison of the animal and human carcinogenicity of several environmental, occupational and therapeutic chemicals, in *Mechanisms and Toxicology of Chemical Carcinogens and Mutagens*, ed. G. Flamm and R. Lorentzen, Princeton Scientific Publishers, Princeton, NJ, 1985, Vol. IX, pp. 207–248.
72. International Agency for Research on Cancer (IARC), Agents classified by the IARC Monographs, Volumes 1-102, IARC, Lyon, France, 2011. http://monographs.iarc.fr/ENG/Classification/ClassificationsGroupOrder.pdf [accessed January 2012].
73. D. E. Williams, G. Orner, K. D. Willard, S. Tilton, J. D. Hendricks, C. Pereira, A. D. Benninghoff and G. S. Bailey, Rainbow trout (oncorhynchus mykiss) and ultra-low dose cancer studies, *Comp. Biochem. Physiol. Toxicol. Pharmacol.*, 2009, **149**, 175–181.
74. D. Schmähl, Combination effects in chemical carcinogenesis, *Arch. Toxicol. Suppl.*, 1980, **4**, 29–40.

75. G. M. Williams, Interactive carcinogenesis in the liver, in *Falk Symposium: Liver Cell Carcinoma*, ed. P. Bannasch, D. Keppler and G. Weber, Academic Press, Boston, 1989, Vol. 51, pp. 197–216.
76. G. M. Williams, S. Katayama and T. Ohmori, Enhancement of hepatocarcinogenesis by sequential administration of chemicals: summation *versus* promotion effects, *Carcinogenesis*, 1981, **2**, 1111–1117.
77. G. M. Williams and K. Furuya, Distinction between liver neoplasm promoting and syncarcinogenic effects demonstrated by exposure to phenobarbital or diethylnitrosamine either before or after *N*-2-fluorenylacetamide, *Carcinogenesis*, 1984, **5**, 171–174.
78. K. Mayura, P. Parker, W. O. Berndt and T. D. Philips, Effect of simultaneous prenatal exposure to ochratoxin A and citrinin in the rat, *J. Toxicol. Environ. Health*, 1984, **13**, 553–561.
79. B. L. Van Duren and B. M. Goldschmidt, Carcinogenic and tumor-promoting agents in tobacco carcinogenesis, *J. Natl. Cancer Inst.*, 1976, **56**, 1237–1242.
80. J. DiGiovanni, J. Rymer, T. J. Slaga and R. K. Boutwell, Anticarcinogenis and cocarcinogenic effects of benzo[e]pyrene and dibenz[a,c]anthracene on skin tumor initiation by polycyclic hydrocarbons, *Carcinogenesis*, 1982, **3**, 371–375.
81. T. A. Smolarek, W. M. Baird, E. P. Fisher and J. DiGiovanni, Benzo(e)-pyrene-induced alterations in the binding of benzo(a)pyrene and 7,12-dimethylbenz(a)anthracene to DNA in Sencar mouse epidermis, *Cancer Res.*, 1987, **47**, 3701–3706.
82. J. E. Rice, T. J. Hosted and E. J. Lavoie, Fluoranthene and pyrene enhance benzo[a]pyrene-DNA adduct formation *in vivo* in mouse skin, *Cancer Lett.*, 1984, **24**, 327–333.
83. G. M. Williams, Food-borne carcinogens, *Prog. Clin. Biol. Res.*, 1986, **206**, 73–81.
84. International Agency for Research on Cancer (IARC), *Some Traditional Herbal Medicines, some Mycotoxins, Naphthalene and Styrene*, IARC Monographs on the Evaluation of Carcinogenic Risks to Humans, Vol. 82, IARC, Lyon, 2002, pp. 169–275.
85. M. Koivusalo and T. Vartiainen, Drinking water chlorination by-products and cancer, *Rev. Environ. Health.*, 1997, **12**, 81–90.
86. J. M. M. Meijers, G. M. H. Swaen and L. J. N. Bloemen, The predictive value of animal data in human cancer risk assessment, *Regul. Toxicol. Pharmacol.*, 1997, **25**, 94–102.
87. S. Felter, R. Conolly, J. Bercu, P. Bolger, A. Boobis, P. Bos, P. Carthew, N. Doerrer, J. Goodman, W. Harrouk, D. Kirkland, S. Lau, G. Llewellyn, R. Preston, R. Schoeny, A. Schnatter, A. Tritscher, F. van Velsen and G. M. Williams, A proposed framework for assessing risk from less-than lifetime exposures to carcinogens, *Crit. Rev. Toxicol.*, 2011, **41**, 507–544.
88. G. M. Williams, T. Tanaka, H. Maruyama, Y. Maeura, J. H. Weisburger and E. Zang, Modulation by butylated hydroxytoluene of liver and bladder carcinogenesis induced by chronic low level exposure to 2-acetylamino-fluorene, *Cancer Res.*, 1991, **51**, 6224–6230.

CHAPTER 1.3

DNA Alkylation and Repair After EMS Exposure: Where Do the Thresholds for Mutagenic/ Clastogenic Effects Arise?

ELMAR GOCKE,* THOMAS SINGER AND
LUTZ MÜLLER

Preclinical Research, F. Hoffmann-La Roche Ltd, Basel, Switzerland
*Email: elmar.gocke@roche.com

1.3.1 Introduction

Mutations and/or clastogenic effects generally arise from exposure to a genotoxic (DNA-damaging) agent. Rarely, does the mutagenic/clastogenic alteration occur directly (*i.e.*, by mispairing); more often, it is related to erroneous (error-prone) or unsuccessful/incomplete repair of the induced DNA lesion. Traditionally, the dose-response relation for mutagenic effects has been thought to have no threshold; that is, even the most miniscule exposure carries a finite (non-zero) chance for eliciting the effect. Recently, several publications have provided convincing experimental evidence that simple alkylating agents, such as methylmethane sulfonate (MMS) and ethylmethane sulfonate (EMS), cause mutational events with a thresholded dose-response relationship.[1,2]

Issues in Toxicology No. 13
The Cellular Response to the Genotoxic Insult: The Question of Threshold for
Genotoxic Carcinogens
Edited by Helmut Greim and Richard J. Albertini
© The Royal Society of Chemistry 2012
Published by the Royal Society of Chemistry, www.rsc.org

When considering possible mechanisms for a thresholded dose-response relationship, two distinctive possibilities, both involving saturation processes, come to mind:

1) a thresholded dose response can arise if a scavenging/detoxification mechanism fully prevents accessibility of the genotoxin to the DNA (and thus averts induction of damages) up to a certain dose level;
2) a thresholded dose response can arise if a repair mechanism is capable of removing all DNA damages practically error free up to a certain damage level—even if damages are induced with clear dose proportionality.

Here we provide evidence that the latter mechanism is responsible for thresholds in the dose response of the induction of micronuclei and gene mutations in the bone marrow of mice treated daily for 7 or 28 days with EMS.

Our studies on EMS were triggered by the need to assess the possible mutational risk of patients accidentally exposed to EMS after ingesting contaminated Viracept (Nelfinavir mesylate) tablets. In early 2007, the impurity had formed over a period of time in a storage tank for methanesulfonic acid after the cleaning fluid (ethanol) was not properly removed prior to refilling.[3] The contaminant remained during the synthesis process and in the finished product at levels reaching approximately 1000 ppm. Retrospective analysis showed that tablets produced prior to the incident had contained EMS at levels several orders of magnitude lower, verifying that only three months of Viracept supply were contaminated.

EMS has long been known for researchers as a standard alkylating agent and is used for its properties as a positive control in many *in vitro* and *in vivo* genetic toxicity assays. Yet, existing *in vitro* and *in vivo* data did not permit a sufficiently convincing risk assessment for the levels at which HIV patients were exposed to EMS at a maximal daily dose of 0.055 mg per kg body weight. Therefore, the marketing authorisation holder of Viracept, F. Hoffmann-La Roche, had to conduct a series of non-clinical investigations to better quantify the risk for adverse effects in the exposed individuals.[2,4] For direct acting, DNA-damaging genotoxins, such as EMS, it has been generally assumed that dose-response relations are linear – or at least non-thresholded – due to their stochastic, all-or-nothing, mode of action. By default, the formation of tumours, heritable birth defects, and teratogenic effects are also accepted to follow a linear dose relation for these genotoxins. However, the publication by Doak *et al.*[1] provided strong evidence for a thresholded dose-response relationship for *in vitro* induction of micronuclei and gene mutations by EMS. Based on this publication, we sought to obtain reassuring evidence for the absence of (geno)toxicological adverse effects in the Viracept patients.

A set of studies was conducted in consultation with the Committee for Medicinal Products for Human Use (CHMP), part of the European Medicines Agency (EMA), the relevant regulatory authority in control of the marketing authorisation of Viracept.[5] In extension of the *in vitro* studies by Doak *et al.*,[1] clastogenic and gene mutation inducing capacities of EMS were evaluated with multiple dosing regimens and complemented by pharmacokinetic investigations and modelling to assess the patient exposures. The full set of non-clinical data was then integrated into a comprehensive risk assessment evaluation.[4] This assessment was discussed with eminent experts in the preclinical areas of genotoxicity/carcinogenicity and metabolism/distribution, as well as with eminent experts on viral infections and clinical care of HIV patients.[6] The studies provided strong evidence for a thresholded dose response regarding both clastogenic and mutagenic activities of EMS. The threshold dose was observed to be 25 mg/kg body weight/day, orders of magnitude higher than the maximal patient dose. This observation also held true when the assessment was based on drug metabolism and pharmacokinetics (DMPK) parameters (AUC, C_{max}). On this basis, it was concluded that the Viracept patients did not experience any additional toxicological adverse effects, which would be attributable to their exposure to EMS.[4] This argument was accepted by the authorities and the initially issued request for registries for birth defects and/or tumorigenic effects in putatively affected patients was retracted.

In the present publication, we focus on the dose-response curves of protein and DNA adduct formation in comparison to the dose-response curves of the clastogenic and mutagenic effects in the bone marrow of mice exposed to EMS.

1.3.2   Material and Methods

The experimental details of the MNT and Muta™Mouse studies is described in Gocke *et al.*[2] In short: The genotoxicity tests were performed under GLP at Covance, Harrogate, UK. Formulation analytics of EMS was conducted at F. Hoffmann-La Roche Ltd, Basel (for MNT test) and RCC Ltd, Itingen, Switzerland (for Muta™Mouse test). Bioanalytics (determination of ethylvaline adduct levels) was performed at Currenta GmbH & Co, Leverkusen, Germany.

1.3.2.1   Test Item

EMS [CAS 62-50-0] and ENU [CAS 62-50-0] were purchased from Sigma-Aldrich, UK. Batches of EMS had purities of 99.4 and 100.4%. Aqueous solutions of the appropriate concentrations were administered by oral gavage.

1.3.2.2   Micronucleus (MNT) Study

The protocol of the MNT was designed to meet the known requirements of the OECD Guideline 474, 1997 and the ICH Tripartite Harmonised Guideline on Genotoxicity: Specific Aspects of Regulatory Tests, 1995. Out-bred Crl:CD-1 (ICR) mice were obtained from Charles River (UK) Ltd, Margate, UK. Animals were treated daily for 7 days (0, 1.25, 2.5, 5, 20, 80, 140, 200, 260 mg/kg/day) and necropsied on day 8. Four thousand PCE per animal were scored for MN (MN-PCE) from 6 animals per dose point.

1.3.2.3 Muta™Mouse Study

The Muta™Mouse study was designed to comply with the recommendations made at the IWGT meeting in 2002.[7] CD$_2$-lacZ80/HazfBR mice were obtained from Harlan, UK. The study consisted of two arms: a 28-day, daily treatment scheme (EMS: 0, 1.56, 3.13, 6.25, 12.5, 25, 50, 100 mg/kg/day, sampling on day 31) and a single treatment scheme (EMS: 0, 350 mg/kg, sampling on day 7 or day 31). Where possible, data were generated for at least 200,000 pfu per tissue for all 7 animals per group.

1.3.2.4 Protein Adduct Determination

Blood was obtained from each animal during necropsy by cardiac puncture and processed for analysis of ethylvaline levels. The determinations were made for all animals treated with EMS.

1.3.2.5 Statistical Assessment

The statistical assessment of the genotoxicity data is described in detail in Gocke and Wall (2009). The threshold software developed by Lutz and Lutz[8] was employed to calculate the threshold dose values including confidence limits.

1.3.3 Results

In the course of our studies[2] on the genotoxic activity of EMS, we investigated the clastogenic action in bone marrow of mice and the mutagenic action in bone marrow, liver and large intestine.

Figure 1.3.1 presents the data on micronuclei induction in the bone marrow of CD1 mice and lacZ mutation induction in the bone marrow of Muta™Mouse mice, and, for both studies, the induction of globin adducts in peripheral blood. The effects are plotted relative to their corresponding control values. It is apparent that low-dose levels do not result in increases of the genotoxic effects but do substantially increase the adduct levels. Indeed, for the clastogenicity endpoint, even a negative (hormesis-like) trend was apparent, which was assessed as being statistically significant.[9] Above the NOEL doses of 25 mg/kg/day for lacZ induction and 80 mg/kg/day for MN induction, clear increases of the genotoxic effects are observed, reaching a factor of 8.7 (MN) or 4.0 (lacZ) above control values. Using the threshold software developed by Lutz and Lutz,[8] we obtained estimates of the threshold values and confidence intervals as shown in Table 1.3.1.

Regarding the adduct levels, Figure 1.3.1 presents our data on the ethyl-adduct levels in terminal valine of globin, which we determined as a measure of exposure. It is evident that no threshold for adduct formation can be derived from the data. In fact, an almost 10 fold increase over background is apparent already at the lowest dose of about 1 mg/kg/day. At the threshold doses for

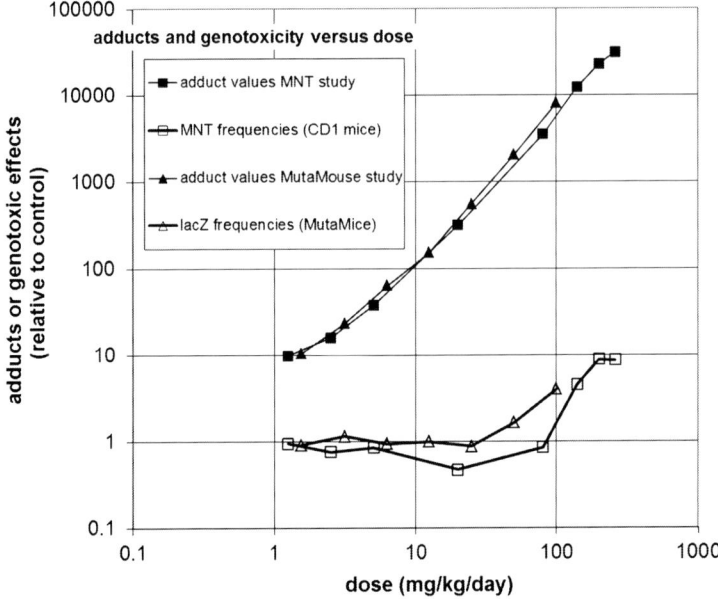

Figure 1.3.1   Induction of protein adducts in globin, micronuclei and lacZ mutations in bone marrow. All values plotted relative to the respective control values.

Table 1.3.1   Threshold analysis.[8,9] (Gocke and Wall 2009, Lutz and Lutz 2009)

Study	Organ	NOEL (mg/kg)	Threshold (td) (mg/kg)	95% Confidence interval of td (mg/kg)
MNT	Bone marrow	80	89.8	56.7 to 118.2
Muta™mouse	Bone marrow	25	35.4	21.5 to 45.7

genotoxicity, the ethylvaline levels surpass the background values by roughly a factor of 1000.

Figure 1.3.2 shows the relative MN and lacZ frequencies (*i.e.*, the biomarkers of effect) plotted against the respective ethylvaline adduct levels (*i.e.*, against the biomarker of exposure) for each animal. Increases above the background level of clastogenic and mutagenic effects are evident only at ≥ 100 nmol ethylvaline per gram of globin compared to a background level of < 0.1 nmol/g.

While it cannot be surmised that the dose relationship of DNA-adduct levels is similar to the dose-response relationship of globin-adduct levels, the investigation of Murthy *et al.*[10] showed a high degree of parallelity for ethylation of both targets as depicted in Figure 1.3.3. Murthy *et al.*[10] used radioactive EMS, enabling them to obtain reliable data down to dose levels below 0.1 mg/kg of EMS because background ethylation does not interfere with the measurements. Figure 1.3.3 shows, at doses below 10 mg/kg, they observed a roughly linear

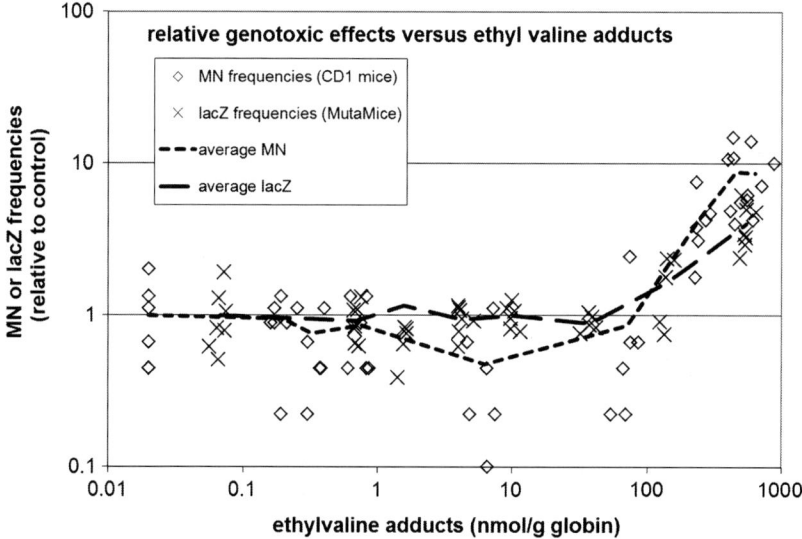

Figure 1.3.2 Scatter plot of individual MN and lacZ frequencies (relative to control mean) plotted against ethylvaline levels of the respective individual. *Note*: The lines connect the means of the relative frequencies of the dose groups plotted against the means of adduct levels of the respective dose groups.

dose relation (*i.e.*, slopes in the log-log plot of about 1), and, at doses above 10 mg/kg, they observed a nearly quadratic relation (*i.e.*, slope close to 2 in log-log plot). When calculating the daily increments of globin- adduct levels in our 7- or 28-day studies, our values compare well with the data reported by Murthy *et al.*[10]

1.3.4 Discussion

EMS has been investigated in numerous *in vitro* and *in vivo* studies for the induction of gene mutations and clastogenic events. *In vivo* rodent studies employing doses of ≥ 50 mg/kg/day were not amenable to the elucidation of dose-response relationships at lower doses, which was of particular relevance for assessing the genotoxic risk of HIV patients exposed to EMS-tainted Viracept. Graphical assessment of these studies, as presented in Gocke *et al.*,[12] suggested sub-linear dose relations for almost all investigations. However, definite conclusions from these studies could not be drawn because of experimental restrictions (*e.g.*, lack of doses below the LOEL, too few animals in controls and treatment groups, and too few cells analysed per animal).

Our studies were designed to give a reliable experimental and statistical basis for determination of a threshold dose, should it exist *in vivo*. Thus, we employed 7 to 8 closely spaced dose levels starting as low as 1.25 mg/kg/day, included several (up to 4) independent control groups, increased the number of

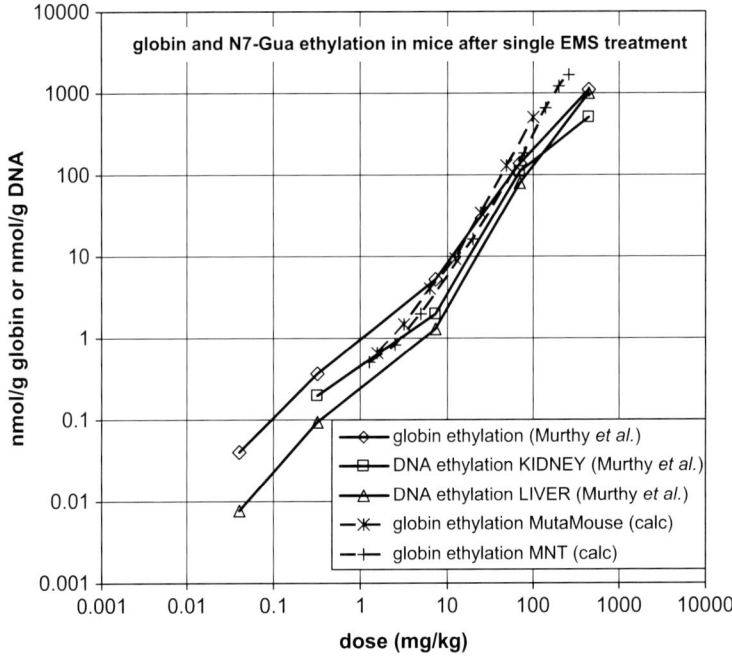

Figure 1.3.3   Plot of adduct levels in DNA and globin after single doses of EMS.
Note: Data are taken from the study by Murthy *et al.*[10] with radioactive
EMS or are calculated from our MN and MutaTMMouse studies using
the formula for calculating the daily increment[11] and the relative reactivity with ethylvaline versus total globin determined by Murthy *et al.*[10]
in whole blood.

animals/dose group and the number of evaluated cells/animal above the levels
recommended in the respective guidelines. This approach provided 5 to 6 dose
levels that were identified to be below the LOEL dose.

Fixation of mutations likely depends on the status of proliferation and differentiation of the exposed cells. Studies with short-term treatments indicated
that the sensitivity of stems cells might be higher or lower than the sensitivity of
transit or differentiated cells, depending on the organ investigated.[13] In our
Muta™Mouse study, we included three organs (liver, intestine, bone marrow)
with different turnover rates, and the treatment period of the study was chosen
to be 28 days, so that the induction of lacZ mutations was measured in
differentiated, transit and stem cells in the different organs. We also included
a single, acute treatment regiment to show that dose fractionation abolishes
the mutagenic effect when the daily dose is below the threshold, while the
acute dose, equal to the total cumulative dose, induces a clear increase.[2]

We believe our data provide unambiguous and reliable support for a
thresholded dose-response relationship that, when applied to human risk
assessment, indicates that there is no increase of the genotoxic effect (clastogenic or mutagenic) above the background level at doses below the threshold.

Based on our pharmacokinetic assessments,[14,15] we were able to estimate the AUC of EMS in mice. For the NOEL dose of 25 mg/kg, an AUC of 350 µM h was calculated. Doak *et al.*[1] determined the lowest effective concentration for EMS to be 1.4 µg/ml in their *in vitro* experiments, and the threshold concentrations for gene mutations and chromosomal aberrations were both between 1.2 to 1.3 µg/ml. In their experiments, the cells were treated for 24 h. For buffer and serum, we have determined[14] the half-life of EMS to be about 11 h. On this basis, we calculate the AUC of EMS for the *in vitro* exposure conditions employed by Doak *et al.*[1] to amount to 147 µM h, in good agreement with the *in vivo* AUC at the threshold dose determined in our study.

While the dose response relations for genotoxic effects were identified to follow a 'hockey-stick' fashion, our data on globin ethylation, as well as the data of Murthy *et al.*[10] on the alkylation of N7-G ethylation, provide convincing evidence that induction of alkylation damage after EMS exposure does not occur in a threshold manner down to dose levels of 0.1 mg/kg. This implies there is no efficient scavenging of EMS prior to reaction with DNA, and, as a consequence, the appearance of thresholds in the dose-response curve for clastogenic and mutagenic effects has to be attributed to error-free removal of DNA damage. In a calculation described in Gocke and Müller,[16] we have derived that in liver cells treated at the NOEL dose of lacZ mutation induction (50 mg/kg/day) that each daily EMS dose induced 380 000 DNA alkylations without any measurable increase in mutation frequency. For bone marrow cells, the calculation yields a total of 78 000 adducts at the NOEL of 25 mg/kg/day, assuming a similar adduct induction in bone marrow DNA as in liver DNA. The difference between the two organs might be due to the different cell turnover (*i.e.*, the liver cells have more time to repair).

In the *in vitro* study by Doak *et al.*,[1] adduct levels were not determined. However, Swenberg *et al.*[17] did measure DNA adduct levels after exposure to MMS and a more or less linear dose response was found. At the threshold concentration of 1 mg/ml, about 350 N7-G adducts per 10^8 nucleotides were observed, corresponding to about 50 000 adducts per diploid human cell. These values show that cells can remove vast numbers of exogenously induced simple alkylations without any measurable consequence to the integrity of their genome.

The absence of thresholds in the dose response relations for ENU and MNU reported by Doak *et al.*,[1] as well as the seemingly linear dose responses for ENU identified in our *in vivo* studies,[2] do not fit well with the fact that these two alkylators induce the same type of damage as EMS and MMS (*i.e.*, ethylation or methylation of DNA) and, thus, should principally show comparable dose response relations. Reassessment of the *in vitro* curves for ENU by Johnson *et al.*[18] suggested that the dose response for HPRT mutations could contain a threshold at 0.32 µg/ml. Based on the analysis of the ethylvaline adduct rates in our studies[2] and the data published by Beranek,[19] we have estimated that an *in vivo* threshold for induction of lacZ mutations in bone marrow could be around 0.4 mg/kg of ENU, which would be compatible with the observed mutation frequencies.[16] More recently, ENU was tested in the pigA mutation assay with rats[20] and a thresholded dose response with a NOEL of approximately 0.88 mg/kg was identified. Thus, we conclude that thresholded dose response relationships also occur following treatment with ENU, albeit at a much lower doses than for EMS.

It is well known[19] that different alkylators target DNA and proteins at various sites, depending on the so called Swain Scott constant (s value). Ethylation/methylation by EMS/MMS occurs predominantly at nitrogen sites (N^7-G in DNA, cystein nitrogens and terminal nitrogens in proteins), while oxygen sites are targeted to a larger extend by ENU/MNU (O^6-G, O^2-T). For removal of the different DNA lesions, various repair mechanisms are available. It has been discussed that clastogenic effects arise from N-ethylations, while gene mutations arise from O-ethylations.[1,21] Beranek[19] reported the relative ratio of N- to O-adducts to be >100 for MMS, about 30 for EMS and about 1 for ENU (absolute values were not reported). If there was a strict channelling of lesion repair, one would expect the thresholds for either endpoint to vary independently from each other according to these relations. In this context, it is notable that Doak *et al.*[1] observed almost the same threshold concentrations for induction of micronuclei versus induction of HPRT mutations with EMS (1.3 µg/ml versus 1.2 µg/ml). In our study, we also found rather similar *in vivo* thresholds (89.8 versus 35.4 mg/kg/day) for EMS in bone marrow. The estimated threshold for gene mutations by ENU is in the range of 0.88 and 0.4 mg/kg for pigA and lacZ mutations. Since the ratio of N- to O-ethylations is about 30 fold lower for ENU than for EMS, one would expect a threshold for clastogenicity of ENU in the order of 10 mg/kg/day. Our data[2] on the clastogenicity of ENU, do not fit to this prediction as already at a dose of 4.5 mg/kg/day the frequency of micronuclei is increased threefold. One likely explanation is that the different repair mechanisms are interrelated to such an extent that the thresholds for either effect (clastogenic or mutagenic) are not independent from one another.

1.3.5   Conclusion

Emerging evidence from studies using alkylating agents, such as EMS and ENU, provide confidence to the conclusion that mammalian cells can faithfully repair high levels of alkylation damage. With cells treated up to the threshold dose, the frequency of mutagenic/clastogenic changes remains at the level observed in untreated cells, and it is only at doses above the threshold that a notable increase of clastogenic and mutagenic effects becomes evident. Critically, the endpoint of DNA alkylation does not show a threshold dose response, indicating that this endpoint can be used to assess the extent of exposure ('biomarker of exposure'), but not the extent of heritable genotoxic consequences. As the induction of cancer can be seen as subsequent to the accumulation of repair errors, the data indicate that thresholds should exist for the tumorigenicity of simple alkylators. Simple alkylation might be considered as an ubiquitous and easily repairable DNA damage for which error-free repair processes have been evolved early on in the history of life. Other, more complex DNA damages might not be repairable without errors, even at very low doses. On this basis, a general change of the paradigm for risk assessment has to await the accumulation of further data on genotoxins inducing other types of DNA damages.

Interestingly, however, a recent study on tumour induction in trout[22] exposed to the genotoxin dibenzo(a,l)pyrene, which induces a host of bulky adducts, has yielded strong evidence for a definitely sub-linear dose response, despite a near linear induction of complex bulky DNA adducts down to very low doses. While a threshold for cancer induction could not be defined in this study, it is clearly evident that linear backextrapolation from TD10 (or TD50) doses to the 'virtual safe' dose (VSD, the dose inducing 1 cancer in a million individuals) yields a value that would be 'too low' by several orders of magnitude compared to the estimate by appropriate non-linear curve fitting.

References

1. S. H. Doak, G. J. Jenkins, G. E. Johnson, E. Quick, E. M. Parry and J. M. Parry, Mechanistic influences for mutation induction curves after exposure to DNA-reactive carcinogens, *Cancer Res.*, 2007, **15**, 3904–3911.
2. E. Gocke, M. Ballantyne, J. Whitwell and L. Müller, MNT and MutaTMmouse studies to define the in vivo dose response relations of the genotoxicity of EMS and ENU, *Toxicol. Lett.*, 2009, **190**, 286–297.
3. C. Gerber and H.-G. Toelle, What happened: The chemistry side of the incident with EMS contamination of Viracept tablets, *Toxicol. Lett.*, 2009, **190**, 248–253.
4. L. Müller, E. Gocke, T. Lavé and T. Pfister, Ethyl methanesulfonate toxicity in Viracept – a comprehensive human risk assessment based on threshold data for genotoxicity, *Toxicol. Lett.*, 2009, **190**, 317–329.
5. L. Müller and T. Singer, EMS in Viracept – a compilation of events in 2007 and 2008 from the non-clinical point of view, *Toxicol. Lett.*, 2009, **190**, 243–247.
6. V. E Walker, D. A. Casciano and D. J. Tweats, The Viracept-EMS case: impact and outlook, *Toxicol. Lett.*, 2009, **190**, 333–339.
7. V. Thybaud, S. Dean, T. Nohmi, J. de Boer, G. R. Douglas, B. W. Glickman, N. J. Gorelick, J. A. Heddle, R. H. Heflich, I. Lambert, H. J. Martus, J. C. Mirsalis, T. Suzuki and N. Yajima, In vivo transgenic mutation assays, *Mutat. Res.*, 2003, **540**, 141–151.
8. R. W. Lutz and W. K. Lutz, Statistical model to estimate a threshold dose and its confidence limitsfor the analysis of a sublinear dose response relationship, exemplified for mutagenicity data, *Mutat. Res.*, 2009, **678**, 118–122.
9. E. Gocke and M. Wall, Statistical assessments for threshold dose response in mice for the induction of micronuclei and lacZ gene mutation by ethyl methanesulfonate, *Toxicol. Lett.*, 2009, **190**, 298–302.
10. M. S. Murthy, C. J. Calleman, S. Osterman-Golkar, D. Segerbäck and K. Svensson, Relationships between ethylation of hemoglobin, ethylation of DNA and administered amount of ethyl methanesulfonate in the mouse, *Mutat. Res.*, 1984, **127**, 1–8.

11. L. Recio, S. Osterman-Golkar, G. A. Csanády, M. J. Turner, B. Myhr, O. Moss and J. A. Bond, Determination of mutagenicity in tissues of transgenic mice following exposure to 1,3-butadiene and N-ethyl-N-nitrosourea, *Toxicol. Appl. Pharmacol.*, 1992, **117**, 58–64.

12. E. Gocke, L. Müller, T. Pfister and H. Buergin, Literature review on the genotoxicity, reproductive toxicity, and carcinogenicity of ethyl methanesulfonate, *Toxicol. Lett.*, 2009, **190**, 254–265.

13. J. A. Heddle, H. J. Martus and G. R. Douglas, Treatment and sampling protocols for transgenic mutation assays, *Environ. Mol. Mutagen.*, 2003, **41**, 1–6.

14. T. Lave, A. Paehler, H. P. Grimm and H. Birnboeck, Pharmacokinetic profile of ethyl methanesulfonate in vitro and in animals, including globin ethylation, *Toxicol. Lett.*, 2009, **190**, 298–302.

15. T. Lavé, A. Paehler, H. P. Grimm and H. Birnboeck, Pharmacokinetic modeling of species differences for prediction in humans, *Toxicol. Lett.*, 2009, **190**, 303–309.

16. E. Gocke and L. Müller, In vivo studies in the mouse to define a threshold fort he genotoxicity of EMS and ENU, *Mutat. Res.*, 2009, **678**, 101–107.

17. J. A. Swenberg, E. Fryar-Tita, Y. C. Jeong, G. Boysen, T. Starr, V. E. Walker and R. J. Albertini, Biomarkers in toxicology and risk assessment: informing critical dose-response relationships, *Chem. Res. Toxicol.*, 2008, **21**, 253–265.

18. G. E. Johnson, S. H. Doak, S. M. Griffiths, E. L. Quick, D. O. Skibinski, Z. M. Zaïr and G. J. Jenkins, Non-linear dose-response of DNA-reactive genotoxins: Recommendations for data analysis, *Mutat. Res.*, 2009, **678**, 95–100.

19. D. T. Beranek, Distribution of methyl and ethyl adducts following alkylation with monofunctional alkylating agents, *Mutat. Res.*, 1990, **231**, 11–30.

20. K. L. Dobo, R. D. Fiedler, W. C. Gunther, C. J. Thiffeault, Z. Cammerer, S. L. Coffing, T. Shutsky and M. Schuler, Defining EMS and ENU dose-response relationships using the Pig-a mutation assay in rats, *Mutat. Res.*, 2011, Jun, 24. [Epub ahead of print]

21. A. T. Natarajan, J. W. Simons, E. W. Vogel and A. A. van Zeeland, Relationship between cell killing, chromosomal aberrations, sister-chromatid exchanges and point mutations induced by monofunctional alkylating agents in Chinese hamster cells. A correlation with different ethylation products in DNA, *Mutat. Res.*, 1984, **128**, 31–40.

22. G. S. Bailey, A. P. Reddy, C. B. Pereira, U. Harttig, W. Baird, J. M. Spitsbergen, J. D. Hendricks, G. A. Orner, D. E. Williams and J. A. Swenberg, Nonlinear cancer response at ultralow dose: a 40800-animal ED(001) tumor and biomarker study, *Chem. Res. Toxicol.*, 2009, **22**, 1264–1276.

2. Metabolic Inactivation of Genotoxic Reactants

CHAPTER 2.1

Enzymic Detoxification of Endogenously Produced Mutagenic Carcinogens Maintaining Cellular Homeostasis

H. M. BOLT

Leibniz Research Centre for Working Environment and Human Factors (IfADo), Ardeystr. 67, D-44139 Dortmund, Germany
Email: bolt@ifado.de

2.1.1 Introduction

There is growing recognition that carcinogenic risk extrapolation to low doses, which is an essential step for setting standards for carcinogenic substances, must consider the mode of action of the carcinogen in question. A continuous scientific discussion on this matter has resulted in new perspectives for standard setting.[1–9] For some genotoxic carcinogens, the existence of 'practical' thresholds has been suggested, for which health-based exposure limits may be deduced.[4,8–10] The scientific support for a practical threshold may differ from compound to compound.[10,11] One aspect is the existence of endogenously produced mutagenic/genotoxic carcinogens. For such compounds, maintaining

Issues in Toxicology No. 13
The Cellular Response to the Genotoxic Insult: The Question of Threshold for Genotoxic Carcinogens
Edited by Helmut Greim and Richard J. Albertini
© The Royal Society of Chemistry 2012
Published by the Royal Society of Chemistry, www.rsc.org

a physiological cellular homeostasis between formation and detoxification is pivotal.

In a key publication for the German MAK Commission in 1998, Neumann *et al.*[1] proposed the derivation of health-based Occupational Exposure Limits (OELs) for a group of genotoxic carcinogens with a very and practically negligible low cancer risk at the respective OEL. Such a categorization must be supported by data on the mode of action, dose–effect dependence and toxicokinetics. As of 2010, the German MAK Commission has listed four compounds in this 'category 5' of carcinogens: acetaldehyde, ethanol, isoprene (2-methyl-1,3-butadiene) and styrene.[12] Whereas the key argument for the inclusion of styrene in this group was a very effective detoxification of its reactive metabolite styrene oxide, a main argument for the other compounds (acetaldehyde, ethanol, isoprene) was that these were at the same time physiological compounds produced within the body. As long as the physiological homeostasis would be effective, no additional cancer risk should be expected if such compounds were exogenously applied.

2.1.2   Discussion of Individual Compounds

The current state of the discussion on endogenously produced genotoxic carcinogens with a physiological homeostasis is presented below. This includes the specific chemicals listed above.

2.1.2.1   Ethylene Oxide

The relevant endpoint for limiting human exposures to ethylene oxide is its carcinogenicity. However, a unique feature for ethylene oxide is that low levels of this chemical are produced endogenously by both human and animal organisms, and that ethylene oxide represents therefore a physiological body constituent.[13] The metabolisc precursor, ethylene, is produced by oxidative attack of biological molecules (lipids and proteins)[14] and is transformed to ethylene oxide.[13] This endogenous production induced the idea of basing regulatory considerations for ethylene oxide on the physiological quantities of ethylene and ethylene oxide in humans.[15]

Ethylene oxide is a weak alkylating agent that is directly mutagenic and carcinogenic. After external exposure, it is distributed within the entire organism and the quantities of DNA akylation are relatively uniform in various tissues. The basic metabolism of ethylene oxide in humans and in experimental animals is shown in Figure 2.1.1.

The carcinogenicity of exogenously administered ethylene oxide is evident from animal experiments. In rats (Fischer 344), the compound has induced brain tumours, mononuclear cell leukaemias and peritoneal mesotheliomas, and in mice (B6C3F1) lung adenomas and carcinomas. In long-term carcinogenicity experiments, significant carcinogenic effects were noted upon repeated daily inhalation of 33 parts per million (ppm) and 100 ppm ethylene oxide.

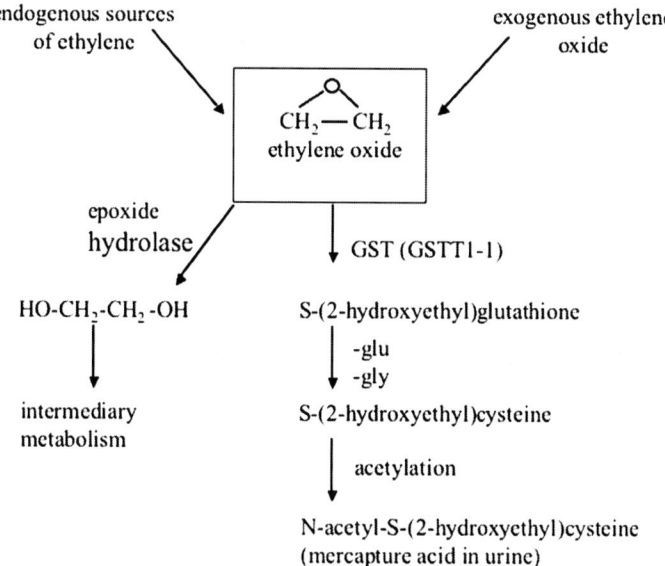

Figure 2.1.1 Metabolic pathways of ethylene oxide (adapted from ref. 13).

The carcinogenicity of ethylene oxide is reasonably connected with DNA alkylation and genotoxicity (see IARC monographs[16,17]). In 2008, IARC[17] confirmed that ethylene oxide is carcinogenic for humans ('group 1')—rating the epidemiological evidence by itself as being limited, but considering further elements of mechanism/mode of action and 'other relevant data'.

Peritoneal mesothelimas, quantitatively a major malignancy observed after ethylene oxide dosing in rats, have no apparent counterpart in humans and there are no epidemiological indications whatsoever of such a target site for ethylene oxide-induced carcinogenesis in humans. However, indications of haematopoietic/lymphatic cancer in humans are to be taken seriously, as pointed out by IARC.[17] No consistent evidence in humans could be found for brain, lung and mammary tumours. The available data on human haematopoietic/lymphatic cancer and leukaemia were starting points of quantitative cancer risk assessments by Kirman *et al.*[18] and by Valdez-Flores *et al.*[19]

2.1.2.1.1 Genotoxicity of Ethylene Oxide

In *Salmonella typhimurium* TA1535 and TA100, ethylene oxide was found to be mutagenic without metabolic activation. Ethylene oxide has mutagenic activity in *Drosophila melanogaster*. Ethylene oxide induces chromosomal aberrations and sister chromatid exchange in mammalian cells, and the substance yields positive results in the micronucleus test and the dominant lethal test. The genotoxic effects of ethylene oxide have been summarized in detail by IARC.[16,17] Ethylene oxide is also effective at inducing genetic damage in germ cells in whole mammals.[20]

Many human studies have been carried out to evaluate the effect of exposure to ethylene oxide on the incidences of chromosomal aberrations (including micronuclei) and sister chromatid exchange in peripheral blood lymphocytes of workers exposed occupationally to ethylene oxide. These studies include workers at hospital and factory sterilization units and those working at ethylene oxide manufacturing and processing plants. The results, as summarized by IARC,[16,17] show that ethylene oxide induces chromosomal damage in exposed humans. In general, the degree of damage is correlated with the level and duration of exposure. The induction of sister chromatid exchange appears to be more sensitive to exposure to ethylene oxide than is the formation of adducts, chromosomal aberrations or micronuclei. Alkali-labile sites and DNA single-strand breaks were not observed in lymphocytes of sterilization workers, but the induction of DNA crosslinking was reported in one study.

Taking these studies together,[16,17] it is clear that occupational ethylene oxide exposure can lead to genotoxic damage in exposed humans. Chromosomal aberrations could be established at exposure levels of 5 ppm and above. At exposures of 1 ppm, no clear cytogenetic changes have so far been established.

2.1.2.1.2 *Individual Differences in Ethylene Oxide Metabolism*

The detection of individual differences in the susceptibility of humans has promoted research into possible genetic factors that have an imprint on the metabolism of ethylene oxide. Fuchs *et al.*[21] distinguished a 'higher sensitive group' as opposed to a 'less sensitive group' when studying single-strand breaks in blood cell DNA using the 'alkaline elution' method of people occupationally exposed to ethylene oxide. Such individual differences in susceptibility to the genotoxicity of ethylene oxide are likely to be linked to differences in ethylene oxide metabolism.

Hallier *et al.*[22] demonstrated that peripheral lymphocytes of persons lacking the glutathione transferase human *(h)GSTT1* gene (homozygote *hGSTT1*0*, addressed as 'non-conjugators') were more susceptible to the sister chromatid exchange (SCE) inducing effect of ethylene oxide than lymphocytes of carriers of the intact *hGSTT1* gene ('conjugators'; Pemble *et al.*[23]). 'Conjugators' were again comprised of two subgroups—heterozygous 'slow conjugators' and homozygous 'high conjugators'.[24] The impact of the glutathione-dependent metabolic pathway for ethylene oxide in humans is visualized by urinary excretion of the respective mercapturic acid, *N*-acetyl-*S*-(2-hydroxyethyl)-L-cysteine,[25] and the background levels of the haemoglobin adduct 2-hydroxy-ethyl-valine.[26] Also, there are differences in the background levels of the hemoglobin adduct 2-hydroxyethyl-valine between GSTT1-positive and GSTT1-negative individuals not occupationally exposed to ethylene oxide.[26]

Thier *et al.*[27,28] studied the metabolism of ethylene oxide and its homolog, propylene oxide, in *S. typhimurium* TA1535 expressing or not expressing human or rat *GSTT1* (*hGSTT1*+/− or *rGSTT1*+/−, respectively); the GSTT1-dependence of metabolism and genotoxicity of both epoxides was clear cut.

Müller *et al.*[29] and Thier *et al.*[26] have studied the influence of the genetic status of the polymorphic glutathione transferases *hGSTT1* and *hGSTM1* on the background hydroxyethylation of *N*-terminal valine (HOEtVal) in haemoglobin. Both studies were in accordance with (endogenous) ethylene oxide being a substrate of hGSTT1 isoenzyme, and homozygote *hGSTT1*0* individuals displayed background HOEtVal levels that were about a third higher than those of people with the intact *hGSTT1* gene or enzyme. Another parameter that influences the background level of HOEtVal in haemoglobin is smoking status, with levels of smokers being about half higher than those of non-smokers.[26,30] In occupational field studies the background HOEtVal level (in non-smokers not exposed to ethylene) was about $20 \, \text{pmol} \, \text{g}^{-1}$ globin.[31]

2.1.2.1.3 DNA Adduct Formation from Ethylene Oxide

In experimental animals, repeated exposures to 10 ppm resulted in statistically elevated DNA adducts, the main adduct being N^7-(2-hydroxyhydroxyl)guanine (HOEtG). In earlier publications, the physiological background of HOEtG in rats was estimated to correspond to a repeated exogenous exposures at 1 ppm ethylene oxide, but during the last 20 years there have been gradual improvements of the methods of adduct detection and quantitation. The experimental study of Marsden *et al.*[32] using liquid chromatography (LC)–tandem mass spectrometry (MS/MS), arrived at the conclusion that, at a single or repeated dose (i.p.) to rats of $0.01 \, \text{mg} \, \text{kg}^{-1}$ ethylene oxide, any DNA adduct increase was negligible compared with the endogenous damage already present. In this case, the DNA damage did not accumulate with repeated ethylene oxide administration.

The formation and persistence of the DNA adduct N^7-(2-hydroxyethyl)guanine (HOEtG) have been studied for a number of years and the methodologies for determination of DNA adducts of ethylene oxide have been continuously refined.[15–17] Levels of HOEtG in target and in non-target tissues were found to be surprisingly similar[33–36] and it was concluded that the tissue/organ specificity for tumour induction by ethylene oxide should primarily be due to factors other than DNA-adduct formation.[33] The diverse picture of tumours induced by ethylene oxide in rodents cannot be interpreted only on toxicokinetic and metabolic grounds, and points to further tissue-specific factors. It is also in accordance with the assumption of a non-linearity in the exposure–carcinogenic response relationship.[18]

The DNA adduct HOEtG is rapidly repaired from the DNA and does therefore not accumulate upon subacute or subchronic exposure.[31,37] This adduct has served as a quantitative indicator of genotoxic damage by ethylene oxide *in vivo* because it is most abundant. Other DNA adducts are also formed, but quantitatively to minor extents.[13]

2.1.2.1.4   Quantitative Aspects of Endogenous DNA Adduct Levels

Taking all available evidence together, and including physiologically based pharmacokinetic (PBPK) modelling, Csanády *et al.*[38] calculated the range of endogenous formation of ethylene oxide in humans to be between 0.4 and 11 nmol h^{-1} per kg body weight (bw). According to Zhao and Hemminki,[39] the mean DNA adduct level in lymphocytes of (non-smoking) humans is 3.0–3.8 adducts per 10^7 nucleotides. In granulocytes of non-exposed persons (five had never smoked, values unadjusted), the mean ± standard error (SE) was 10.8 ± 7.0 HOEtG adducts per 10^7 nucleotides.[40] These values are basically consistent with data of others.[16,17] In rats (different organs examined), Marsden *et al.*[32] found a physiological background of 0.11–0.35 HOEtG adducts per 10^7 nucleotides in rat tissues, which is lower than that observed in humans.

In earlier publications it was noted that the observed background HOEtG levels in rats would correspond to exogenous airborne ethylene oxide exposures of about 1 ppm.[36,41] However, the analytical techniques for adduct detection have been gradually refined. Using the technically advanced LC-MS/MS technique for the quantitation of ethylene oxide-induced DNA adducts, Marsden *et al.*[32] established the above mentioned background adduct levels of HOEtG in tissues of non-exposed rats. The HOEtG adduct levels gradually increased with administration of a single or of three daily doses (i.p.) of 0.1 and 1.0 mg kg^{-1} bw ethylene oxide; at the lowest dose of 0.01 mg kg^{-1}, there was no difference of total HOEtG adduct levels to those detected in control animals. It was concluded that, at this lowest dose, any DNA adduct increase was negligible compared with the endogenous damage already present, and the DNA damage did not accumulate with repeated ethylene oxide administration.[32]

Using a similarly sensitive gas chromatography–electron capture–mass spectrometry (GC-EC-MS) technique,[42] Yong *et al.*[40] studied a group of sterilizer operators exposed to less that 1 ppm ethylene oxide and non-exposed controls from ten hospitals. From an earlier study[43] it was anticipated that the levels of ethylene-oxide induced haemoglobin adducts would be associated with ethylene oxide exposures in these groups. The study of Yong *et al.*[40] examined HOEtG adduct levels in granulocytes in relation to actual (eight-hour time-weighted average; TWA) exposures and to cumulative exposures (ppm-hour) during a four-month period before sample collection. There was considerable inter-individual variability in HOEtG adduct levels, with a range of 1.6–241.3 adducts per 10^7 nucleotides, the most prominent confounder being cigarette smoking. However, differences between the exposure groups (non-exposed; lower exposure: ≤ppm-h cumulative or 0.03 ± 0.05 ppm eight-hour TWA; higher exposure: >32 ppm-h cumulative or 0.36 ± 0.31 ppm eight-hour TWA) were not statistically significant. The unadjusted HOEtG levels per 10^7 nucleotides for those who had never smoked within these groups were as follows: for non-exposed (*n* = 5) 10.8 ± 7.0, for the lower exposure group (*n* = 21) 18.2 ± 6.6, for the higher exposure group (*n* = 8) 11.1 ± 4.4. Thus, it was concluded that human workplace exposures within the range that was studied (higher dose: 0.36 ± 0.31 ppm eight-hour TWA) did not lead to any significant elevation of the HOEtG adduct levels in granulocytes over the endogenous background.

On this basis, it was suggested by the EU Scientific Committee on Occupational Exposure Limits (SCOEL) (public consultation period in 2009–2010) that the genotoxic/carcinogenic risk of human occupational ethylene exposures at levels (eight-hour TWA) of 0.1 ppm is practically negligible. This view was backed by a quantitative risk assessment.[18]

2.1.2.1.5 Investigations unto the Human Cancer Risk Induced by Ethylene Oxide

At higher airborne concentrations of several ppm, ethylene oxide exposures have led to genotoxic damage in occupationally exposed humans. Cytogenetic signs of genotoxicity were clearly noted at exposure levels of 5 ppm and above. At exposures of 1 ppm, no genotoxic changes could be directly established in exposed humans.

In a British cancer mortality study, conducted on 2876 ethylene oxide exposed persons, Coggon *et al.*[43] found no significant associations of any tumour category with ethylene oxide exposures that had occurred in the UK in recent decades (at exposures < 5 ppm TWA).

Van Sittert *et al.*,[41] applying the 'radiation-dose equivalent' approach of Ehrenberg *et al.*,[44] concluded that exposure to 1 ppm ethylene oxide per year corresponded to a radiation dose of about 5 rad and that this low risk of occupational exposure would be consistent with the outcome of epidemiological investigations. Considering the genotoxic effects of ethylene oxide in animals and humans, the DNA binding of other related carcinogens, the natural DNA adduct background and the genotoxicity of low energy transfer radiation, they postulated that long-term occupational exposure to airborne ethylene oxide at or below 1 ppm did not 'produce an unacceptable increased risk in man'.

The IARC[16] assessment of a possible association between ethylene oxide exposure and lympho/hematopoietic cancer mortality was the starting point of a human risk assessment by Kirman *et al.*[18] Based on a meta-analysis by Teta *et al.*,[45] the authors selected two epidemiology studies of sufficient size and length of follow-up which were considered to provide adequate exposure information to quantify an exposure–response relationship for ethylene oxide. These were (i) the leukaemia/lymphoid tumour mortality data from a Union Carbide cohort[46] and (ii) the leukaemia/lymphoid tumour data from a US National Institute for Occupational Safety and Health study.[47] These data were used for a quantitative cancer potency estimate. It was concluded that the weight of evidence supported the use of a nonlinear assessment and a unit risk value of 4.5×10^{-8} was derived for ethylene oxide, with a range between 1.4×10^{-8} and 1.4×10^{-7} reflecting the given uncertainty.

In total, several lines of argument support the definition of exposure conditions under which a carcinogenic risk of ethylene oxide appears to be practically negligible. According to SCOEL (see above), these conditions correspond to an eight-hour TWA of 0.1 ppm (0.183 mg m^{-3}).

2.1.2.1.6   Conclusions on Ethylene Oxide

Ethylene oxide represents a paradigmatic case of a DNA-alkylating genotoxic carcinogen that is also produced endogenously. It is derived from endogenous ethylene and represents a physiological body constituent. In consequence, there is an endogenous background of ethylene oxide-induced DNA adducts. According to the available evidence, this background is not increased at low exogenous exposure levels to ethylene oxide, even at repeated dosing (*i.e.* airborne exposures to 0.1 ppm). This level of exposure to ethylene oxide can therefore be regarded as being within the boundaries of physiological homeostasis.

This view is supported by a quantitative risk assessment with the derivation of a 'unit risk' for ethylene oxide-induced haemato/lymphopoietic malignancies,[18] which indicates that an exposure to 0.1 ppm ethylene oxide for (working) lifetime would lead to a very low theoretical cancer risk, which in practice cannot be distinguished from zero.

2.1.2.2   Isoprene (2-methyl-1,3-butadiene)

Isoprene, in its biologically activated form of isopentenyl pyrophosphate, is an important physiological compound of intermediary metabolism in both plants and animals. It is the physiological C_5 monomer from which carotinoids and terpenes are formed in plants. In mammals and humans, it serves to build up cholesterin, gall acids, hormonal steroids and the side-chain of vitamin K compounds. There is a constant spontaneous formation of isoprene from its energetically activated form. Hence, its physiological origin is much different from that of ethylene/ethylene oxide. Among the hydrocarbons exhaled by animals and humans, isoprene is a quantitatively important single constituent and species differences have been noted.[48,49]

A first concern about possible carcinogenicity was raised by analogy to 1,3-butadiene, as it is metabolized by monooxygenase (*e.g.* CYP2E1) to monoepoxides and diepoxide.[50,51] Interestingly, there is some stereoselectivity between the optical isoprene isomers, as the C-2 atom of isoprene is a chiral centre.[52] Isoprene itself appears not to be mutagenic, but there is clear direct mutagenicity of its diepoxide, 2-methyl-2,2,3,4-diepoxybutane.[53] The metabolic pathways are shown in Figure 2.1.2.

2.1.2.2.1   Carcinogenicity of Isoprene

Isoprene has been categorized by IARC in Group 2B as a possible human carcinogen, based on positive animal bioassays conducted by the US National Toxicology Program (NTP).[54,55]

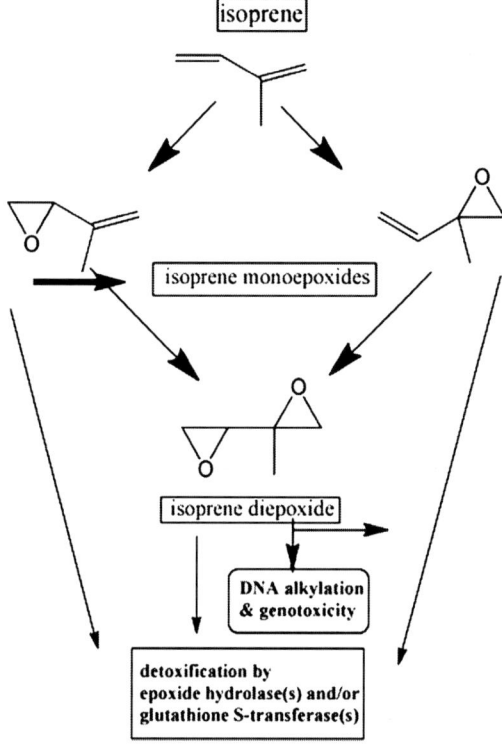

Figure 2.1.2 Metabolic pathways of isoprene.[50-52] Chiral centres are marked (*).

In long-term inhalation carcinogenicity studies in mice, isoprene led to increased incidences of ademomas of the Harderian gland and to pituitary tumours, starting from 70 ppm isoprene. Higher concentrations of isoprene led to hepatocellular adenomas and carcinomas, alveolar, bronchiolar adenomas and carcinomas, forestomach adenomas and carcinomas, haemangiosarcomas and histiocytic sarcomas.[56] In male rats, the lowest tested inhalation concentration of 220 ppm induced increased incidences of mammary fibroadenomas, adenomas of renal tubuli and interstitial cell adenomas of the testes. Female rats developed mammary fibroadenomas.[57] There are no data on the carcinogenicity of isoprene in humans.[58]

By analogy to 1,3-butadiene, it appears plausible that the di-epoxide of isoprene, 2-methyl-2,2,3,4-diepoxybutane, is the ultimate genotoxic metabolite that is responsible for the experimental tumour formation upon exposure to isoprene.[53,54]

2.1.2.2.2 Toxicokinetic Studies

The toxicokinetics of isoprene has been the subject of several reports.[49,59,60] In order to quantitate and explain obvious species differences, a physiological toxicokinetic model has been developed[59] for inhaled isoprene in mouse, rat and man. Partition coefficients (blood:air and tissue:blood) were determined *in vitro* by a headspace method. Parameters of a saturable isoprene metabolism in B6C3F1 mice, Sprague–Dawley rats and volunteers were obtained from gas uptake experiments in closed systems, analysed by means of a two-compartment model. Incorporation of these parameters into the model revealed that isoprene was metabolized in the liver and in extrahepatic organs. Endogenous production of isoprene in humans was quantified from experiments with volunteers breathing into a closed system. The model was validated across species for mice, rats and humans.[59]

In a second step, a physiologically based toxicokinetic model was developed for isoprene in mouse, rat and human. Metabolic parameters were obtained from gas uptake experiments. The measured data could be described by introducing hepatic and extrahepatic metabolism into the model. At exposure concentrations up to 50 ppm, the rate of metabolism at steady-state was 14 times faster in mice and about eight times faster in rats than in humans (2.5 micromol $h^{-1} kg^{-1}$ at 50 ppm isoprene in air). Isoprene accumulated only barely owing to its fast metabolism and a low thermodynamic partition coefficient (body:air). The endogenous production was low in rodents compared with humans. About 90% of isoprene produced endogenously in humans was metabolized and 10% was exhaled unchanged. The blood concentration of isoprene in non-exposed humans was predicted to be 9.5 nmol L^{-1}. The area under the blood concentration-time curve (AUC) following exposure over eight hours to 10 ppm isoprene was about four times higher than the AUC resulting from the unavoidable endogenous isoprene production over 24 h.[60] These data provided the quantitative basis for the derivation of a health-based exposure limit for isoprene.

2.1.2.2.3   Considerations of a Health-Based Occupational Exposure Limit

The above mentioned data were used to derive a health-based Occupational Exposure Limit (MAK value).[58] The formation rate of endogenous isoprene in humans was anticipated to be $13.1 \pm 10.0 \, \mu mol \, h^{-1}$, leading to an isoprene concentration in venous blood of $5.2 \pm 4.0 \, nmol \, L^{-1}$ and, within a lifetime (80 years), an AUC of $3.6 \pm 2.8 \, mmol \times h \, L^{-1}$. For an occupational exposure to 10 ppm airborne isoprene, the blood concentration at the end of an eight-hour shift was calculated to be 133 nmol L^{-1}, and the AUC of a 40-year working life exposure (eight hours per day, five days per week, 48 weeks per year) was calculated to be $9.84 \, mmol \times h \, L^{-1}$. An occupational exposure for working lifetime to 3 ppm isoprene would therefore result in an exogenous isoprene burden that is about equal to the standard deviation of the endogenous burden. With this reasoning, the German MAK Commission justified a health-based Occupational Exposure Limit (MAK) for isoprene of 3 ppm.

2.1.2.3 Products of Intermediary C_1 and C_2 Metabolism

The intermediary C_1 and C_2 metabolism involves some compounds that are also relevant in terms of occupational exposure and which have proven to be genotoxic and/or carcinogenic if exogenously administered. In these cases, the maintenance of the physiological homeostasis is key. This refers to formaldehyde (and formaldehyde generating precursors), acetaldehyde and ethanol.

2.1.2.3.1 *Formaldehyde*

The setting of occupational[61] and environmental[62] standards for formaldehyde has been the subject of recent debates.[63] For the derivation of safe exposure limits for formaldehyde, with regard to carcinogenic risk, the avoidance of cell proliferation is critical. Upon inhalation, the critical target for formaldehyde is the upper respiratory tract and the cause of cell proliferation appears closely related to the irritant effect of formaldehyde.[61,62] Formaldehyde is mutagenic *in vitro*, but has apparently no genotoxic effect under relevant *in vivo* conditions. In this respect the discussion of a balance of endogenous formaldehyde is important.

Formaldehyde, bound to tetrahydrofolic acid, is the active C_1-compound in intermediary metabolism that is required for the synthesis of amino acids and nucleic acid bases. It may be released from its activated form and is then rapidly inactivated. The molecule may re-enter the C_1 pool of cell metabolism and there is effective glutathione (GSH) dependent oxidation by formaldehyde dehydrogenase.[64,65] The concentration of endogenous formaldehyde in human blood is about $2–3\,\text{mg L}^{-1}$; similar concentrations are found in other species such as monkeys and rats. Exposure of humans, monkeys or rats to formaldehyde by inhalation has not been found to alter the concentration of endogenous formaldehyde in the blood. The average level of formate in the urine of persons not occupationally exposed to formaldehyde is $12.5\,\text{mg L}^{-1}$ and varies considerably both within and between individuals. No significant changes of urinary formate were detected after exposure to 0.4 ppm formaldehyde for up to three weeks in humans.[66] This points to the essential maintenance of cellular homeostasis by rapid detoxification of free formaldehyde.

Several mechanisms are involved in the local inactivation of formaldehyde. The inhaled hydrophilic gas dissolves first in the layer of mucus covering the nasal epithelium; reactions with components of the mucus and mechanical clearance of the mucus represent the first barrier.[67] From a certain exposure concentration, mucociliary clearance is impaired. In inhalation studies with rats exposed to 15 ppm, the mucociliary function in the frontal nasal region was inhibited and marked mucostasis was observed. After 6 ppm, only some areas were affected. After 2 ppm, minimal changes in the mucus flow rate were observed, whilst 0.5 ppm had no effect.[68] With sufficiently high exposure concentrations, a concentration gradient of free formaldehyde could be established within the layers of the nasal epithelium. Under these circumstances, in the fully differentiated cells near the surface, the actual concentration is higher than in the lower lying proliferating stem cells. In the rostral third of the respiratory epithelium, however, the epithelium consists of only two cell layers with few basal cells.[69] In the epithelial cells there are several ways inactivation can take place. Direct reactions with protein and RNA in the cytosol may remove a large amount of free formaldehyde.[70]

Such considerations of an endogenous homeostasis have been one important point in deriving health-based occupational and environmental exposure limits for formaldehyde.[61,62]

2.1.2.3.2   *Acetaldehyde and Ethanol*

Both acetaldehyde and ethanol, which are linked in an equilibrium by alcohol dehydrogenase, are physiological compounds. Acetaldehyde is genotoxic, mainly inducing DNA crosslinks and chromosomal aberrations.[71] Experimentally following inhalation it has produced carcinomas of the nasal mucosa in rats[71,72] and laryngeal carcinomas in hamsters.[72] Consequently, acetaldehyde has been categorized by IARC into its Group 2B based on sufficient evidence for carcinogenicity to animals.[72] The argumentation for ethanol is more complicated because of the consumption of alcoholic beverages, which have been categorized by IARC in Group 1 as being carcinogenic to humans.[73]

In the human organism, possible sources of acetaldehyde (and ethanol) are intestinal bacteria[74] and pathways of intermediary metabolism.[75] The physiological processes lead, in sober persons, to blood ethanol levels of $0.27 \pm 0.17 \, mg \, L^{-1}$.[74] The physiological blood levels of acetaldehyde are very similar, and recent published data range between 0.1 and $0.3 \, mg \, L^{-1}$.[76,77] The mean endogenous acetaldehyde production of a human adult has been calculated as $5.6 \, \mu mol$ (260 µg) per minute.[78]

Among others, these fundamentals were decisive for the German MAK Commission when it established Occupational Exposure Limits (MAK values) for acetetaldehyde and ethanol of 50 ppm and 500 ppm, respectively.[78,79]

A related case is vinyl acetate, also being an irritant and carcinogenic compound to the upper respiratory tract upon inhalation. This compound is locally split, at the site of first contact with the organism, by ubiquitous esterases into acetaldehyde and acetic acid. Considerations for setting a health-based Occupational Exposure Limit for this compound also included the maintenance of cellular homeostasis of acetaldehyde.[80]

2.1.3   Oxidative Stress Leading to Secondary Genotoxicity

In the discussion of possible thresholds for carcinogenic compounds the induction of oxidative stress plays an important role.[9–11] In general, oxidative stress (*i.e.* biological actions/consequences of reactive oxygen species, ROS) is viewed as an important source of indirect genotoxicity that can be triggered by exposure to exogenous primary factors such as ultraviolet (UV) light, ionizing radiation, anoxia or hyperoxia. Oxidative stress is also mediated by chemicals producing reactive oxygen species; paraquat and oxidants such as potassium bromate or hydrogen peroxide are classical examples in this respect. Other exogenous sources of ROS are tobacco smoke, fatty acids, transition metals, ethanol and redox cycling compounds. ROS interact with critical molecules within cells and with intracellular signalling, leading to cell death, mutagenesis and to toxicity associated with lipid peroxidation. Increased oxidative stress (excessive ROS production) also causes damage to DNA, modifying bases and altering DNA strands, and may contribute to cancer.[9,81]

It has been argued that ROS-mediated processes in carcinogenesis should have at least practical thresholds.[6,9,80] Hence, the homeostasis of oxidative stress and the reactive compounds involved are of key interest.

The endogenous oxidative stress-related oxidant that has been studied best is hydrogen peroxide. Hydrogen peroxide is again a physiological component of the body which originates from biochemical pathways involving molecular oxygen. The organism protects itself against the toxic effects of reactive oxygen species very efficiently by detoxifying enzymes like catalases, glutathione peroxidases and superoxide dismutase, and by antioxidants.[82] The specific adverse health effects of hydrogen peroxide in humans and animals have been compiled and assessed by the European Centre for Ecotoxicology and Toxicology of Chemicals (ECETOC)[83] and by the European Chemicals Bureau under the Existing Chemicals Directive.[84] Reference can be made to such existing documentation.

2.1.3.1 Physiological Homeostasis of Oxidative Stress as Indicated by Hydrogen Peroxide Levels in Human Blood

Hydrogen peroxide is the only reactive oxygen compound of which quantitative determinations are easily possible in the intact human organism. Therefore, it serves as a quantitative indicator of oxidative stress in physiological and pathophysiological studies. From such studies, data on its physiological homeostasis can be deduced, which mirror the general homeostasis of oxidative stress in the intact human organism. Key data are summarized below.

The rate of hydrogen peroxide production through physiological pathways in the liver has been estimated to be $3.8 \, \text{g kg}^{-1}$ liver weight per day. This is equivalent to a total production of about 6.8 g hydrogen peroxide each day in a 70 kg man whose liver weight is 1800 g, and equals $4.7 \, \text{mg min}^{-1}$ per person.[85,86]

Human blood plasma has hydrogen peroxide generating capacity by itself, mainly due to the activity of xantine oxidase. In the presence of sodium azide, a catalase inhibitor, a mean plasma concentration of $36 \, \mu\text{M}$ ($1.22 \, \mu\text{g mL}^{-1}$) can be reached *in vitro*.[87] Within the blood compartments, the concentration of hydrogen peroxide is several times higher in erythrocytes than in plasma, consistent with the role of erythrocytes in the physiological oxygen transport.[88]

Earlier literature data on the physiological range of hydrogen peroxide plasma concentrations have been widely variable, ranging from 0.25 µM to about 50 µM.[82,88,89] This was obviously due to the use of non-standardized measurement methodologies. Recent data from independent research groups are much more uniform. Thus, Lacy *et al.*[87] reported plasma hydrogen peroxide levels of 2.5 ± 0.16 µM in 29 normotensive and 3.16 ± 0.14 mM in 21 hypertensive subjects. Cakmak *et al.*[90] evaluated the 'total oxidative status' (TOS) in the plasma of healthy ($n = 32$) and asthmatic ($n = 42$) children; healthy children had a TOS of 9.0 ± 3.5 µM and asthmatic children of 13.4 ± 7.0 µM H_2O_2 equivalents. This indicates no basic difference between children and adults.

The physiological homeostasis of hydrogen peroxide in blood has been intensively studied by Bloomer *et al.*[91-93] Modulating factors influencing plasma hydrogen peroxide levels and other parameters of oxidative stress were sex and postprandial status, and women experienced a lower oxidative stress response to nutrients compared to men.[91] An oral supplementation with glycine propionyl-L-carnitine did not attenuate the increase in oxidative biomarkers, including plasma hydrogen peroxide.[93]

With this background, the effect on oxidative biomarkers, including plasma hydrogen peroxide, of a standard high-fat diet was assessed in 14 obese and 16 non-obese women. The test meal, consisting of an 'American milkshake' of whole milk, ice cream and whipped cream, was to be consumed within 15 minutes. The dietary energy was dependent on the individual body mass and equalled 1.2 g of each fat and carbohydrate, and 0.25 g of protein per kg. It provided approximately 17 kilocalories (71 kilojoules) per kg. Results, in terms of the plasma hydrogen peroxide response, are shown in Figure 2.1.3. In non-obese women, plasma hydrogen peroxide levels increased from 5 µM before the challenge to 13 µM after four hours, with subsequent reversal; in obese women

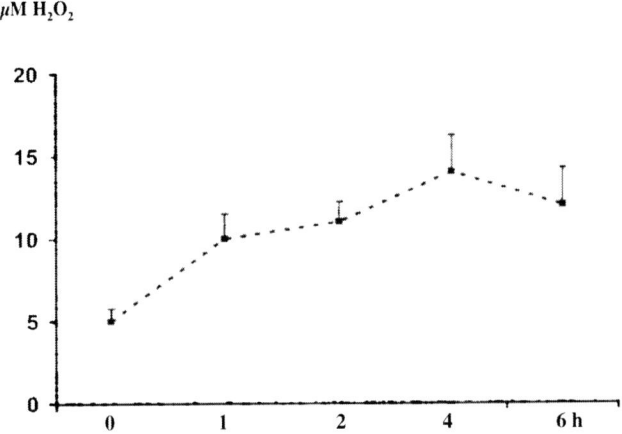

Figure 2.1.3 Homeostasis of hydrogen peroxide: Plasma levels of hydrogen peroxide in non-obese women before and after intake of a high fat meal (data according to ref. 92, with modifications).

the response was from $7\,\mu M$ to $20\,\mu M$ (six hours after the challenge). The differential increase in plasma hydrogen peroxide concentration, evoked as a physiological response to the high-fat meal, amounted to $8\,\mu M$ in non-obese and $13\,\mu M$ in obese women.[92]

The decay of hydrogen peroxide added to human blood plasma *in vitro* is very rapid and has been determined to be about $1\,min.$[94] This rapid detoxification is explained both by the action of antioxidants and enzymic degradation in the plasma.

Taking these data together, there is clear evidence that hydrogen peroxide is a physiological body constituent that is constantly produced and at the same time rapidly eliminated (detoxified) from the body. Concentrations of hydrogen peroxide are indicative of oxidative stress and vary in response to physiological challenges. The concentration range of physiological homeostasis of hydrogen peroxide, derived from the present data,[92] is $5–20\,\mu M$ $(170–680\,\mu g\,L^{-1})$ hydrogen peroxide in human plasma. This also visualizes the quantitative frame of the variations of oxidative stress, when remaining within the boundaries of physiological homeostasis.

2.1.3.2 Oxidative Stress and Endogenous Reactive Carbonyls

Reactive carbonyls are endogenous electrophiles formed during the oxidative fragmentation of unsaturated membrane lipids, sugars, amino acids and nucleic acids. Quite a number of compounds have been mentioned in this respect, including malondialdehyde, 4-hydroxynonenal, acrolein, crotonaldehyde, dehydroascorbate and 3-deoxyglucosone.[95] Besides cancer, there have been speculations about connections of reactive carbonyls with other chronic diseases such as Alzheimer's disease, atherosclerosis, diabetes type II and obesity.[95] The umbrella term 'carbonyl stress' has been proposed to denote cellular conditions displaying either heightened formation and/or diminished detoxication of endogenous carbonyls.[95] 'Carbonyl stress' is tightly linked to oxidative stress.

Much effort has been devoted to characterizing reactions between endogenous electrophiles and cellular targets, partly in the expectation that chemically modified amino acids or DNA bases may serve as biomarkers of exposure to these reactive species.[95] There are quantitative differences between human tissues in endogenous aldehydic DNA lesions; levels found were similar in liver, lung, kidney and white matter of the cerebrum, higher in colon and in grey matter of the cerebrum, and lower in the cerebellum. An interindividual variability in the quantity of such DNA damage was noted, which was partly ascribed to differences in repair capacity.[96]

As far as exogenous factors are concerned, it has been postulated that environmental air pollution may alter endogenous oxidative DNA damage levels in humans.[97] Genetic polymorphisms of genes involved in metabolism and detoxification, as well as differences in DNA repair capacity and antioxidant status have also been taken to explain variations in the levels of endogenous oxidative DNA damage between different populations.[97]

2.1.3.3 Genotoxic 'Etheno' DNA Adducts Resulting from Products of Oxidative Stress

Mechanistic and clinical studies have indicated that some inherited and acquired human risk factors increase the levels of promutagenic 'etheno' DNA adducts in target organs, in which tumours later develop as a consequence of increased oxidative stress and consecutive lipid peroxidation. Such factors include metal storage diseases, thalassaemia, overproduction of nitric oxides in consequence of chronic inflammation, chronic inflammatory bowel disease (Crohn's disease), pancreatitis, persistent hepatitis B infection, liver fluke, atherosclerosis, high intake of omega-6 polyunsaturated fatty acids and alcohol abuse.[98–103]

'Etheno' DNA adducts were first noted as a consequence of exposure to the human carcinogen vinyl chloride.[104] The structural formulae of the most intensively investigated adducts ($1,N^6$-ethenodeoxyadenosine, $N^2,3$-ethenodeoxyguanosine, $3,N^4$-ethenodeoxycytidine) are shown in Figure 2.1.4. These specific DNA lesions are clearly pro-mutagenic, as demonstrated in site-directed mutagenesis experiments, and cause single base changes.[104,105]

Low concentrations of such cyclic DNA adducts have consistently been detected in tissues of non-exposed animals and humans, and background levels reported in human tissues have been compiled.[104] Interestingly the background levels in the liver of rats appear to be lower than those in the liver of humans.[106] For instance, the mean endogenous $N^2,3$-ethenoguanine level measured in adult rats was 0.46 per 10^7 guanine moieties,[107,108] whilst 1.7 adducts per 10^7 guanine moieties (*i.e.* 0.33 $N^2,3$-ethenoguanine adducts per 10^7 nucleotides) were measured in six samples of human liver from donors with no known exposure to vinyl chloride.[108] The background adduct figures in humans show little variability.[109]

The endogenous cyclic DNA adducts arise from processes of lipid peroxidation, induced by oxidative stress. Figure 2.1.5 shows the reaction sequence as proposed by Nair *et al.*[100] Redox cycling (in this case *o*-semiquinone/-quinone metabolites of environmental polycyclic aromatic hydrocarbons; PAH) leads to reactive oxygen production (ROS) and subsequent lipid peroxidation. One product of this oxidative lipid attack is trans-4-hydroxy-2-nonenal, which is epoxidized to form etheno-bridged DNA base modifications. The same mechanism has been proposed to be triggered by *o*-semiquinone/quinone metabolites of endogenous oestrogens, thus providing an explanation of sex differences, which was demonstrated in rats.[110]

The endogenous formation of 'etheno' DNA adducts under the influence of oxidative stress provides another example for the induction of a secondary genotoxicity, with a link between oxidative stress and carcinogenicity.[111,112] The link between oxidative stress and 'etheno' adduct formation is also clinically supported. Frank *et al.*[113] investigated the levels of $1,N^6$-ethenoadenine in DNA of liver samples from patients diagnosed with diseases predisposing to hapatocarcinogenesis such as hepatitis, fatty liver, fibrosis and cirrhosis, primary haemochromnatosis and Wilson's disease. Using a specific immunohistochemical detection method, the relative prevalence of this specific adduct

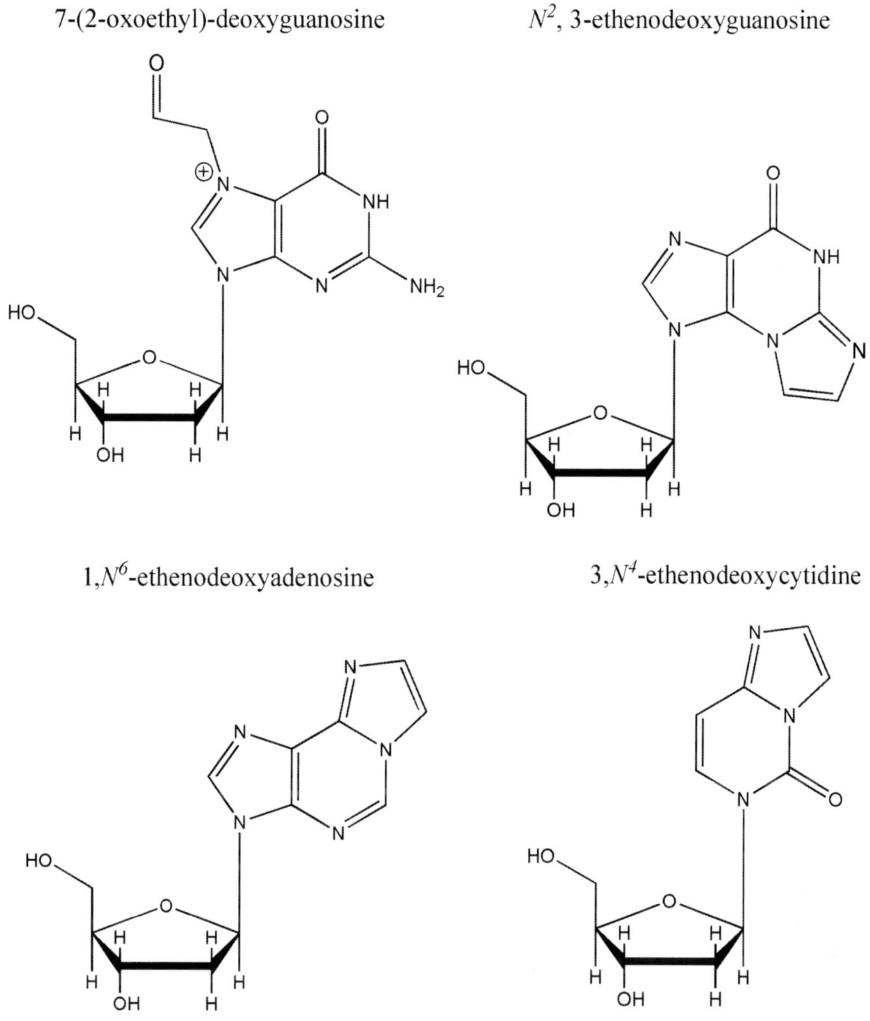

7-(2-oxoethyl)-deoxyguanosine N^2, 3-ethenodeoxyguanosine

1,N^6-ethenodeoxyadenosine   3,N^4-ethenodeoxycytidine

Figure 2.1.4 Structural formulae of 'etheno' DNA adducts: 1,N^6-ethenodeoxy-adenosine, N^2,3-ethenodeoxyguanosine, 3,N^4-ethenodeoxycytidine.

was assessed. Compared with normal livers, the adduct prevalence was significantly higher in alcoholic fatty liver, fibrosis, Wilson's disease and haemochromatosis.[113]

2.1.3.4   Implications for Cancer Risk Assessment

From the data reviewed above, it follows that oxidative stress is a physiological feature in response to quite a number of exogenous challenges. There is a high impact of nutrition. The reactive oxygen species involved in oxidative stress are

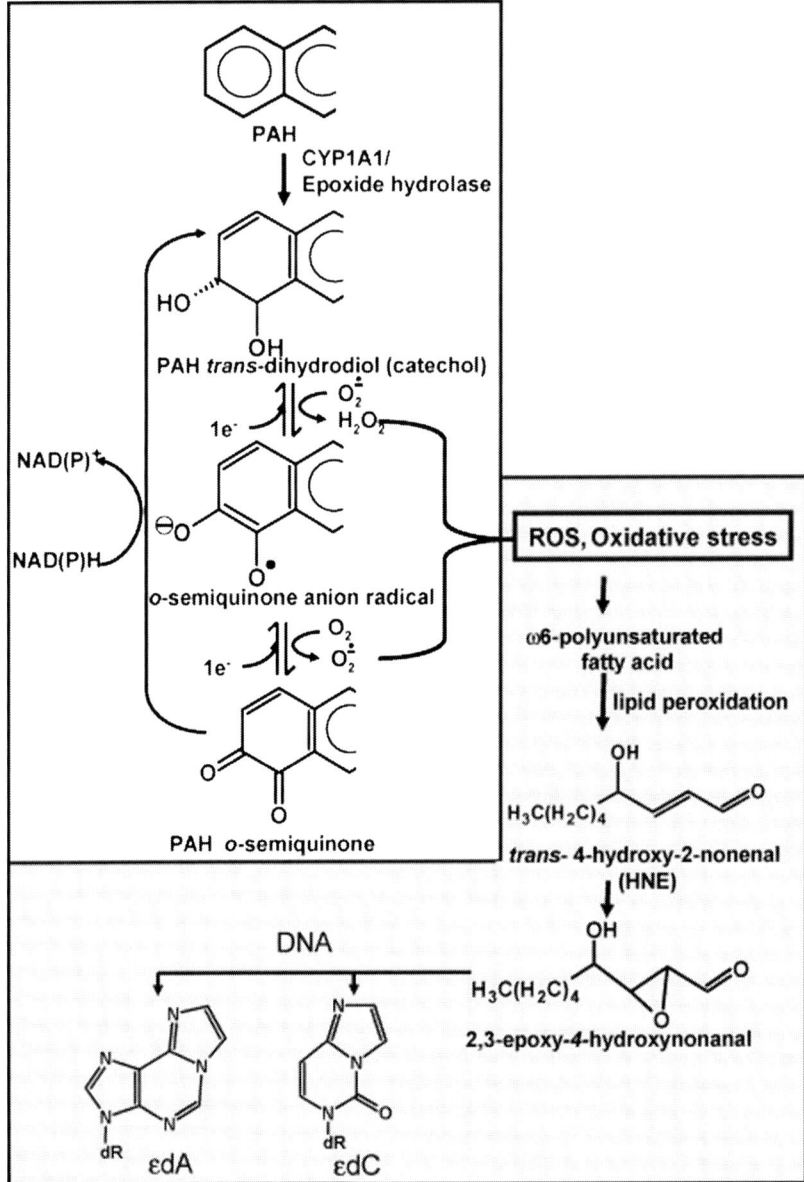

Figure 2.1.5   Proposed reaction sequence for 'etheno' DNA adduct formation in consequence of oxidative stress, elicited by redox cycling. Reproduced from ref. 100, with permission.

rapidly detoxified by enzymatic routes, leading to a physiological homeostasis. Processes consecutive to oxidative stress are linked with oxidative attacks of body constituents, such as the liberation of reactive secondary products by lipid peroxidation, which may lead to further consequences. One of these consequences is the formation of etheno-bridged DNA adducts, which are clearly pro-mutagenic and identical to the adducts formed from the well-known human carcinogen, vinyl chloride.

It has been stressed that the presence of endogenous 'etheno' DNA adducts in humans has implications for contemporary cancer risk assessment.[109] Thus, it was argued that repeated inhalation exposure of rats to 10 ppm vinyl chloride for four weeks resulted in a five-fold increase over the endogenous concentration of N^2,3-ethenodeoxyguanosine in DNA, whilst a linear interpolation suggested that exposure to 1 ppm vinyl chloride would result in only a 5% increase over the endogenous N^2,3-ethenodeogyguanine background. This was taken as an argument why no new cases of human hepatic angiosarcoma (the typical cancer induced by vinyl chloride) were reported in workers exposed only after the lowering of the threshold limit value (TLV) of vinyl chloride to 1 ppm.[109] By contrast, a linear non-threshold (LNT) approach has resulted in a theoretical angiosarcoma risk of 3×10^{-4} for working lifetime exposure to vinyl chloride.[114]

2.1.4 Conclusions

The cases discussed in this chapter show that a number of very different mutagenic carcinogens are endogenously produced by different mechanisms. As a consequence of continuous production, the examples show that the enzymic detoxification of such endogenous carcinogens is generally rapid and effective, so that a reasonable cellular homeostasis is maintained. This maintenance of a physiological homeostasis has consequences for the assessment of cancer risks induced, when the organisms is exogenously exposed to low levels of such compounds. In such cases, a practical cancer threshold may be reasonably assumed, which then allows the derivation of health-based exposure limits.

References

1. H. G. Neumann, S. Vamvakas, H. W. Thielmann, H. P. Gelbke, J. G. Filser, U. Reuter, H. Kappus, K. H. Norpoth, P. Wardenbach and H. E. Wichmann, Changes in the classification of carcinogenic chemicals in the work area, *Int. Arch. Occup. Environ. Health*, 1998, **71**, 566–574.
2. M. R. Seeley, L. E. Tonner-Navarro, B. D. Beck, R. Deskin, V. J. Feron, G. Johanson and H. M. Bolt, Procedures of health risk assessment in Europe, *Regul. Toxicol. Pharmacol.*, 2001, **34**, 153–169.
3. S. M. Cohen, M. E. Meek, J. E. Klaunig, D. E. Patton and P. A. Fenner-Crisp, The human relevance of information on carcinogenic modes of action: overview, *Crit. Rev. Toxicol.*, 2003, **33**, 581–589.
4. H. M. Bolt, Genotoxicity—threshold or not? Introduction of cases of industrial chemicals, *Toxicol. Lett.*, 2003, **140/141**, 43–51.
5. M. Kirsch-Volders, M. Aardema and A. Elhajouji, Concept of threshold in mutagenesis and carcinogenesis, *Mutat. Res.*, 2000, **464**, 3–11.

6. I. S. Pratt and T. Barron, Regulatory recognition of indirect genotoxicity mechanisms in the European Union, *Toxicol. Lett.*, 2003, **140/141**, 53–62.

7. R. Thier, D. Bonacker, T. Stoiber, K. J. Böhm, M. Wang, E. Unger, H. M. Bolt and G. H. Degen, Interaction of metal salts with cytoskeletal motor protein systems, *Toxicol. Lett.*, 2003, **140/141**, 75–81.

8. C. Streffer, H. M. Bolt, D. Føllesdal, P. Hall, J. G. Hengstler, P. Jacob, D. Oughton, K. Priess, E. Rehbinder and E. Swaton, *Environmental Standards: Dose-Effect Relations in the Low Dose Range and Risk Evaluation*, Springer-Verlag, Berlin, Heidelberg, New York, 2004.

9. H. Foth, G. H. Degen and H. M. Bolt, New aspects in the classification of carcinogens, *Arh. Hig. Rada Toksikol.*, 2005, **56**, 165–173.

10. H. M. Bolt and A. Huici-Montagud, Strategy of the Scientific Committee on Occupational Exposure Limits (SCOEL) in the derivation of occupational exposure limits for carcinogens and mutagens, *Arch. Toxicol.*, 2008, **82**(1), 61–64.

11. H. M. Bolt, The concept of 'practical thresholds' in the derivation of occupational exposure limits for carcinogens by the Scientific Committee on Occupational Exposure Limits (SCOEL) of the European Union, *Genes Environ.*, 2008, **30**, 114–119.

12. Deutsche Forschungsgemeinschaft, *MAK-und BAT-Werte Liste 2010*, Senatskommission zur Prüfung gesundheitsschädlicher Arbeitsstoffe, Wiley-VCH, Weinheim, 46th edn, 2010.

13. R. Thier and H. M. Bolt, Carcinogenicity and genotoxicity of ethylene oxide: new aspects and recent advances, *Crit. Rev. Toxicol.*, 2000, **30**, 595–608.

14. W. Kessler and H. Remmer, Generation of volatile hydrocarbons from amino acids and proteins by an iron/ascorbate/GSH system, *Biochem. Pharmacol.*, 1990, **39**, 1347–1351.

15. J. G. Filser, P. E. Kreuzer, H. Greim and H. M. Bolt, New scientific arguments for regulation of ethylene oxide residues in skin-care products, *Arch. Toxicol.*, 1992, **68**(7), 401–405.

16. IARC, Ethylene oxide, *IARC Monogr. Eval. Carcinog. Risks Hum.*, 1994, **60**, 73–159.

17. IARC, Ethylene oxide. *IARC Monogr Eval. Carcinog. Risks Hum.*, 2008, **97**, 185–309.

18. C. R. Kirman, L. M. Sweeney, M. J. Teta, R. L. Sielken, C. Valdez-Flores, R. J. Albertini and M L. Gargas, Addressing nonlinearity in the exposure-response relationship for a genotoxic carcinogen: Cancer potency estimates for ethylene oxide, *Risk Anal.*, 2004, **24**, 1165–1183.

19. C. Valdez-Flores, R. L. Sielken jr and M. J. Teta, Quantitative cancer risk assessment based on NIOSH and UCC epidemiological data for workers exposed to ethylene oxide, *Regul. Toxicol. Pharmacol.*, 2010, **56**, 312–320.

20. V. Dellarco, W. Generoso, G. A. Sega, J. R. Fowle, D. Jacobson-Kram and H. E. Brockman, Review of the mutagenicity of ethylene oxide, *Environ. Mol. Mutagen.*, 1990, **16**, 85–103.

21. J. Fuchs, U. Wullenweber, J. G. Hengstler, H. G. Bienfait, G. Hiltl and F. Oesch, Genotoxic risk for humans due to work place exposure to ethylene oxide: remarkable individual differences in susceptibility, *Arch. Toxicol.*, 1994, **68**, 343–348.

22. E. Hallier, T. Langhof, D. Dannappel, M. Leutbecher, K. Schröder, H. W. Goergens, A. Müller and H. M. Bolt, Polymorphism of glutathione conjugation of methyl bromide, ethylene oxide and dichloromethane in human blood: influence on the induction of sister chromatid exchange (SCE) in lymphocytes, *Arch. Toxicol.*, 1993, **67**, 173–178.

23. S. Pemble, K. R. Schröder, S. R. Spencer, D. J. Meyer, E. Hallier, H. M. Bolt, B. Ketterer and J. B. Taylor, Human glutathione S-transferase theta (GSTT1): cDNA cloning and the characterization of a genetic polymorphism, *Biochem. J.*, 1994, **300**, 271–276.

24. F. A. Wiebel, A. Dommermuth and R. Thier, The heriditary transmission of the glutathione transferase hGST1-1 conjugator phenotype in a large family, *Pharmacogenetics*, 1999, **9**, 251–256.

25. D. B. Barr and D. L. Ashley, A rapid, sensitive method for the quantitation of N-acetyl-S-(2-hydroxyethyl)-L-cysteine in human urine using isotope dilution HPLC-MS-MS, *J. Anal. Toxicol.*, 1998, **22**, 96–104.

26. R. Thier, J. Lewalter, M. Kempkes, S. Selinski, T. Brüning and H. M. Bolt, Haemoglobin adducts of acrylonitrile and ethylene oxide in acrylonitrile workers, dependent on polymorphisms of the glutathione transferases GSTT1 and GSTM1, *Arch. Toxicol.*, 1999, **73**, 197–202.

27. R. Thier, S. E. Pemble, J. B. Taylor, W. G. Humphreys, M. Persmark, B. Ketterer and F. P. Guengerich, Expression of mammalian glutathione S-transferase 5-5 in *Salmonella thyphimurium* TA1535 leads to base-pair mutations upon exposure to dihalomethanes, *Proc. Nat. Acad. Sci. U. S. A.*, 1993, **90**, 8576–8580.

28. R. Thier, S. E. Pemble, H. Kramer, J. B. Taylor, F. P. Guengerich and B. Ketterer, Human glutathione S-transferase T1-1 enhances mutagenicity of 1,2-dibromoethane, dibromomethane and 1,2,3,4-diepoxybutane, *Carcinogenesis*, 1996, **17**, 163–167.

29. M. Müller, A. Krämer, J. Angerer and E. Hallier, Ethylene oxide-protein adduct formation in humans: influence of glutathione-S-transferase polymorphisms, *Int. Arch. Occup. Environ. Health*, 1998, **71**, 499–502.

30. R. Bono, M. Vincenti, V. Meineri, C. Pignata, U. Saglia, O. Giachino and E. Scursatone, Formation of N-(2-hydroxyethyl)valine due to exposure to ethylene oxide *via* tobacco smoke: a risk factor for onset of cancer, *Environ. Res.*, 1999, **81**, 62–71.

31. P. J. Boogaard, P. S. I. Rocchi and N. J. van Sittert, Biomonitoring of exposure to ethylene oxide and propylene oxide by determination of hemoglobin adducts: correlations between airborne exposure and adduct levels, *Int. Arch. Occup. Environ. Health*, 1999, **72**, 142–150.

32. D. A. Marsden, D. J. L. Jones, J. H. Lamb, E. Tompkins and P. B. Farmer, Determination of endogenous and exogenously derived

     N7-(2-hydroxyethyl)guanine adducts in ethylene oxide treated rats, *Chem. Res. Toxicol.*, 2007, **20**, 290–299.

33. V. E. Walker, T. R. Fennell, J. A. Boucheron, N. Fedtke, F. Ciroussel and J. A. Swenberg, Macromolecular adducts of ethylene oxide: a literature review and a time-course study on the formation of 7-(2-hydroxyethyl)guanine following exposures of rats by inhalation, *Mutat. Res.*, 1990, **233**, 151–164.

34. V. E. Walker, T. R. Fennell, P. B. Upton, T. R. Skopek, V. Prevost, D. E. G. Shuker and J. A. Swenberg, Molecular dosimetry of ethylene oxide: formation and persistence of 7-(2-hydroxyethyl)guanine in DNA following repeated exposures of rats and mice, *Cancer Res.*, 1992, **52**, 4328–4334.

35. H. M. Bolt, M. Leutbecher and K. Golka, A note on the physiological background of the ethylene oxide adduct 7-(2-hydroxyethyl)guanine in DNA from human blood, *Arch. Toxicol.*, 1997, **71**, 719–721.

36. K. Y. Wu, A. Ranasinghe, P. B. Upton, V. E. Walker and J. A. Swenberg, Molecular dosimetry of endogenous and ethylene oxide-induced N7-(2-hydroxyethyl)guanine formation in tissues of rodents, *Carcinogenesis*, 1999, **20**, 1787–1792.

37. I. Rusyn, S. Asakura, Y. Li, O. Kosyk, H. Koc, J. Nakamura, P. B. Upton and J. A. Swenberg, Effects of ethylene oxide and ethylene inhalation on DNA adducts, apurinic/apyrimidinic sites and expression of base excision DNA repair in rat brain, spleen and liver, *DNA Repair*, 2005, **4**, 1099–1110.

38. G. Y. Csanády, B. Denk, C. Pütz, P. E. Kreuzer, W. Kessler, C. Baur, M. L. Gargas and J. G. Filser, A physiological toxicokinetic model for exogenous and endogenous ethylene and ethylene oxide in rat, mouse and human: formation of 2-hydroxyethyl adducts with hemoglobin and DNA, *Toxicol. Appl. Pharmacol.*, 2000, **165**, 1–26.

39. C. Zhao and K. Hemminki, The *in vivo* levels of DNA alkylation products in human lymphocytes are not age dependent. An assay of 7-methyl and 7-(2-hydroxyethyl)guanine DNA adducts, *Carcinogenesis*, 2002, **23**, 307–310.

40. L. Yong, P. Schulte, C. Y. Kao, R. W. Giese, M. F. Boeniger, G. H. S. Strauss, M. R. Petersen and L. K. Wiencke, DNA adducts in granulucytes of hospital workers exposed to ethylene oxide, *Am. J. Ind. Med.*, 2007, **50**, 293–302.

41. N. J. Van Sittert, P. J. Boogaard, A. T. Natarajan, A. D. Tate, L. E. Ehrenberg and M. A. Törnqvist, Formation of DNA adducts and induction of mutagenic effects in rats following 4 weeks inhalation exposure to ethylene oxide as a basis for cancer risk assessment, *Mutat. Res.*, 2000, **447**, 27–48.

42. C. Kao and R. W. Giese, Measurement of N^7-2′-hydroxyethyl)guanine in human DNA by gas chromatography electron capture mass spectrometry, *Chem. Res. Toxicol.*, 2005, **18**, 70–75.

43. D. Coggon, E. C. Harris, J. Poole and K. T. Palmer, Mortality of workers exposed to ethylene oxide: extended follow up of a British cohort, *Occup. Environ. Med.*, 2004, **61**, 358–362.

44. L. Ehrenberg, E. Moustacchi, S. Osterman-Golkar and G. Ekman, Dosimetry of genotoxic agents and dose-response relationships of their effects, *Mutat. Res.*, 1983, **123**, 121–183.
45. M. J. Teta, R. L. Sielken Jr. and C. Valdez-Flores, Ethylene oxide cancer risk assessment based on epidemiological data: application of revised regulatory guidelines, *Risk Anal.*, 1999, **19**, 1135–1155.
46. M. J. Teta, L. O. Benson and J. N. Vitale, Mortality study of ethylene oxide workers in chemical manufacturing: a 10 year update, *Br. J. Ind. Med.*, 1993, **50**, 704–709.
47. L. Stayner, K. Steenland, A. Greife, R. Hornun, R. B. Hayes, S. Nowlin, J. Morawetz, V. Ringenburg, L. Elliot and W. Halperin, *Am. J. Epidemiol.*, 1993, **138**, 787–798.
48. E. D. DeMaster and H. T. Nagasawa, Isoprene, an endogenous constituent of human alveolar air with a diurnal pattern of excretion, *Life Sci.*, 1978, **22**, 91–98.
49. H. Peter, H. J. Wiegand, H. M. Bolt, H. Greim, G. Walter, M. Berg and J. G. Filser, Pharmacokinetics of isoprene in mice and rats, *Toxicol. Lett.*, 1987, **36**, 9–14.
50. M. Del Monte, L. Citti and P. G. Gervasi, Isoprene metabolism by liver microsomal monooexygenases, *Xenobiotica*, 1985, **15**, 591–597.
51. V. Lingo, L. Citti and P. G. Gervasi, Hepatic microsomal metabolism of isoprene by various rodents, *Toxicol. Lett.*, 1985, **29**, 33–37.
52. D. Wistuba, K. Weigand and H. Peter, Stereoselectivity of in vitro isoprene metabolism, *Chem. Res. Toxicol.*, 1994, **7**(3), 336–343.
53. P. G. Gervasi, L. Citti, M. Del Monte, V. Longo and D. Benetti, Mutagenicity and chemical reactivity od epoxidic intermediates of the isoprene metabolism and other structurally related compounds, *Mutat. Res.*, 1985, **156**, 77–82.
54. IARC, Isoprene, *IARC Monogr. Eval. Carcin. Risks Hum.*, 1994, **60**, 215–232.
55. IARC, Isoprene, *IARC Monogr. Eval. Carcin. Risks Hum.*, 1999, **71**, 1015–1025.
56. NTP, Technical report on toxicity studies of isoprene (CAS No 78-79-5) administered by inhalation to F344/N rats and B6C3F1 mice, *NTP Toxicity Rep. Ser.*, 1995, **31**.
57. NTP, Toxicology and carconogenesis studies of isoprene (CAS No 78-79-5) in F33/N rats (inhalation studies), *NTP Tech. Rep. Ser.* 1999, **486**.
58. Deutsche Forschungsgemeinschaft (DFG), Isopren (2-Methyl-1,3-butadien), in *Toxikologisch-arbeitsmedizinische Begründungen von MAK-Werten*, Wiley-VCH, Weinheim, 49th edn, 2009.
59. J. G. Filser, G. A. Csanády, B. Denk, M. Hartmann, A. Kauffmann, W. Kessler, P. E. Kreuter, C. Pütz, J. H. Shen and P. Stei, Toxicokinetics of isoprene in rodents and humans, *Toxicology*, 1996, **113**(1–3), 278–287.
60. G. A. Csanády and J. H. Filser, Toxicokinetics of inhaled and endogenous isoprene in mice, rats and humans, *Chem.-Biol. Interact.*, 2001, **135/136**, 679–685.

61. Scientific Committee on Occupational Exposure Limits (SCOEL), *Recommendation from the Scientific Committee on Occupational Exposure Limits on Occupational Exposure Limits for formaldehyde*, European Commission, Directorate-General for Employment, Social Affairs and Inclusion, Brussels, 2008.

62. G. D. Nielsen and P. Wolkoff, Cancer effects of formaldehyde: a proposal for an indoor air guideline value, *Arch. Toxicol.*, 2010, **84**, 423–446.

63. H. M. Bolt, G. H. Degen and J. G. Hengstler, The carcinogenicity debate on formaldehyde: how to derivesafe exposure limits?, *Arch. Toxicol.*, 2010, **84**, 421–422.

64. H. d'A. Heck and M. Casanova-Schmitz, Biochemical toxicology of formaldehyde, *Rev. Biochem. Toxicol.*, 1984, **6**, 155–189.

65. H. d'A. Heck and M. Casanova, The implausibility of leucemia induction by formaldehyde: a critical review of the biological evidence on distant-site toxicity, *Regul. Toxicol. Pharmacol.*, 2004, **40**, 92–106.

66. IARC, Formaldehyde, IARC Monogr. Eval. Carcinog. Risks Hum., 2006, **88**.

67. M. S. Bogdanffy, P. H. Morgan, T. B. Starr and K. T. Morgan, Binding of formaldehyde to human and rat nasal mucus and bovine serum albumin, *Toxicol. Lett.*, 1987, **38**, 145–154.

68. K. T. Morgan, E. A. Gross and D. L. Patterson, Responses of the nasal mucociliary apparatus of F-344 rats to formaldehyde gas, *Toxicol. Appl. Pharmacol.*, 1986, **82**, 1–13.

69. R. Hermann, *Karzinogenese der Nasenschleimhaut durch N-Nitrosodi-methylamin—Inhalation bei der Ratte, Inaugural Dissertation, University of Heidelberg*, Germany, 1997.

70. M. Casanova-Schmitz, T. B. Starr and H. d'A. Heck, Differentiation between metabolic incorporation and covalent binding in the labeling of macromolecules in the rat nasal mucosa and bone marrow by inhaled [^{14}C]- and [^{3}H]-formaldehyde, *Toxicol. Appl. Pharmacol.*, 1984, **76**, 26–44.

71. R. A. Woutersen, L. M. Appelman, A. van Garderen-Hoetmer and V. Feron, Inhalation toxicity of acetaldehyde in rats. III. Carcinogenicity study, *Toxicology*, 1986, **41**, 213–231.

72. IARC, Acetaldehyde, *IARC Monogr. Carcinog. Risks Hum.*, 1985, **36**, 101–132 and 1987, Suppl. **7**, 77–78.

73. IARC, Alcohol drinking, *IARC Monogr. Carcinog. Risks Hum.*, 1988, **44**.

74. P. Geertinger, J. Bodenhoff, K. Helweg-Larsen and A. Lund, Endogenous alcohol production by intestinal fermentation in sudden infant death, *Z. Rechtsmed.*, 1982, **89**, 167–172.

75. Y. M. Ostrovsky, Endogenous ethanol—its metabolic, behavioural and biomedical significance, *Alcohol*, 1986, **3**, 239–247.

76. T. Fukunaga, P. Sillanaukee and C. J. P. Eriksson, Problems involved in the determination of endogenous acetaldehyde in human blood, *Alcohol Alcohol. (Oxford, U. K.)*, 1993, **28**, 535–541.

77. M. R. Halvorson, J. K. Noffsinger and C. M. Peterson, Studies on whole blood associated acetaldehyde levels in teetotallers, *Alcohol*, 1993, **10**, 409–413.
78. Deutsche Forschungsgemeinschaft (DFG), Acetaldehyd, in *Toxikologisch-arbeitsmedizinische Begründungen von MAK-Werten*, Wiley-VCH, Weinheim, 44th edn, 2008.
79. Deutsche Forschungsgemeinschaft (DFG), Ethanol, in *Toxikologisch-arbeitsmedizinische Begründungen von MAK-Werten*, Wiley-VCH, Weinheim, 26th edn, 1998.
80. J. G. Hengstler, M. S. Bogdanffy, H. M. Bolt and F. Oesch, Challenging dogma: Thresholds for genotoxic carcinogens? The case of vinyl acetate, *Annu. Rev. Pharmacol. Toxicol.*, 2003, **43**, 485–520.
81. A. Collins, Oxidative DNA damage, antioxidants and cancer, *Bioessays*, 1999, **21**, 238–246.
82. B. Chance, B. H. Sies and A. Boveries, Hydroperoxide metabolism in mammalian organs, *Physiol. Rev.*, 1979, **59**, 527–605.
83. European Centre for Ecotoxicology and Toxicology of Chemicals (ECETOC), *Hydrogen peroxide*, Special Report No. 10, ECETOC, Brussels, 1996.
84. European Chemicals Bureau, European Risk Assessment Report, 2nd Priority List, Vol. 38. Hydrogen peroxide, European Commission, Brussels, 2003.
85. A. Boveris, N. Oshino N and B. Chance, The cellular production of hydrogen peroxide, *Biochem. J.*, 1972, **128**, 617–630.
86. J. M. Desesso, A. L. Lavin and S. M. Hsia, Assessment of the carcinogenicity associated with oral exposures to hydrogen peroxide, *Food Chem. Toxicol.*, 2000, **38**, 1021–1041.
87. F. Lacy, D. T. O'Connor and G. W. Schmid-Schönbein, Plasma hydrogen peroxide production in hypertensives and normotensive subjects at genetic risk of hypertension, *J. Hypertension*, 1998, **16**, 291–303.
88. S. D. Varma and P. S. Devamanoharan, Hydrogen peroxide in human blood, *Free Rad. Res. Commun.*, 1991, **14**, 125–131.
89. B. Frei, Y. Yamamoto, D. Niclas and B. N. Ames, Evaluation of an isoluminol chemolunimiscence assay for the detection of hydroperoxides in human blood plasma, *Anal.. Biochem.*, 1988, **175**, 120–130.
90. A. Cakmak, D. Zeyrek, A. Atas, S. Selek and O. Erel, Oxidative status and paraoxonase activity in children with asthma, *Clin. Invest. Med.*, 2009, **32**, E327–E334.
91. R. J. Bloomer, D. E. Ferebee, K. H. Fisher-Wellman, J. C. Quindry and B. K. Schilling, Postprandial oxidative stress: Influence of sex and exercise training status, *Med. Sci. Sports Exercise*, 2009, **41**, 2111–2119.
92. R. J. Bloomer and K. H. Fisher-Wellman, Systematic oxidative stress is increased to a greater degree in young, obese women following consumption of a high fat meal, *Oxid. Med. Cell. Longevity*, 2009, **2**, 19–25.

93. R. J. Bloomer, W. A. Smith and K. H. Fisher-Wellman, Oxidative stress in response to forearm ischemia-reperfusion with and without carnitine administration, *Int. J. Vitam. Nutr. Res.*, 2010, **80**, 12–23.

94. V. Bocci, L. Aldinucci and L. Bianchi, The use of hydrogen peroxide as a medical drug, *Rivista Italiana di Ossigeno-Ozonoterapia*, 2005, **4**, 30–39.

95. P. C. Burcham, Potentialities and pitfalls accompanying chemico-pharmacological strategies against endogenous electrophiles and carbonyl stress, *Chem. Res. Toxicol.*, 2008, **21**, 779–786.

96. A. Barbin, H. Ohgaki, J. Nakamura, M. Kurrer, P. Kleihues and J. A. Swenberg, Endogenous deoxyribonucleic acid (DNA) damage in human tissues: a comparison of etheno bases with aldehydic DNA lesions, *Cancer Epidemiol. Biomarkers Prev.*, 2003, **12**(11 Pt 1), 1241–1247.

97. R. Singh, B. Kaur, I. Kalina, T. A. Popov, T. Georgieva, S. Garte, B. Binkova, R. J. Sram, E. Taioli and P. B. Farmer, Effects of environmental air pollution on endogenous oxidative DNA damage in humans, *Mutat. Res.*, 2007, **620**, 71–82.

98. H. Bartsch, Dr Jagadeesan Nair, senior scientist at the German Cancer Research Center (DKFZ) 1953–2007 [obituary], *Carcinogenesis*, 2008, **29**, 887–888.

99. J. Nair, F. Gansauge, H. Beger, P. Dolara, G. Winde and H. Bartsch, Increased etheno-DNA adducts in affected tissues of patients suffering from Crohn's disease, ulcerative colitis and chronic pancreatitis, *Antioxid. Redox Signal.*, 2006, **8**, 1003–1010.

100. J. Nair, S. De Flora, A. Izotti and H. Bartsch, Lipid peroxidation-derived etheno-DNA adducts in human atherosclerotic lesions, *Mutat. Res.*, 2007, **621**, 95–105.

101. M. Meerang, J. Nair, P. Sarankapracha, C. Thepinlap, S. Srichairatanakool, S. Fucharoen and H. Bartsch, Increased urinary 1,N6-ethenodeoxyadenosine and 3,N4-ethenodeoxycytidine excretion in thalassemia patients: markers for lipid peroxidation-induced DNA damage, *Free Radical Biol. Med.*, 2008, **44**, 1863–1868.

102. S. Dechakhamphu, P. Yongvanit, J. Nair, S. Pinlaor, P. Sitthithaworn and H. Bartsch, High excretion of etheno adducts in liver fluke-infected patients: protection by praziquantel against DNA damage, *Cancer Epidemiol. Biomarkers Prev.*, 2008, **17**, 1658–1664.

103. A. Frank, H. K. Seitz, H. Bartsch, N. Frank and J. Nair, Immunohistochemical detection of 1,N6-ethenodeoxyadenosine in nuclei of human liver affected by diseases predisposing to hepato-carcinogenesis, *Carcinogenesis*, 2004, **25**, 1027–1031.

104. H. M. Bolt, Vinyl chloride—a classical industrial toxicant of new interest, *Crit. Rev. Toxicol.*, 2005, **35**, 307–323.

105. A. Barbin, Etheno-adduct-forming chemicals: From mutagenicity testing to tumor mutation spectra, *Mutat. Res.*, 2000, **462**, 55–69.

106. J. Nair, A. Barbin, Y. Guichard and H. Bartsch, 1,N^6-Ethenodeoxyadenosine and 3,N^4-ethenodeoxycytidine in liver DNA from humans and

untreated rodents detected by immunoaffinity/^{32}P-postlabelling, *Carcinogenesis*, 1995, **16**, 613–617.

107. E. J. Morinello, H. Koc, A. Ranasinghe and J. A. Swenberg, Differential induction of N^2,3-ethenoguanine in rat brain and liver after exposure to vinyl chloride, *Cancer Res.*, 2002, **62**, 5183–5188.

108. E. J. Morinello, A. J. Ham, A. Ranasinghe, J. Nakamura, P. B. Upton and J. A. Swenberg, Molecular dosimetry and repair of N(2),3-ethenoguanine in rats exposed to vinyl chloride, *Cancer Res.*, 2002, **62**, 5189–5195.

109. A. Albertini, H. Clewell, M. W. Himmelstein, E. Morinello, S. Olin, J. Preston, S. Scarano, M. T. Smith, J. Swenberg, R. Tice and C. C. Travis, The use of non-tumor data in risk assessment: reflections on butadiene, vinyl chloride, and benzene, *Regul. Toxicol. Pharmacol.*, 2003, **37**, 105–132.

110. Q. Fang, J. Nair, X. Sub, D. Hadjiolov and H. Bartsch, Etheno-DNA adduct formation in rats gavaged with linoleic acid, oleic acid and coconut oil is organ- and gender specific, *Mutat. Res.*, 2007, **624**, 71–79.

111. J. G. Hengstler and H. M. Bolt, Induction and control of oxidative stress, *Arch. Toxicol.*, 2007, **81**, 823–824.

112. H. M. Bolt and J. G. Hengstler, Oxidative stress and hepatic carcinogenesis: New insights and applications, *Arch. Toxicol.*, 2010, **84**, 87–88.

113. A. Frank, H. K. Seitz, H. Bartsch, N. Frank and J. Nair, Immunohistochemical detection of 1,N^6-ethenodeoxyadenosine in nuclei of human liver affected by diseases predisposing to hepatocarcinogenesis, *Carcinogenesis*, 2004, **25**, 1027–1031.

114. Scientific Committee on Occupational Exposure Limits (SCOEL), *Recommendation from the Scientific Committee on Occupational Exposure Limits: Risk Assessment for Vinyl Chloride.* European Commission, Directorate-General for Employment, Social Affairs and Inclusion, Brussels, 2002.

Phase 2 Detoxifying Enzymes and Anti-oxidant Defense Mechanisms in the Inactivation of Genotoxic Carcinogens

WOLFGANG DEKANT* AND ANGELA MALLY

University of Würzburg, Department of Toxicology, Versbacher Str. 9, 97078 Würzburg, Germany
*Email: dekant@toxi.uni-wuerzburg.de

2.2.1 Biotransformation

Most xenobiotics are lipophilic and easily penetrate lipid membranes. However, the major excretory mechanisms of the mammalian organism require the solubility of the xenobiotic in aqueous media and thus a certain degree of water solubility. In the absence of efficient excretion of non-volatile chemicals, exposure to a lipophilic chemical could result in accumulation of the xenobiotic.[1–4]

Therefore, mammals have developed a number of biochemical processes that convert lipophilic chemicals to hydrophilic chemicals. These enzymatic processes are termed biotransformation. The enzymes of biotransformation differ from most other enzymes active in organisms by having a broad substrate specificity and by catalyzing reactions at comparably low rates. The low rates of biotransformation reactions are often compensated by high concentrations of biotransformation enzymes in the organ responsible for biotransformation.

Issues in Toxicology No. 13
The Cellular Response to the Genotoxic Insult: The Question of Threshold for Genotoxic Carcinogens
Edited by Helmut Greim and Richard J. Albertini
© The Royal Society of Chemistry 2012
Published by the Royal Society of Chemistry, www.rsc.org

Biotransformation is most often the sum of several processes by which the structure of a chemical is changed during passage through the organism. The metabolites formed from the parent chemical are usually soluble in water; the increased water solubility thus facilitates excretion with urine or bile. Besides biotransformation, mammalians have also developed specific mechanisms to excrete xenobiotic conjugates by active transport across cell membranes.[5,6]

2.2.1.1 Phase I and Phase II Reactions

Xenobiotic metabolism is catalyzed by a number of different enzymes. For solely operational purposes, the enzymes of biotransformation are separated into two phases. Phase I reactions involve oxidation, reduction and hydrolysis, while phase II reactions are biosynthetic and link the metabolite formed by phase I reactions to a polar endogenous molecule producing a conjugate. Various endogenous molecules with high polarity and thus even more water solubility are utilized for conjugation. Moreover, phase II enzymes play an important role in the detoxification of reactive intermediates formed during phase I biotransformation and in the detoxification of reactive oxygen molecules whose formation may be induced by interference of xenobiotics with oxygen utilization processes in the organisms.[7]

2.2.1.2 The Role of Biotransformation in Detoxification and Bioactivation

The purpose of biotransformation reactions is detoxification and facilitation of excretion. However, depending on the structure of the chemical and the enzyme catalyzing the biotransformation reaction, metabolites with a higher potential for toxicity than the parent compound are often formed. This process is termed bioactivation and is the basis for the toxicity and carcinogenicity of many xenobiotics. The interaction of the toxic metabolite with proteins, lipids or nucleic acids initiates events that may ultimately result in cell death, cancer or other manifestations of toxicity.

Since reactive metabolite formation has been observed with many naturally occurring chemicals, capacity for detoxification of such metabolites has evolved in parallel with the general capacity of the organism for biotransformation and detoxification is an integral part of biotransformation mechanisms.[8–12]

2.2.1.2.1 Bioactivation of Xenobiotics

Many xenobiotics with low chemical reactivity (for example the solvent carbon tetrachloride, the environmental contaminant hexachlorobutadiene, and the heat exchanger fluid tri-*o*-cresyl phosphate) cause toxic effects. These toxic effects are initiated by the covalent binding to macromolecules of metabolites formed in the organisms by the enzymes of biotransformation. This process is termed bioactivation.[8] With many chemicals, reactive metabolites formed

Table 2.2.1 Basic mechanisms involved in the bioactivation of xenobiotics based on the chemical reactivity of the intermediates formed.

Mechanism	Structure and reactivity of the intermediate	Examples
Biotransformation to stable but toxic metabolites	Different structures, selective interaction of formed metabolite with specific acceptors or disruption of specific biochemical pathways	Dichlormethane Acetonitrile Parathion
Biotransformation to electrophiles	Reactive electrophiles	Dimethylnitrosamine Acetaminophen Bromobenzene
Biotransformation to free radicals	Radicals	Carbon tetrachloride
Formation of reactive oxygen metabolites	Radicals	Paraquat Aromatic nitro-compounds

during bioactivation may be efficiently detoxified; thus, toxic effects only occur when the balance between the production of reactive metabolites and their detoxification is disrupted. For example, toxic effects may be observed with a certain chemical only when the formation of reactive intermediates is enhanced or when the capacity for detoxification is diminished.[13]

The mechanisms of bioactivation of xenobiotics may be classified into four categories describing the basic types of reactive intermediates formed and their potential reactivity (Table 2.2.1).[8,14,15]

2.2.1.2.2 Biotransformation to Reactive Electrophiles

This is the most common pathway of bioactivation. The carcinogenicity of many chemicals is associated with the formation of electrophiles and the ensuing alkylation of DNA. Reactive intermediates include chemically diverse functionalities such as epoxides, quinones, acyl halides, carbocations and nitrenium ions. The metabolic formation of electrophiles may be catalyzed by many different enzymes, although the majority of example elucidated up to date involve cytochrome P-450 mediated oxidations.[10,16–19] Covalent binding of reactive metabolites may also be involved in other forms of toxic responses such as idiosyncratic drug reactions.[20–22]

Cytochrome P-450 catalyzes the transformation of olefins to reactive and electrophilic oxiranes. For example, the carcinogenicity of the industrial intermediate vinyl chloride and the fungal toxin aflatoxin$_{B1}$ (Figure 2.2.1) are dependent on their transformation to electrophilic oxiranes.[12,23]

2.2.1.2.3 Biotransformation of Xenobiotics to Radicals

Free radicals are chemical species which may be formed by a one-electron oxidation to yield a cation radical:

$$R \rightarrow R^{\bullet} + e^{-} \qquad (2.2.1)$$

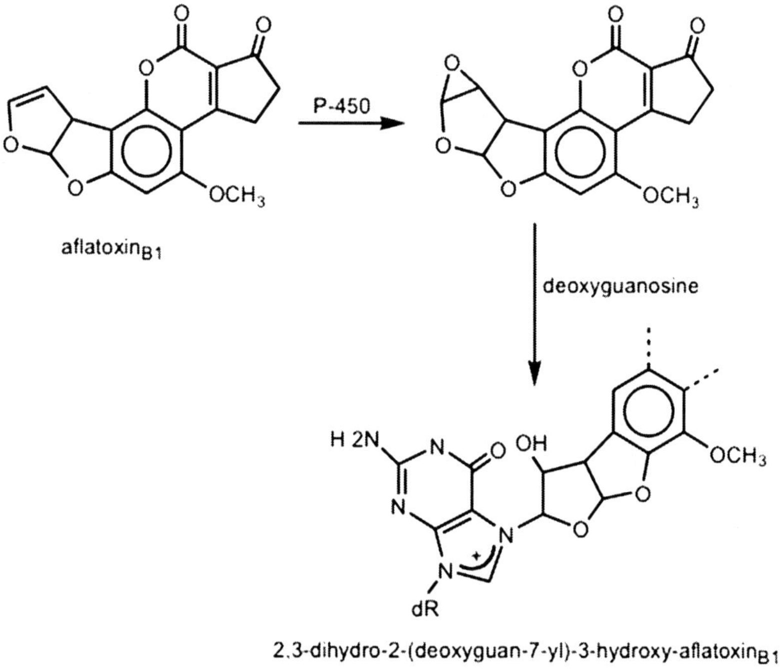

aflatoxin$_{B1}$

P-450

deoxyguanosine

2.3-dihydro-2-(deoxyguan-7-yl)-3-hydroxy-aflatoxin$_{B1}$

Figure 2.2.1   Bioactivation of aflatoxin$_{B1}$ to an electrophilic oxirane. This reaction results in the formation of DNA adducts and is believed to initiate tumor induction by aflatoxin$_{B1}$ in the liver.

or by a one electron reduction to yield an anion radical:

$$R + e^- \rightarrow R^{\bullet -} \qquad (2.2.2)$$

or by homolytic fission of a σ-bond to give a neutral radical:

$$R–R \rightarrow 2R^{\bullet} \qquad (2.2.3)$$

Free radicals are highly reactive and, when formed in biological systems, are expected to react with a variety of tissue molecules. Radicals may abstract hydrogen atoms, undergo oxidation–reduction reactions, dimerization and disproportionation reactions. Radicals may also participate in a chain mechanism, which is initiated by a reaction causing a free radical and propagated by a subsequence of reactions causing further radicals as products.

The toxic solvent carbon tetrachloride is the outstanding example of a bioactivation reaction to a free radical. Carbon tetrachloride is biotransformed by a one-electron reduction to yield the trichloromethyl radical and chloride (eqn 2.2.4).

$$Cl_4C + e^- \rightarrow Cl_3C^{\bullet} + Cl^- \qquad (2.2.4)$$

The trichloromethyl radical may abstract hydrogen atoms from tissue macromolecules to give chloroform or may dimerize to give hexachloroethane. Toxic effects of the formation of radicals during biotransformation reactions are lipid peroxidation and oxidative modification of proteins and nucleic acids (see below). Radicals may be formed by the NADPH-dependent cytochrome P-450 reductase, by nitroreductases or by one-electron oxidations catalyzed by peroxidases such as prostaglandin synthetases. Formation of free radicals from tissue constituents also plays an important role in the toxic effects of ionizing radiation.[24–27]

2.2.1.2.4 Formation of Reactive Oxygen Metabolites by Xenobiotics

Xenobiotic-induced formation of reduced oxygen metabolites such as the superoxide radical anion, hydrogen peroxide and the hydroxyl radical have been implicated as a mechanism of producing cell damage—the so-called oxidative stress.[28–31] The biotransformation of certain xenobiotics that are able to undergo redox cycling or enzyme-catalyzed oxidation–reduction reactions may be associated with the production of reduced oxygen metabolites. 2-Methyl-naphtoquinone (menadione) has been intensively used to study the formation and cellular reactions of reduced oxygen metabolites. Menadione and other quinones undergo enzymatic redox cycling; these one-electron oxidation reactions are associated with the formation of the superoxide radical anion $(O^{2-\bullet})$ by a one electron reduction of triplet oxygen. In aqueous solution, superoxide is not impressively reactive; the dismutation or further reduction of superoxide may result in hydrogen peroxide (Figure 2.2.2).

Hydrogen peroxide is also a poor oxidant in biological systems, but sufficiently stable to cross biological membranes. The toxicity of hydrogen peroxide is attributed to the formation of the hydroxyl radical by a Fenton reaction catalyzed by metal ions such as Fe^{2+} (eqn 2.2.5, M = transition metal).

$$M^n + H_2O_2 \rightarrow M^{(n+1)} + HO^\bullet + HO^- \tag{2.2.5}$$

The highly reactive hydroxyl radical may then initiate cellular damage by radical-based mechanisms. Besides menadione, oxidative stress may also be initiated by other xenobiotics such as the *bis*-pyridinium herbicide paraquat and nitroheterocycles (Table 2.2.1). Moreover, the formation of reduced oxygen metabolites plays an important role in host defense against infectious agents and in the initiation and propagation of certain diseases such as artheriosclerosis and polyarthritis.[32]

Since oxygen radicals are also formed in low concentrations during cellular respiration, efficient mechanisms for their detoxification exist (see below). Oxidative stress is thus only observed when the equilibrium between oxidants and reductants is disturbed and detoxification mechanisms are overwhelmed.[33,34]

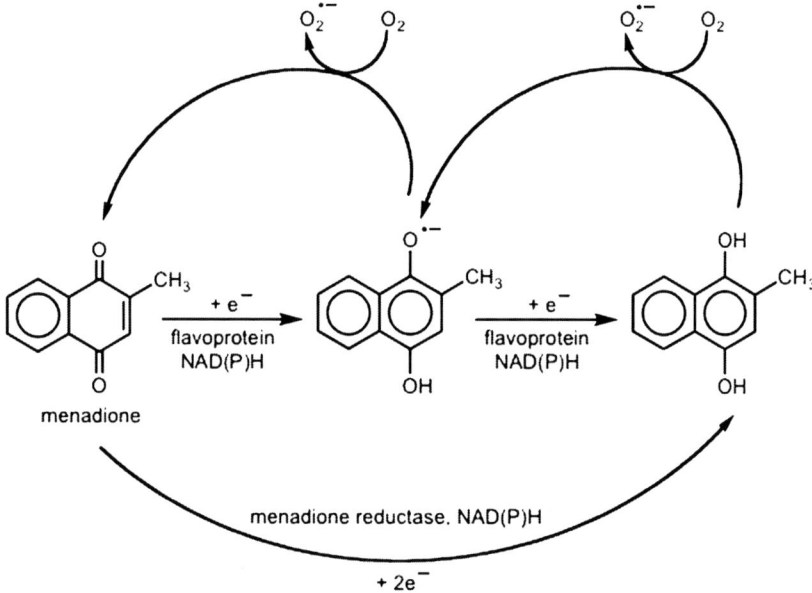

Figure 2.2.2 Biotransformation of menadione and the induction of oxidative stress by reduction of triplet oxygen to the superoxide radical anion.

2.2.1.2.5 Detoxification Mechanisms

Since multicellular organisms had to deal with reactive metabolites and reactive oxygen species, a number of efficient systems to detoxify such molecules developed during evolution. The enzymes of detoxification are present in all organs and usually have a high capacity in organs responsible for bio-transformation of xenobiotics such as the liver, the kidney, and the lung; enzymatic systems to detoxify reactive oxygen species are present in high activity in tissue and subcellular fractions with high oxygen turnover.

The tripeptide glutathione plays an important role in the detoxification reaction for reactive electophiles and for reactive oxygen species. Glutathione consists of the amino acids cysteine, glutamic acid and glycine (γ-glutamylcys-teinyl-glycine) and is present in many cells in high concentrations (up to 10 mM in liver cells).[35] Glutathione conjugation transforms reactive electrophiles to stable thioethers and glutathione peroxidase reduces reactive oxygen species with formation of glutathione disulfide, which may be regenerated to give two molecules of glutathione by glutathione reductase; therefore, glutathione conjugates protects cells from the toxic effects of such reactive metabolites.

2.2.1.2.6 Glutathione Conjugate Formation and Mercapturic Acid Excretion

Glutathione conjugation is catalyzed by a family of enzymes termed glutathione *S*-transferases, which are present in high concentration in the liver,

Table 2.2.2 Substrates for mammalian glutathione S-transferases.

Type of reaction	Examples
Aryl transferase	(1-chloro-2,4-dinitrobenzene) + GSH $\longrightarrow$ (2,4-dinitrophenyl-SG) + HCl
Arylalkyl-transferase	C$_6$H$_5$CH$_2$Cl + GSH $\longrightarrow$ C$_6$H$_5$CH$_2$SG + HCl
Alkene transferase	$\begin{array}{l}CHCOOC_2H_5\\ \parallel \\ CHCOOC_2H_5\end{array}$ + GSH $\longrightarrow$ $\begin{array}{l}CH_2COOC_2H_5\\ \mid \\ GS-CHCOOC_2H_5\end{array}$
Epoxide transferase	$C_6H_5O-CH_2-CH(O)CH_2$ (epoxide) + GSH $\longrightarrow$ $C_6H_5O-CH_2-CH(OH)-CH_2-SG$

kidney, testes and lung.[36] Glutathione S-transferases are both membrane-bound and soluble enzymes.[37,38] Soluble glutathione S-transferases are present as numerous enzymes, each species being a dimer differing in subunit composition.[36,39] The glutathione S-transferase gene family exists of many different families. In contrast to the multiple forms of soluble glutathione S-transferases, only one form of the membrane-bound enzyme is known. The glutathione S-transferases catalyze the reaction of the sulfhydryl group of glutathione with chemicals containing electrophilic carbon atoms; for examples of reactions, see Table 2.2.2.

Thioethers are formed by the reaction of the thiolate anion of glutathione with the electrophile; usually, a spontaneous reaction of the electrophile with glutathione without assistance by glutathione S-transferases occurs at low rates. Glutathione thioethers formed in the organism are further processed to excretable mercapturic acids. Mercapturic acids are thioethers derived from N-acetyl-L-cysteine; their formation is initiated by conjugation of the xenobiotic or an electrophilic metabolite with glutathione (Figure 2.2.3).

This is followed by cleavage of the glutamate catalyzed by γ-glutamyl-transpeptidase, an enzyme specifically recognizing γ-glutamyl peptides and found in high concentrations in the kidney and other excretory organs. Dipeptidases then catalyze the loss of glycine from the intermediary

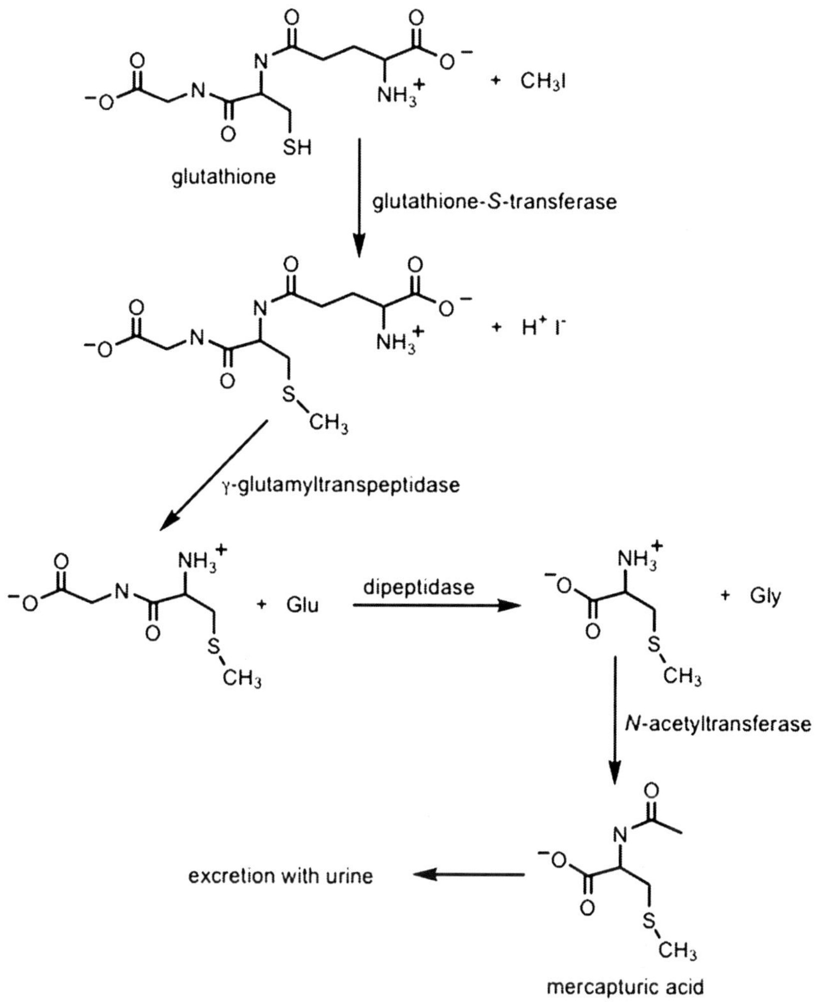

Figure 2.2.3 Formation of mercapturic acids by processing of glutathione *S*-conjugates as exemplified by the metabolism of methyl iodide.

cysteinylglycine *S*-conjugate to give the cysteine *S*-conjugate which, in the final step of mercapturic acid formation, is *N*-acetylated by a cysteine conjugate specific *N*-acetyltransferase using acetyl coenzyme A as cofactor. The mercapturic acids formed are readily excreted into urine by active transport mechanisms in the kidney.[40,41]

Glutathione conjugation is one of the most important detoxification reactions for reactive intermediates formed in the organisms. Often, reactive intermediates are efficiently detoxified; however, under specific circumstances, glutathione conjugation may be overwhelmed by high concentrations of electrophiles resulting in covalent binding of intermediates causing disruption of important cellular functions resulting in cell death (see below).

2.2.1.2.7   *Detoxification and Interactions of Reactive Metabolites with Cellular Macromolecules*

Reactive intermediates formed inside cells may react with low and high molecular weight cellular constituents. These interactions may result in formation of less reactive chemicals and thus in detoxification, or they may perturb important cellular functions resulting in acute and/or chronic toxic effects such as necrosis or cancer. Usually, the interaction with low molecular weight constituents in the cell results in detoxification, whereas the irreversible interaction with cellular macromolecules results in adverse effects.[40,42–44]

Detoxification of reactive intermediates may be due to hydrolysis, glutathione conjugation or interactions with cellular antioxidants. The reaction of electrophilic xenobiotics with the nucleophile water, present in high concentrations in all cells, is the simplest detoxification. Many of the products thus formed are of low reactivity and may be rapidly excreted. For example, acyl halides formed by the oxidation of olefins such as perchloroethene are hydrolyzed rapidly to halogenated carboxylic acids; only minor amounts of the intermediary acyl halide reacts with protein and lipids (Figure 2.2.4).

Glutathione-dependent detoxifications represent an important detoxification mechanism for metabolically formed electrophiles, free radicals and reduced oxygen metabolites.[41,46–49] Electrophiles react with the nucleophilic sulfur atom of glutathione; this reaction may be spontaneous or enzyme catalyzed. Spontaneous reactions are only observed in appreciable rates with soft electrophiles (glutathione is a soft nucleophile); the conjugation of hard electrophiles with glutathione requires enzymatic catalysis; usually, the glutathione S-transferase catalyzed rates of conjugation differs between hard and soft electrophiles, soft electrophiles being conjugated more efficiently. For example, the hard electrophile aflatoxin$_{B1}$-8,9-oxide does not spontaneously react with glutathione; only in the presence of a certain glutathione S-transferase enzyme is a

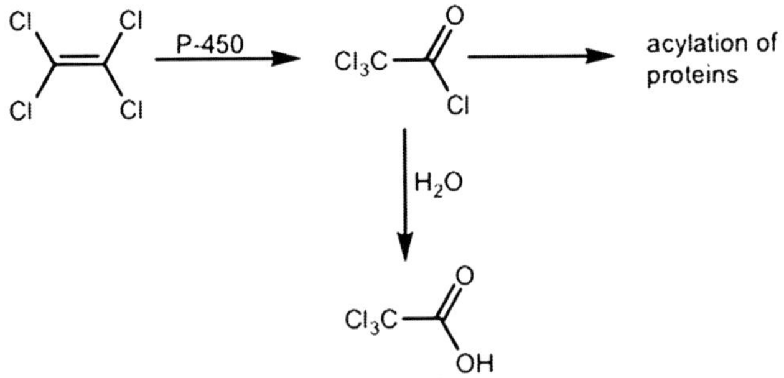

Figure 2.2.4   Perchloroethene biotransformation to trichloroacetyl chloride followed by hydrolysis to trichloroacetic acid, the major urinary metabolite formed from perchloroethene. Only minor amounts of the acyl halide formed react with proteins.[45]

glutathione *S*-conjugate of aflatoxin$_{B1}$-8,9-oxide formed (Figure 2.2.1). Species differences in the tumorigenesis of aflatoxin$_{B1}$ may serve to illustrate the important role of glutathione *S*-transferases in the expression of toxicity and carcinogenicity. Aflatoxin$_{B1}$ is a potent liver carcinogen in rats but is only weakly carcinogenic in mice. The liver of mice contains a glutathione *S*-transferase which efficiently detoxifies aflatoxin$_{B1}$-8,9-oxide. This glutathione *S*-transferase enzyme is not present in rat liver; thus, the binding of aflatoxin$_{B1}$-8,9-oxide to rat liver DNA and liver carcinogenicity of aflatoxin$_{B1}$ is much higher in rats as compared to mice.

Glutathione also plays a major role in the detoxification of reactive oxygen metabolites and radicals. Selenium-dependent glutathione peroxidases are important enzymes catalyzing the detoxification of hydrogen peroxide. In the glutathione peroxidase catalyzed reaction, two moles of glutathione are oxidized to glutathione disulfide (eqn 2.2.6).

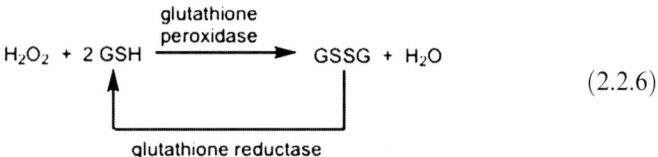

$$\tag{2.2.6}$$

Glutathione may be recycled by the reduction of glutathione disulfide by glutathione reductase. The copper and zinc dependent cytosolic and the manganese dependent mitochondrial superoxide dismutase detoxify superoxide radical anions. Hydrogen peroxide formed by dismutation of superoxide is transformed to water and oxygen and is thus detoxified by catalase (eqn 2.2.7).

$$2H_2O_2 \xrightarrow{\text{catalase}} 2H_2O + \tfrac{1}{2}O_2 \tag{2.2.7}$$

Several cellular antioxidants also play a role in the detoxification of radicals. α-Tocopherol is an important lipophilic antioxidant whose presence in lipid membranes prevents damage to lipid constituents (*e.g.* unsaturated fatty acids) by radicals. The hydroxyl radical, the superoxide radical anion and peroxy radicals react with α-tocopherol to yield water, hydrogen peroxide and hydroperoxides, which may be detoxified further by catalase and glutathione peroxidase. α-Tocopherol is transformed during these reactions to give a stable radical of comparatively low reactivity. Ascorbic acid is an important antioxidant present in the cytoplasm of the cell and may also participate in the detoxification of radicals.

Oxidative stress produces mixed disulfides of proteins with low molecular weight thiols such as glutathione also altering protein structure and function. In addition, oxidants and radicals promote the oxidation of amino acids in proteins which may increase the susceptibility of these proteins to proteolysis.[50,51] Increased protein oxidation has been implicated in cellular aging and in the mechanisms of toxicity of several redox active transition metals.

Radicals formed during the biotransformation of xenobiotics may abstract hydrogen atoms from cellular components.[24] The abstraction of hydrogen atoms from polyunsaturated fatty acids of lipids results in a process termed lipid peroxidation. The fatty acid radicals thus formed may react with molecular oxygen to give peroxy radicals and further to hydroperoxides. The initiated radical chain reactions cause the cleavage of carbon-carbon bonds in the fatty acids to short fragments such as α,ß-unsaturated carbonyl compounds.[52,53]

The disruption of membranes and the formation of toxic hydroperoxides and α,ß-unsaturated carbonyl compounds may cause disruptions in cellular calcium homeostasis and thus cause biochemical changes ultimately leading to cell death.[54,55] The reaction of electrophilic metabolites with DNA constituents results in the formation of altered purine and pyrimidine bases, or other DNA damage such as DNA strand breaks or the loss of single bases from the double helix. Many of these modifications are 'pre-mutagenic lesions'. After gene expression, these lesions may be translated into mutations.[56] Mutations in certain genes are considered as the basis for the evolution of neoplastic cells and cancer, and thus play a major role in chemical carcinogenesis. Other types of DNA damage may result in the activation of genes important for cellular differentiation or other regulatory functions. Electrophilic intermediates alkylate the nitrogen and oxygen atoms of the purines and pyrimidines in DNA; deoxyguanosine is often preferentially alkylated. The site of alkylation of a certain base in DNA is again dependent on the electrophilicity of the alkylation agent; hard electrophiles preferentially react with the oxygen atoms of guanosine, soft electrophiles alkylate the exocyclic amino groups (Figure 2.2.5).

8-Hydroxy-deoxyguanosine, a pre-mutagenic modification, is considered one of the more important lesions induced by oxidative DNA damage; due to sensitive methods for quantitation, it may serve as a marker for the extent of oxidative DNA modification caused by a xenobiotic or by other processes. DNA oxidation has also been implicated in aging; an increase in oxidative DNA modifications may occur with age due the degreased availability of antioxidants in cells of aging mammals. Several theories suggest a correlation between increased oxidative DNA damage and the increased tumor incidence in the aged population.[57,58]

In addition to cell death, disturbances in the regulation of cell division induced by toxic xenobiotics may have harmful long-term consequences for the organism affected. It has been understood for many years that non-lethal alterations in the genome of somatic cells may result in mutations and may lead to malignant transformation and tumor formation.[59] More recently, it has become clear that compounds not directly interacting with the genomic DNA can also produce cancer by so-called epigenetic

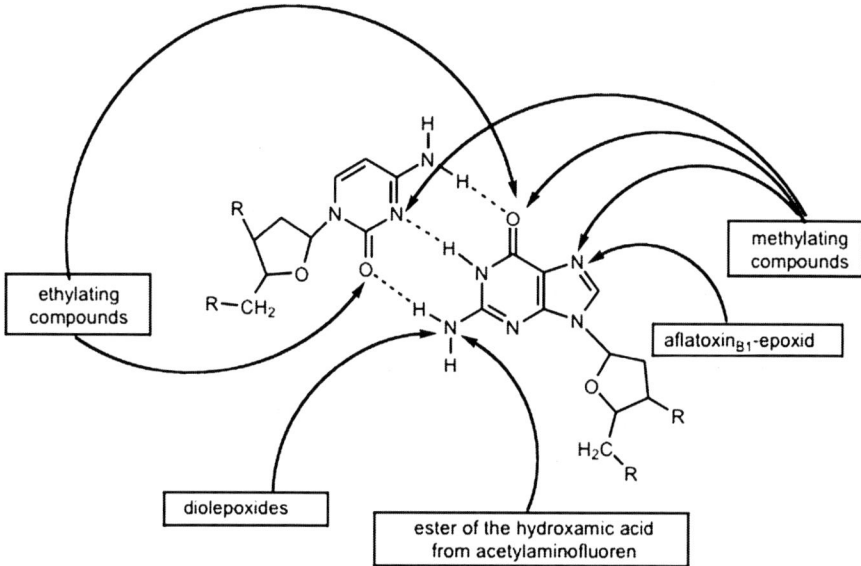

Figure 2.2.5 Regioselectivity of DNA alkylation by different electrophiles.

mechanisms.[60] These may involve a proliferative response of epithelial cells to cytotoxicity, as suggested to occur with high-dose carcinogens such as allyl isothiocyanate and chloroform, or a more direct action enhancing the rate of cell division in the absence of cytotoxicity seen with 2,3,7,8-tetra-chlorodibenzo-*p*-dioxin.

Increased cell replication, whatever the cause, is accompanied by increased chance of unrepaired DNA lesions that may be fixed as mutations. It has long been suspected that hyperplasia precedes neoplasia, but the inevitability of such a progression has never been established on a pathological basis alone. Even in the absence of foreign compounds, there is a considerable extent of damage to DNA, *i.e.* by reactive oxygen species formed during different biochemical processes in the cell, it may thus be speculated that the nongenotoxic carcinogens act by enhancing the likelihood of this normal DNA damage being fixed as a mutation, leading to cancer.[61]

The complex interaction of biotransformation, bioactivation and detoxication can be exemplified by the interactions of acetaminophen and ensuing liver toxicity (Figure 2.2.6). After intake of low doses, most of the parent acetaminophen is transformed by conjugation reactions with activated sulfate and glucuronide to give conjugates. After higher doses, a part of the administered dose is oxidized to the electrophilic quinone imine. The quinone imine can be detoxified by reaction with glutathione as long as glutathione levels are sufficient. Covalent binding resulting in toxicity only occurs when this detoxication pathway is overwhelmed.[62–68]

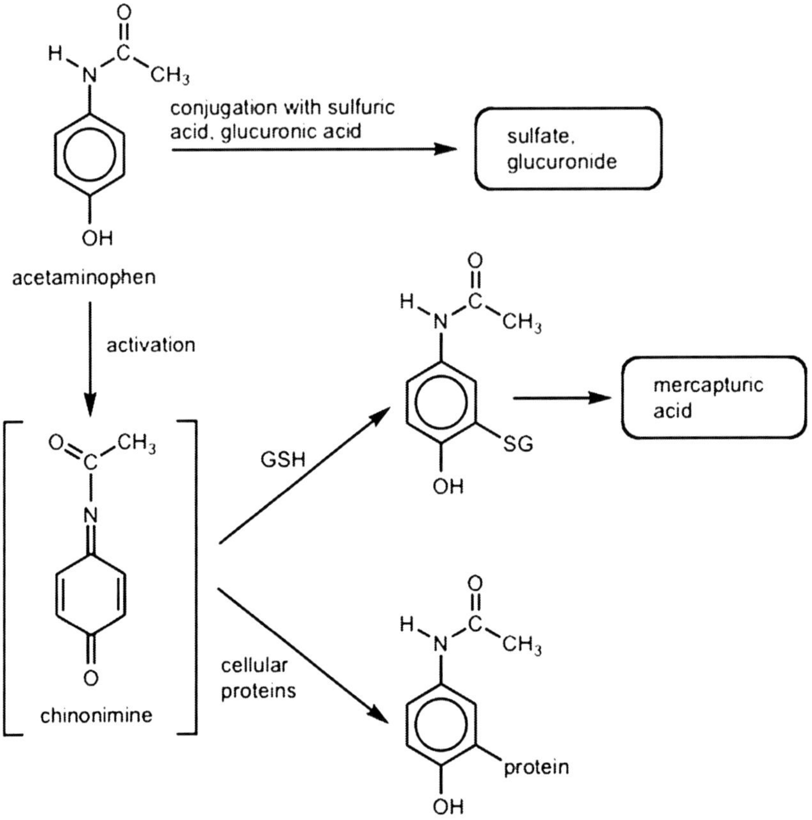

Figure 2.2.6 Interplay between bioactivation, detoxication and covalent binding as initiation of toxicity by the liver toxicant acetaminophen.

2.2.1.2.8 *Mechanisms of Cellular Defense: Receptor and Non-Receptor Mediated Regulation of Detoxifying Enzymes*

Exposure to xenobiotics can activate homeostatic control mechanisms that serve to protect cells from harmful effects of xenobiotics, electrophiles and reactive oxygen species through increasing the cell's capacity for detoxification. This is achieved by induction of xenobiotic metabolizing enzymes, transporters and antioxidant enzymes, resulting in enhanced elimination and clearance of foreign compounds and other stressors from the body. Key mechanisms of cellular defense that regulate the expression of detoxifying enzymes under conditions of xenobiotic stress include the xenobiotic-sensing nuclear receptors constitutive androstane receptor (CAR) and pregnane X receptor (PXR), and the nuclear factor-erythroid 2 p45-related factor 2 (Nrf2).[69]

CAR and PXR are nuclear receptors that are expressed predominantly in the enterohepatic axis. They belong to a family of orphan nuclear receptors which lack previously identified physiological ligands but bind a remarkable diversity of xenobiotic compounds due to their relatively large and flexible hydrophobic ligand pocket, although binding affinities vary widely across species. The broad substrate specificity enables CAR and PXR to serve as xenobiotic sensors that translocate to the nucleus and activate transcription of xenobiotic metabolizing enzymes and transporters upon ligand binding (Figure 2.2.7).[70] Ligands for CAR include the classic rodent tumor promoter phenobarbital, drugs such as clotrimazol and chlorpromazine, pesticides such as polychorinated biphenyls (PCBs), dieldrin and DDT, and the pesticide contaminant 1,4-*bis*[2-(3,5 dichlorpyridyloxy)] benzene (TCPOBOP).[70,71] On exposure to a ligand, CAR dissociates from its co-chaperones cytoplasmic CAR retention protein (CCRP) and HSP90 by a phosphorylation-dependent mechanism involving protein phosphatase PP2A.[72] Dissociation from its cytoplasmic binding partners triggers translocation into the nucleus, where CAR associates with the retinoid X receptor (RXR) and binds to phenobarbital responsive elements (PBREM) within promoter regions of CAR target genes, leading to activation of gene expression (Figure 2.2.7). The battery of genes induced by CAR includes various cytochrome P450s (CYPs) (*e.g.* CYP2B6, CYP2C8, CYP2C9, CYP2C19, CYP3A4), phase II enzymes, *i.e.* γ-glutamyltransferases (GSTs), UDP-glucuronyltransferases (UGTs) and sulfotransferases (SULTs), and membrane bound efflux transporters (MDRs, MRPs),[71,72] which serve to limit absorption and to increase elimination of xenobiotics or their metabolites. Importantly, however, activation of CAR (or PXR) can also contribute to toxicity by allowing some xenobiotics to induce their own activation to a reactive metabolite, as appears to be the case with acetaminophen.

Similar to CAR, activation of PXR involves ligand binding to PXR, dissociation of PXR from cytoplasmic retention proteins, nuclear transport and dimerization with RXR (Figure 2.2.7). Binding of the PXR/RXR heterodimer to xenobiotic responsive elements (XRE) then causes transcriptional activation of CYPs (*e.g.* CYP2B6, CYP2C8, CYP2C9, CYP2C19, CYP3A4), conjugating enzymes, and drug transporters involved in phase I–III metabolism, as well as a number of other enzymes important in cellular defense against xenobiotic stress, *e.g.* superoxide dismutase (SOD) and inducible nitric oxide synthase (iNOS).[71,72] Although CAR and PXR were originally identified as xenobiotic receptors that mediate induction of CYP2B and CYP3A, respectively, they share a significant overlap with respect to their gene targets. CAR and PXR also respond to an overlapping set of xenobiotics, although the multiplicity of ligands that activate PXR, which include the antibiotic macrolide rifampicin, the antineoplastic drug paclitaxel, dexamethasone and various environmental chemicals (*e.g.* PCBs, bisphenol A, chlordane) is much larger than the number of known CAR ligands.[70,71]

(A) Receptor mediated regulation of detoxication

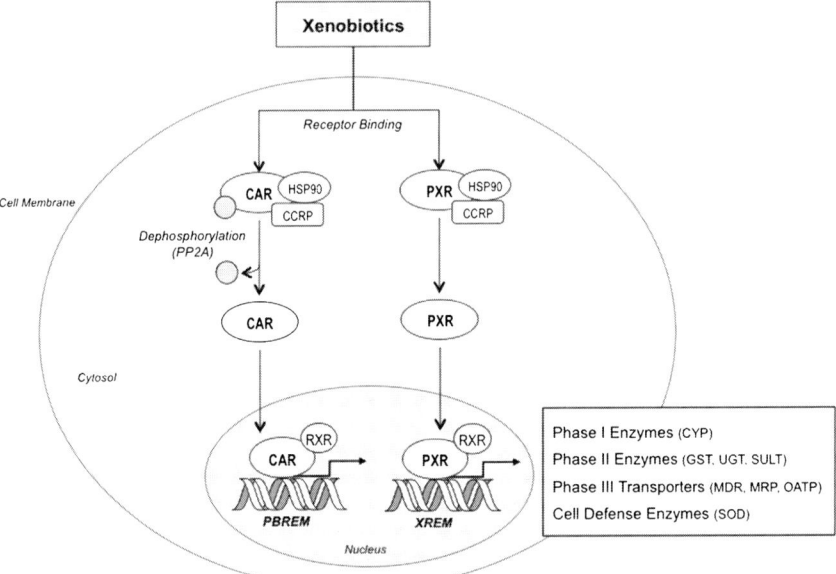

(B) Non-receptor mediated regulation of detoxication

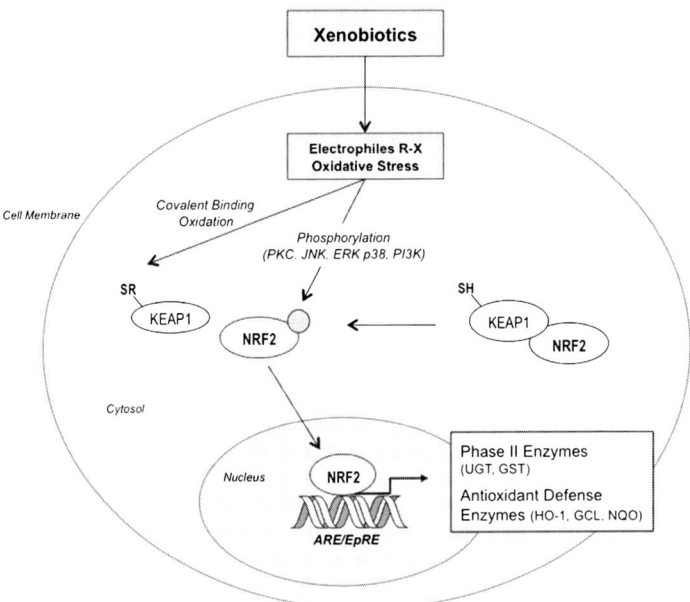

The nuclear factor-erythroid 2 p45-related factor 2 (Nrf2) is a nuclear transcription factor and key mediator of the antioxidant response which controls the coordinate induction of conjugating enzymes and antioxidant proteins upon sensing oxidative and electrophilic stress. In its inactive form, Nrf2 is sequestered in the cytoplasm by its inhibitor Keap1 (Kelch-like ECH-associating protein 1), a cysteine-rich cytosolic protein. Distinct cysteine residues in Keap1 mediate binding to Nrf2 and subsequent ubiquitin-dependent proteolysis of Nrf2. Exposure to pro-oxidants and electrophiles can induce Nrf2 activation by two distinct mechanisms (Figure 2.2.7) Oxidation or covalent modification of sensor cysteine residues by ROS or electrophiles can induce conformational changes in Keap1, resulting in stabilization of Nrf2.[69,73–75] Direct oxidation of cysteine thiol groups in Keap1, for instance, appears to be a critical step in activation of the Nrf2 pathway by reactive quinone derivatives. Alternatively, xenobiotic stress can lead to activation of intracellular signaling cascades involving protein kinases such as protein kinase C (PKC), c-Jun *N*-terminal kinase (JNK), extracellular signal regulated kinase (ERK), p38 mitogen-activated protein kinase and phosphatidylinositol 3-kinase (PI3K), which mediate dissociation of the Nrf2/Keap1 complex through phosphorylation of Nrf2 at specific serine or threonine residues.[69,73–75] Release from its repressor Keap1 allows for nuclear translocation of Nrf2 and subsequent induction of Nrf2 target genes via binding to antioxidant/electrophile response elements (ARE/ EpER). Nrf2-responsive genes encode phase II enzymes (*e.g.* GSTs, UGTs) and proteins involved in antioxidant defense and cellular stress response, such as NAD(P)H-quinone oxidoreductase (NQO), heme oxygenase 1 (HO-1) and γ-glutamylcysteine ligase (GCL).[69,73–75] Consequently, up-regulation of cytoprotective enzymes through activation of Nrf2 signaling enhances the ability of cells to cope with oxidative/electrophilic stress induced by exposure to xenobiotics, thereby providing protection from DNA damage or epigenetic change induced by reactive intermediates.

Figure 2.2.7 Cellular protection from xenobiotic stress through (A) receptor- and (B) non-receptor mediated induction of xenobiotic metabolizing enzymes and enzymes involved in cellular stress response. (A) Ligand binding to CAR and PXR causes nuclear translocation and transcriptional activation of genes encoding phase I–III enzymes and other cytoprotective proteins through binding of CAR/RXR and PXR/RXR heterodimers to specific DNA response elements. (B) Enhanced detoxification of electrophiles and ROS through Nrf2 mediated transcription of phase II enzymes and enzymes involved in cellular defense following dissociation of Nrf2 from its cytoplasmic repressor Keap1. CYP = cytochrome P450, GCL = γ-glutamylcysteine ligase, GSt = γ-glutamyltransferase, HO-1 = heme oxygenase 1, NQO = NAD(P)H-quinone oxidoreductase, SOD = superoxide dismutase, SULT = sulfotransferase, UGT = UDP-glucuronyltransferase.

2.2.2   Conclusions

The mammalian organism has developed an impressive array of systems to metabolize and deactivate toxic chemicals, and to react to chemical-induced stress. Many of these systems have a high capacity for detoxication and therefore low doses of reactive and toxic chemicals or toxically induced cellular stress may be efficiently counteracted resulting in tolerable exposures to such chemicals in low levels.

References

1. K. S. Pang, X. Xu and M. V. St-Pierre, Determinants of metabolite disposition, *Annu. Rev. Pharmacol. Toxicol.*, 1992, **32**, 623–669.
2. M. Medinsky and J. L. Valentine, Toxicokinetics, in *Casarett and Doull's Toxicology. The Basic Science of Poisons*, ed. C. D. Klassen, McGraw-Hill,New York, 2001, pp. 225–237.
3. G. C. Gibson and P. Skett, *Introduction to Drug Metabolism*, Nelson Thornes Publishers, Cheltenham, 3rd edn, 2001.
4. J. T. M. Buters, Phase I metabolism, in *Toxicology and Risk Assessment*, ed. H. Greim and R. Snyder, John Wiley & Sons, Chichester, 2008, pp. 49–74.
5. J. Muntane, Regulation of drug metabolism and transporters, *Curr. Drug Metab.*, 2009, **10**, 932–945.
6. M. S. Benedetti, R. Whomsley, I. Poggesi, W. Cawello, F. X. Mathy, M. L. Delporte, P. Papeleu and J. B. Watelet, Drug metabolism and pharmacokinetics, *Drug Metab. Rev.*, 2009, **41**, 344–390.
7. B. Testa and S. D. Kramer, The biochemistry of drug metabolism-an introduction: Part 5. Metabolism and bioactivity, *Chem. Biodivers.*, 2009, **6**, 591–684.
8. M. W. Anders, ed., *Bioactivation of Foreign Compounds*, Academic Press, Orlando, FL, 1985.
9. J. D. deBethizy and J. R. Hayes, Metabolism, a determinant of toxicity, in *Principles and Methods of Toxicology*, ed. A. W. Hayes, Raven Press, New York, 1994, pp. 59–100.
10. F. P. Guengerich, Cytochrome p450 and chemical toxicology, *Chem. Res. Toxicol.*, 2008, **21**, 70–83.
11. F. P. Guengerich, The 1992 Bernard B. Brodie award lecture, Bioactivation and detoxication of toxic and carcinogenic chemicals, *Drug Metab. Dispos.*, 1993, **21**, 1–6.
12. F. P. Guengerich and D. C. Liebler, Enzymatic activation of chemicals to toxic metabolites, *CRC Crit. Rev. Toxicol.*, 1985, **14**, 259–307.
13. W. W. Johnson, Many drugs and phytochemicals can be activated to biological reactive intermediates, *Curr. Drug Metab.*, 2008, **9**, 344–351.
14. M. W. Anders, Bioactivation mechanisms and hepatocellular damage, in *The Liver: Biology and Pathology,* ed. I. M. Arias, W. B. Jakoby, H. Popper, D. Schachter and D. A. Shafritz, Raven Press, New York, 2nd edn, 1988, pp. 389–400.

15. J. G. Bessems and N. P. Vermeulen, Paracetamol (acetaminophen)-induced toxicity: molecular and biochemical mechanisms, analogues and protective approaches, *Crit. Rev. Toxicol.*, 2001, **31**, 55–138.
16. M. W. Anders, W. Dekant and S. Vamvakas, Glutathione-dependent toxicity, *Xenobiotica*, 1992, **22**, 1135–1145.
17. M. W. Anders, Glutathione-dependent bioactivation of haloalkanes and haloalkenes, *Drug Metab. Rev.*, 2004, **36**, 583–594.
18. F. P. Guengerich, Cytochrome P450s and other enzymes in drug metabolism and toxicity, *AAPS J.*, 2006, **8**, E101–E111.
19. F. P. Guengerich, Mechanisms of cytochrome P450 substrate oxidation: MiniReview, *J. Biochem. Mol. Toxicol.*, 2007, **21**, 163–168.
20. H. Takakusa, H. Masumoto, H. Yukinaga, C. Makino, S. Nakayama, O. Okazaki and K. Sudo, Covalent binding and tissue distribution/retention assessment of drugs associated with idiosyncratic drug toxicity, *Drug Metab. Dispos.*, 2008, **36**, 1770–1779.
21. B. K. Park, D. J. Naisbitt, S. F. Gordon, N. R. Kitteringham and M. Pirmohamed, Metabolic activation in drug allergies, *Toxicology*, 2001, **158**, 11–23.
22. A. Srivastava, J. L. Maggs, D. J. Antoine, D. P. Williams, D. A. Smith and B. K. Park, Role of reactive metabolites in drug-induced hepatotoxicity, *Handb. Exp. Pharmacol.*, 2010, **196**, 165–194.
23. D. C. Liebler and F. P. Guengerich, Olefin oxidation by cytochrome P-450: evidence for group migration in catalytic intermediates formed with vinylidene chloride and *trans*-1-phenyl-1-butene, *Biochemistry*, 1983, **22**, 5482–5489.
24. S. D. Aust, C. F. Chignell, T. M. Bray, B. Kalyanaraman and R. P. Mason, Free radicals in toxicology, *Toxicol. Appl. Pharmacol.*, 1993, **120**, 168–178.
25. B. D. Goldstein, B. Czerniecki and G. Witz, The role of free radicals in tumor promotion, *Environ. Health Perspect.*, 1989, **81**, 55–57.
26. B. D. Goldstein and G. Witz, Free radicals and carcinogenesis, *Free Radical Res. Commun.*, 1990, **11**, 3–10.
27. J. P. Kehrer, B. T. Mossman, A. Sevanian, M. A. Trush and M. T. Smith, Free radical mechanisms in chemical pathogenesis. Summary of the symposium presented at the 1988 annual meeting of the Society of Toxicology, *Toxicol. Appl. Pharmacol.*, 1988. **95**, 349–362.
28. P. A. Cerutti and B. F. Trump, Inflamation and oxidative stress in carcinogenesis, *Cancer Cell*, 1991, **3**, 1–7.
29. E. Chacon and D. Acosta, Mitochondrial regulation of superoxide by Ca^{2+}: An alternate mechanism for the cardiotoxicity of doxorubicin, *Toxicol. Appl. Pharmacol.*, 1991, **107**, 117–128.
30. R. H. Burdon, V. Gill and C. Rice-Evans, Oxidative stress and tumour cell proliferation, *Free Rad. Res. Commun.*, 1990, **11**, 65–76.
31. R. W. Pero, G. C. Roush, M. M. Markowitz and D. G. Miller, Oxidative stress, DNA repair, and cancer susceptibility, *Cancer Detect. Prev.*, 1990, **14**, 551–561.

32. D. B. Kell, Towards a unifying, systems biology understanding of large-scale cellular death and destruction caused by poorly liganded iron: Parkinson's, Huntington's, Alzheimer's, prions, bactericides, chemical toxicology and others as examples, *Arch. Toxicol.*, 2010, **84**, 825–889.

33. D. Ziech, R. Franco, A. G. Georgakilas, S. Georgakila, V. Malamou-Mitsi, O. Schoneveld, A. Pappa and M. I. Panayiotidis, The role of reactive oxygen species and oxidative stress in environmental carcinogenesis and biomarker development, *Chem. Biol. Interact.*, 2010, **188**, 334–339.

34. D. Ziech, R. Franco, A. Pappa and M. I. Panayiotidis, Reactive oxygen species (ROS)–induced genetic and epigenetic alterations in human carcinogenesis, *Mutat. Res.*, 2011, **711**, 167–173.

35. I. M. Arias and W. B. Jakoby, ed. *Glutathione: Metabolism and Function*, Raven Press, New York, 1976.

36. B. Ketterer, D. J. Meyer and A. G. Clark, Soluble glutathione transferase isoenzymes, in *Glutathione Conjugation: Mechanisms and Biological Significance*, ed. H. Sies and B. Ketterer, Academic Press, London, 1988, pp. 73–135.

37. R. Morgenstern, G. Lundqvist, G. Andersson, L. Balk and J. W. Depierre, The distribution of microsomal glutathione transferase among different organelles, different organs, and different organisms, *Biochem. Pharmacol.*, 1984, **33**, 3609–3614.

38. J. D. Hayes, J. U. Flanagan and I. R. Jowsey, Glutathione transferases, *Annu. Rev. Pharmacol. Toxicol.*, 2005, **45**, 51–88.

39. J. L. Cmarik, P. B. Inskeep, M. J. Meredith, D. J. Meyer, B. Ketterer and F. P. Guengerich, Selectivity of rat and human glutathione *S*-transferases in activation of ethylene dibromide by glutathione conjugation and DNA binding and induction of unscheduled DNA synthesis in human hepatocytes, *Cancer Res.*, 1990, **50**, 2747–2752.

40. J. Caldwell and W. B. Jakoby, ed., *Biological Basis of Detoxification*, Academic Press, New York, 1983.

41. E. Boyland and L. F. Chasseaud, Role of glutathione and glutathione S-transferases in mercapturic acid biosynthesis, *Adv. Enzymol.*, 1969, **32**, 173–177.

42. W. B. Jakoby and D. M. Ziegler, The enzymes of detoxication, *J. Biol. Chem.*, 1990, **265**, 20715–20719.

43. H. Sies, Zur Biochemie der Thiolgruppe: Bedeutung des Glutathions, *Naturwissenschaften*, 1989, **76**, 57–64.

44. D. M. Ziegler, Microsomal flavin-containing monooxygenase: oxygenation of nucleophilic nitrogen and sulfur compounds, in *Enzymic Basis of Detoxification*, ed. W. B. Jacoby, Academic Press, New York, 1980, pp. 201–227.

45. W. Völkel, M. Friedewald, E. Lederer, A. Pähler, J. Parker and W. Dekant, Biotransformation of perchloroethene: Dose-dependent excretion of trichloroacetic acid, dichloroacetic acid and *N*-acetyl-*S*-(trichlorovinyl)-L-cysteine in rats and humans after inhalation, *Toxicol. Appl. Pharmacol.*, 1998, **153**, 20–27.

46. T. A. Fjellstedt, R. H. Allen, B. K. Duncan and W. B. Jakoby, Enzymatic conjugation of epoxides with glutathione, *J. Biol. Chem.*, 1973, **248**, 3702–3707.
47. R. C. James and R. D. Harbison, Hepatic glutathione and hepatotoxicity, *Biochem. Pharmacol.*, 1982, **31**, 1829–1835.
48. L. I. McLellan, C. R. Wolf and J. D. Hayes, Human microsomal glutathione S-transferase. Its involvement in the conjugation of hexachlorobuta-1,3-diene with glutathione, *Biochem. J.*, 1989, **258**, 87–93.
49. A. Oakley, Glutathione transferases: A structural perspective, *Drug Metab. Rev.*, 2011, **43**, 138–151.
50. E. R. Stadtman and C. N. Oliver, Metal-catalyzed oxidation of proteins, *J. Biol. Chem.*, 1991, **266**, 2005–2008.
51. E. R. Stadtman, Protein oxidation and aging, *Science*, 1992, **257**, 1220–1224.
52. H. Esterbauer, H. Zollner and R. J. Schaur, Hydroxyalkenals: cytotoxic products of lipid peroxidation, *ISI Atlas of Sci.: Biochem.*, 1988, **1**, 311–317.
53. H. Kappus, Overview of enzyme systems involved in bioreduction of drugs and in redox cycling, *Biochem. Pharmacol.*, 1986, **35**, 1–6.
54. J. L. Farber, The role of calcium in lethal cell injury, *Chem. Res. Toxicol.*, 1990, **3**, 503–508.
55. J. P. Kehrer, D. P. Jones, J. J. Lemasters, J. L. Farber and H. Jaeschke, Mechanisms of hypoxic cell injury, *Toxicol. Appl. Pharmacol.*, 1990, **106**, 165–178.
56. K. Hemminki, DNA adducts, mutations and cancer, *Carcinogenesis*, 1993, **14**, 2007–2012.
57. B. N. Ames, Endogenous DNA damage as related to cancer and aging, *Mutat. Res.*, 1989, **214**, 41–46.
58. R. Adelman, R. L. Saul and B. N. Ames, Oxidative damage to DNA: Relation to species metabolic rate and life span, *Proc. Natl. Acad. Sci. U. S. A.*, 1988, **85**, 2706–2708.
59. S. M. Cohen and L. B. Ellwein, Cell proliferation in carcinogenesis, *Science*, 1990, **249**, 1007–1011.
60. B. Butterworth, Consideration of both genotoxic and non–genotoxic mechanisms in predicting carcinogenic potential, *Mutat. Res.*, 1990, **239**, 117–132.
61. Q. Chen, J. marsh, B. Ames and B. Mossman, Detection of 8-oxo-2'-deoxyguanosine, a marker of oxidative DNA damage, in culture medium from human mesothelial cells exposed to crocidolite asbestos, *Carcinogenesis*, 1996, **17**, 2525–2527.
62. J. R. Mitchell, S. S. Thorgeirsson, W. Z. Potter, D. J. Jollow and H. Keiser, Acetaminophen-induced hepatic injury: protective role of glutathione in man and rationale for therapy, *Clin. Pharmacol. Ther.*, 1974, **16**, 676–84.
63. D. J. Jollow, S. S. Thorgeirsson, W. Z. Potter, M. Hashimoto and J. R. Mitchell, Acetaminophen-induced hepatic necrosis. VI. Metabolic disposition of toxic and nontoxic doses of acetaminophen, *Pharmacology*, 1974, **12**, 251–271.

64. W. Z. Potter, S. S. Thorgeirsson, D. J. Jollow and J. R. Mitchell, Acetaminophen-induced hepatic necrosis. V. Correlation of hepatic necrosis, covalent binding and glutathione depletion in hamsters, *Pharmacology*, 1974, **12**, 129–143.

65. J. R. Mitchell, D. J. Jollow, W. Z. Potter, J. R. Gillette and B. B. Brodie, Acetaminophen-induced hepatic necrosis. IV. Protective role of glutathione, *J. Pharmacol. Exp. Ther.*, 1973, **187**, 211–217.

66. D. J. Jollow, J. R. Mitchell, W. Z. Potter, D. C. Davis, J. R. Gillette and B. B. Brodie, Acetaminophen-induced hepatic necrosis. II. Role of covalent binding *in vivo*, *J. Pharmacol. Exp. Ther.*, 1973, **187**, 195–202.

67. J. R. Mitchell, D. J. Jollow, W. Z. Potter, D. C. Davis, J. R. Gillette and B. B. Brodie, Acetaminophen-induced hepatic necrosis. I. Role of drug metabolism, *J. Pharmacol. Exp. Ther.*, 1973, **187**, 185–194.

68. W. Z. Potter, D. C. Davis, J. R. Mitchell, D. J. Jollow, J. R. Gillette and B. B. Brodie, Acetaminophen-induced hepatic necrosis. 3. Cytochrome P-450-mediated covalent binding *in vitro*, *J. Pharmacol. Exp. Ther.*, 1973, **187**, 203–210.

69. C. Xu, C. Y. Li and A. N. Kong, Induction of phase I, II and III drug metabolism/transport by xenobiotics, *Arch. Pharm. Res.*, 2005, **28**, 249–268.

70. Y. E. Timsit and M. Negishi, CAR and PXR: the xenobiotic-sensing receptors, *Steroids*, 2007, **72**, 231–246.

71. A. di Masi, E. De Marinis, P. Ascenzi and M. Marino, Nuclear receptors CAR and PXR: Molecular, functional, and biomedical aspects, *Mol. Aspects Med.*, 2009, **30**, 297–343.

72. S. Kakizaki, Y. Yamazaki, D. Takizawa and M. Negishi, New insights on the xenobiotic-sensing nuclear receptors in liver diseases-CAR and PXR, *Curr. Drug Metab.*, 2008, **9**, 614–621.

73. Y. J. Surh, J. K. Kundu and H. K. Na, Nrf2 as a master redox switch in turning on the cellular signaling involved in the induction of cytoprotective genes by some chemopreventive phytochemicals, *Planta Med.*, 2008, **74**, 1526–1539.

74. J. W. Kaspar, S. K. Niture and A. K. Jaiswal, Nrf2:INrf2 (Keap1) signaling in oxidative stress, *Free Radical Biol. Med.*, 2009, **47**, 1304–1309.

75. S. K. Niture, J. W. Kaspar, J. Shen and A. K. Jaiswal, Nrf2 signaling and cell survival, *Toxicol. Appl. Pharmacol.*, 2010, **244**, 37–42.

3. DNA Repair

CHAPTER 3.1

Consequences and Repair of Oxidative DNA Damage

STÉPHANIE DUCLOS, SYLVIE DOUBLIÉ AND
SUSAN S. WALLACE*

Department of Microbiology and Molecular Genetics, The Markey Center
for Molecular Genetics, University of Vermont, Stafford Hall, 95 Carrigan
Drive, Burlington, VT 05405-0068, USA
*Email: susan.wallace@uvm.edu

3.1.1 Introduction

Cellular DNA is constantly subjected to damaging agents, predominantly reactive oxygen species (ROS) (for reviews see refs. 1 and 2). The most common source of ROS is the mitochondria where approximately 1% of the oxygen intake leads to the production of the superoxide anion radical, $O_2^{\bullet-}$, which can either be hydrated to highly reactive hydroxyl radicals, $^{\bullet}OH$, or converted by superoxide dismutase to hydrogen peroxide. In the presence of transition metals such as Fe^{2+}, H_2O_2 can form $^{\bullet}OH$ radicals in a Fenton reaction. In addition, $O_2^{\bullet-}$ can react with $^{\bullet}NO$ to produce peroxynitrite, $OONO^-$; in the presence of carbon dioxide and other reactants, $OONO^-$ can also lead to the formation of $^{\bullet}OH$ radicals (for reviews see refs. 3 and 4). Both H_2O_2 and $^{\bullet}NO$ are readily permeable and diffuse throughout the cell. In fact, the steady-state cellular concentration of hydrogen peroxide is 10 nM.[5] Outside of the mitochondria, ROS are formed by cytochrome P450-dependent reactions,[6,7]

Issues in Toxicology No. 13
The Cellular Response to the Genotoxic Insult: The Question of Threshold for
Genotoxic Carcinogens
Edited by Helmut Greim and Richard J. Albertini
© The Royal Society of Chemistry 2012
Published by the Royal Society of Chemistry, www.rsc.org

by P450 peroxisomes,[8] by NADPH oxidase in macrophages, neutrophils, and endothelial cells and by lipoperoxidation.[9,10] Reactive oxygen species can also result from the intracellular metabolism of xenobiotics and drugs. The steady-state levels of ROS and H_2O_2 are maintained by detoxifying mechanisms involving enzymes such as glutathione reductase and catalase. Since H_2O_2 and ROS are used in a variety of cell signaling pathways (for reviews see refs. 11–16), a basic steady state level must be maintained.

Common exogenous sources that result in free radical ROS production are inflammation, ionizing radiation and ultraviolet (UV) light (for reviews see refs. 17 and 18). Inflammation has been well-documented to occur during exposure to a number of exogenous chemical agents such as cigarette smoke as well as during infection, for example *Helicobacter pylori* (for reviews see refs. 19 and 20). Upon exposure to UV light, UVB produces cellular reactive oxygen species. However, the most-well studied source of •OH radical production is ionizing radiation; •OH are produced during the radiolysis of water and, if this occurs in the vicinity of the DNA molecule, direct damage to DNA results.

3.1.2   Oxidative DNA Lesions

3.1.2.1   Types of Oxidative DNA Lesions

Although many types of ROS can be produced by the processes described above, the most highly reactive species is the •OH radical, which is responsible for the majority of oxidative DNA damages (reviewed in refs. 21–23). Hydroxyl radicals can damage DNA bases and sugar moieties, and can generate DNA strand breaks (for a review see ref. 24). The •OH radical adds to the C5 and C6 positions of thymine and cytosine generating the $C5\text{-OH}^-$ and $C6\text{-OH}^-$ adduct radicals, respectively. Oxidation of the $C5\text{-OH}^-$ adduct radicals of cytosine and thymine leads to cytosine glycol and thymine glycol, which are the most common products (see Figure 3.1.1). Oxidation of the allyl radical of thymine generates 5-hydroxymethyluracil and 5-formyluracil, whereas oxidation of the $C6\text{-OH}^-$ adduct radical of thymine gives rise to 5-hydroxy-5-methyl-hydantoin. The cytosine products are unique insofar as they can both dehydrate and deaminate.[22,23,25–27] Cytosine glycol is unstable and deaminates to uracil glycol, which can dehydrate to 5-hydroxyuracil while the direct dehydration product of cytosine glycol is 5-hydroxycytosine (Figure 3.1.1).

Guanine has the lowest redox potential of the four DNA bases, and hence is the most readily oxidized, giving rise to the $C8\text{-OH}^-$ adduct radical. Under oxidizing conditions 8-oxoguanine is formed, whereas 2,6-diamino-4-hydroxy-5-formamidopyrimidine (FapyG) is formed under reducing conditions (Figure 3.1.2) (for reviews see refs. 22, 23, 28 and 29). 8-Oxoguanine in DNA also has a lower redox potential than the four normal bases and can be further oxidized to two major products, spiroiminodihydantoin (Sp) and guanidino-hydantoin (Gh) (Figure 3.1.2) (for review see ref. 30). As with guanine, the $C8\text{-OH}^-$ radical adduct of adenine gives rise to either 8-oxoadenine or

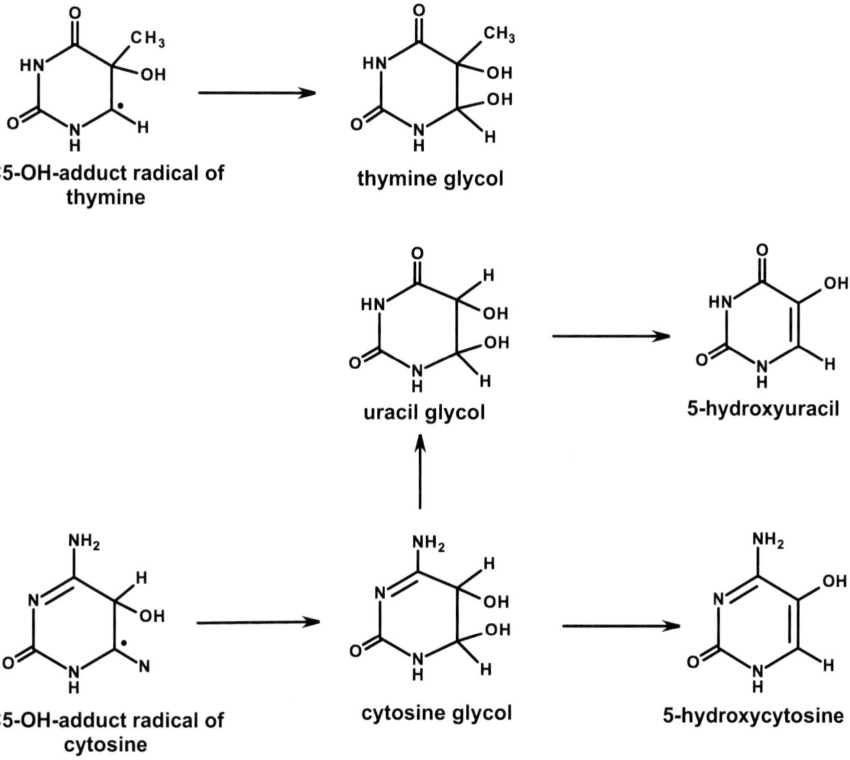

Figure 3.1.1 Structures of the common oxidation products of DNA pyrimidines.

4,6-diamino-5-formamidopyrimidine (FapyA) (Figure 3.1.2). 8-Oxopurines and formamidopyrimidines are unique in that they are formed in DNA both in the presence and absence of oxygen, although the presence of oxygen or the presence of reducing agents can affect yields. 2-Hydroxyadenine can also be formed by hydroxyl radical attack at the C2 position of adenine and α deoxyadenosine is produced by ionizing radiation.[31]

In addition, hydroxyl radicals can extract an H atom from each of the carbon atoms of the sugar moiety in DNA, resulting either in strand breaks or sites of base loss. A substantial number of sites of base loss, about 18 000 per human cell per day, are formed endogenously by spontaneous depurination.[32] Single-strand breaks are the hallmarks of agents that act *via* free radical formation such as ionizing radiation and hydrogen peroxide. Strand breaks formed in this manner usually have 3′-blocked termini with phosphate or phosphoglycolate groups derived from the remnants of the deoxyribose moiety.[21,33,34] In the case where the DNA backbone is not cleaved and the DNA base is released, the resulting sugar moiety can adopt a number of different structures depending on which carbon was attacked and the subsequent chemical reactions (for a review see ref. 35).

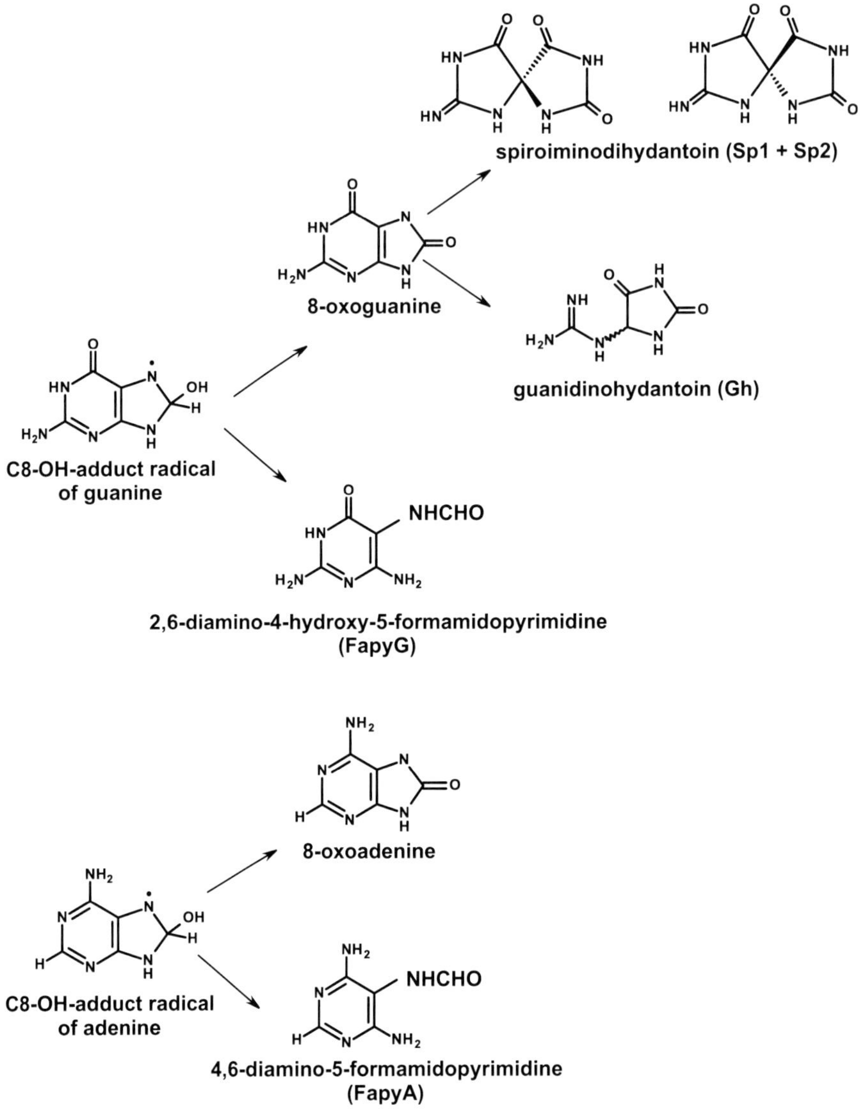

Figure 3.1.2   Structures of the common oxidation products of DNA purines.

3.1.2.2   Consequences of Unrepaired Oxidative DNA Damages

The biological consequences of oxidative DNA lesions have been predicted by studying the interactions of individual lesions with DNA polymerases *in vitro* and by assessing the potential lethality and/or mutagenicity of these same lesions in biologically active DNA transfected into bacterial or mammalian cells. If a lesion blocks a replicative DNA polymerase, it is potentially lethal although it may be bypassed by the plethora of cellular translesion DNA polymerases (for reviews see refs. 36–40). If a lesion does not block the DNA polymerase and if misinsertion of a non-cognate base occurs, the lesion is potentially mutagenic (for reviews see refs. 30 and 41–47).

The most common stable oxidized cytosine bases found in DNA (*i.e.* uracil glycol, 5-hydroxyuracil and 5-hydroxycytosine) do not distort the DNA molecule and thus are readily bypassed by the replicative DNA polymerases.[48,49] In the case of uracil glycol and 5-hydroxyuracil, A is always inserted opposite.[48,49] Since both damages are derived from DNA cytosine and miscode for A, they are potentially mutagenic. 5-Hydroxycytosine has also been shown *in vitro* to miscode for adenine.[48] When single-stranded biologically active DNA containing the oxidized cytosine residues was transfected into *Escherichia coli*, the *in vivo* studies mirror the *in vitro* work showing that 5-hydroxyuracil and uracil glycol are not cytotoxic but induce C to T transitions at very high efficiencies.[50] 5-Hydroxycytosine, which was also not cytotoxic, was much less mutagenic than the uracil residues, presumably because it often codes for cognate G. 5-Hydroxycytosine primarily gives rise to C toT transitions with some C to G transversions.[50,51]

In contrast to the cytosine oxidation products, thymine glycol distorts the DNA duplex, although the base pairing properties of thymine glycol remain intact.[52–54] *In vitro* primer extension studies with proofreading DNA polymerases show progress up to the site of the thymine glycol lesion and insertion of an A opposite, but no further extension occurs;[55–58] G is inserted several orders of magnitude less efficiently.[59] There are, however, certain sequence contexts that facilitate polymerase bypass of the thymine glycol lesion.[58,60,61] Recent X-ray crystallographic studies have confirmed the earlier *in silico* studies which predicted that the methyl group at the 5-position of thymine glycol would hinder the proper stacking of the neighboring 5′ template base into the helix.[53,54,62] The displaced 5′ template base is then unable to pair with an incoming deoxynucleoside triphosphate[63] (Figure 3.1.3). When introduced into either single-stranded or double-stranded biologically active DNA molecules, and transfected into *E. coli*, thymine glycol is lethal.[64–67] On the other hand, thymine glycol is almost 100% bypassed under SOS conditions in *E. coli*.[67,68] A number of the Y- and A-family DNA polymerases have been shown to readily bypass the otherwise blocking thymine glycol lesion, inserting the correct base opposite.[69–72] As expected, thymine glycol is not mutagenic.[66] Like thymine glycol, 5-hydroxy-5,6-dihydrothymine is also a strong block to Klenow fragment in template DNA.[73,74]

The thymine oxidation product, 5-hydroxymethyluracil, is non blocking, pairs correctly *in vitro*, and is therefore a poor premutagenic lesion.[75,76] 5-Formyluracil is another thymine oxidation product which does not block DNA polymerases, although misincorporation of guanine and cytosine is observed *in vitro* opposite the lesion.[77,78] When 5-formyluracil is placed in a plasmid vector and assayed in *E. coli*, T to C transitions and T to A transversions are observed,[79] and in mammalian cells both types of transversions are found.[80] The consequences of 5-hydroxy-5-methylhydantoin have been measured *in vitro* using primer extension assays and the lesion was shown to be strongly

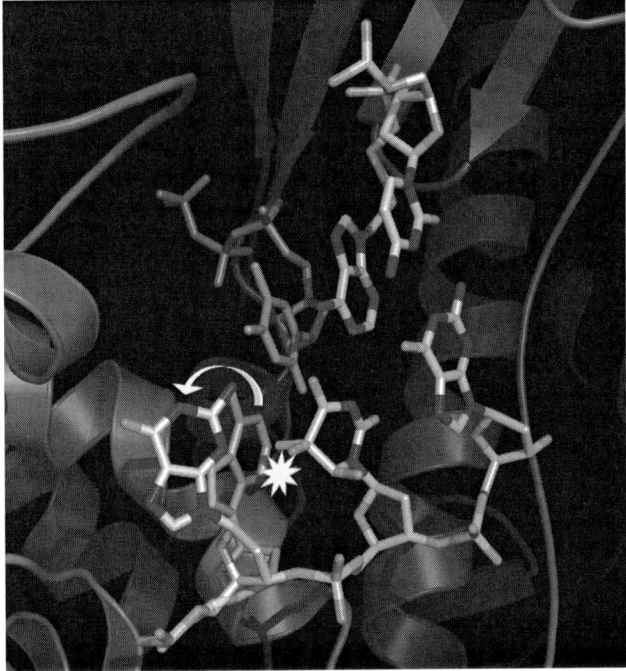

Figure 3.1.3   Effect of a templating thymine glycol on the 5′ templating guanine in the active site of a replicating DNA polymerase.

blocking with adenine being the preferred base inserted.[81] In *E. coli nth nei* mutants, defective in the removal of oxidized thymines and cytosines, all of the spontaneous mutations observed are C to T transitions supporting both the *in vitro* data as well as the data with transfecting DNA demonstrating that thymine oxidation products are poor premutagenic lesions.[82]

Of all the oxidative DNA lesions, 8-oxoguanine is the most well-studied and is often taken as the benchmark for oxidative DNA damage. 8-Oxoguanine does not distort the DNA molecule and is readily bypassed by the replicative polymerases.[83–86] In addition to pairing with cognate C, 8-oxoG readily forms Hoogsteen base pairs with A, which is often inserted by high-fidelity DNA polymerases (for reviews see refs. 87 and 88). When site specifically introduced into biologically active DNA and transfected into either bacterial or mammalian cells, 8-oxoguanine is mutagenic leading to G to T transversions.[83,85,89,90]

Although FapyG is formed in similar amounts in cellular DNA as 8-oxoG, it has only recently been chemically synthesized[91–93] and therefore methyl-FapyG has been used as a surrogate. Methyl-FapyG, produced by alkylating agents, is a strong block to DNA polymerases and is lethal.[46,94–97] In contrast FapyG, like 8-oxoguanine, can be bypassed by Klenow exo⁻ DNA polymerase which inserts adenine opposite[98] and in *E. coli* leads to G-T transversions. In a single sequence context, FapyG is not as mutagenic in *E. coli* as it is in simian kidney cells;[99,100] however, recent studies suggest that FapyG is as potent a premutagenic lesion in *E. coli* as 8-oxoguanine.[101]

Similar to guanine, the 8-position on the imidazole ring of adenine is also susceptible to oxidation resulting in 8-oxoadenine. 8-Oxoadenine is neither a distorting nor a blocking lesion to DNA polymerases *in vitro* and thymine is exclusively incorporated opposite 8-oxoA.[102] In *E. coli*, 8-oxoA is neither cytotoxic nor mutagenic.[103] With mammalian polymerases, however, some dGTP is incorporated opposite the 8-oxoA and in mammalian cells, A to G and A to C mutations have been observed.[104–106] Like FapyG, FapyA is also formed and is a weak block to DNA synthesis *in vitro*[107] and chemically synthesized FapyA causes preferential insertion of A opposite.[108] In a mammalian cell line FapyA was at best weakly mutagenic.[99] 2-Hydroxyadenine is bypassed *in vitro* and all bases appear to be misincorporated opposite.[109,110] With both single- and double-stranded vectors containing 2-hydroxyadenine, no cytotoxicity was observed in *E. coli*; however, A to T and A to G substitutions were observed.[111] In mammalian Cos-7 cells, a −1 deletion was the most frequent mutation found followed by A to T transversions.[112] α-Deoxyadenosine is a moderate block to DNA synthesis *in vitro*[113] and in *E. coli*, one-base deletions are found.[114]

Spiroiminodihydantoin (Sp) and guanidinohydantoin (Gh), the further oxidation products of 8-oxoG, are both strong blocks to Klenow fragment and Klenow fragment exo⁻ with A and G being primarily inserted opposite the lesions.[115,116] A crystal structure of Gh with a replicative polymerase shows Gh to be extrahelical and rotated towards the major groove and no longer in a position to serve as a template for an incoming base.[117] When specifically inserted into single-stranded DNA and introduced into *E. coli*, Gh is more readily bypassed than Sp. In the case of Gh, G to C transversions were almost exclusively observed, whereas for Sp, depending on the isomer, G to C transversions predominated over G to T transversions.[118,119] Additional secondary lesions are oxaluric acid, oxazalone and cyanuric acid which are readily bypassed in *E. coli* with insertion of A opposite.[120]

The consequences of sites of base loss or abasic (AP) sites in DNA have been primarily studied using two types of AP sites, an abasic site produced by the action of uracil glycosylase on a uracil-containing substrate and a chemically synthesized tetrahydrofuran—the latter being significantly more stable than a standard AP site. Abasic sites are strong blocks to replicative polymerases[121–125] and X-ray crystallographic analysis has shown that the DNA containing an abasic site in the polymerase active site distorts the DNA, which does not translocate after incorporation of dAMP.[126] In contrast, some Y family polymerases are able to bypass abasic sites with crystal structures showing that the active site is able to accommodate looping out of the template strand containing the abasic site with pairing of the incoming base occurring with the upstream base on the template strand (for reviews see refs. 127–129). Subsequently, the abasic site repositions itself opposite the newly incorporated base. In *E. coli*, in the absence of SOS, abasic sites are lethal lesions,[65,67,130–132] when bypassed, abasic sites are mutagenic.[130,131,133–135]

Biologically active DNA containing single-strand breaks has also been investigated and single-strand breaks have been shown to be lethal in *E. coli*.[136] Single-strand breaks block replication forks and need to be processed by homology directed repair (for reviews see refs. 137 and 138).

3.1.3 Repair of Oxidative DNA Damages

3.1.3.1 The Base Excision Repair (BER) Pathway

As discussed above, the plethora of DNA lesions arising from oxidative stress constantly jeopardize the integrity of the genome. In order to maintain genomic stability and prevent the initiation and progression of many diseases, the cell must be able to detect and repair the damage sustained by its DNA. The vast majority of oxidative lesions are repaired by the base excision repair (BER) system, a highly conserved process among organisms.[139–143]

The BER system discovered in *E.coli* 35 years ago by Tomas Lindahl as the mechanism to remove uracil from DNA[144] turns out to be the primary pathway to repair oxidized bases, alkylated bases, abasic sites (AP site) and single-strand breaks in most organisms. The core of this versatile and ubiquitous DNA repair pathway can be characterized by five distinct enzymatic reactions (for reviews see refs. 145–148 and Figure 3.1.4). The recognition of the lesion by the appropriate DNA glycosylase is a crucial step in this process since it initiates the BER enzymatic cascade. In the wake of lesion recognition, the DNA glycosylase catalyzes cleavage of the *N*-glycosylic bond that links the damaged base to the deoxyribose, generating an AP site. The second step corresponds to the cleavage of the DNA backbone at the abasic site by a DNA AP endonuclease (APE). APE creates a nick in the DNA 5′ of the AP site, resulting in 3′OH and 5′deoxyribose phosphate (5′dRP) termini (for a review see ref. 149). The removal of the 5′dRP blocking end is carried out by the intrinsic 5′dRP lyase activity of DNA polymerase β (Polβ). Polβ then fills in the gap with the appropriate nucleotide[150–153] providing the correct substrate for an efficient ligation. Finally, the remaining nick is sealed by the formation of a phosphodiester bond catalyzed by the DNA ligase III (LigIII) (for a review see ref. 154).

Alternatively, the second step of the core BER process, cleavage of the DNA backbone, can be carried out by the DNA AP lyase activity present in some DNA glycosylases.[155,156] These bifunctional DNA glycosylases not only excise the lesion base but also cleave the DNA backbone 3′ of the AP site, creating 3′ phopho α/β-unsaturated aldehyde (3′PUA) and 5′phosphate termini *via* a β-elimination reaction. In some cases, the β-elimination reaction is followed by δ-elimination reaction that cleaves PUA leaving a 3′P terminus at the strand break. The 3′PUA product of the β-elimination reaction, and the 3′P product of β,δ-elimination are further processed by APE and polynucleotide kinase (PNK),[157] respectively, in order to generate a 3′OH substrate for Polβ. The repair process is then completed by a subsequent gap filling and ligation as described above.

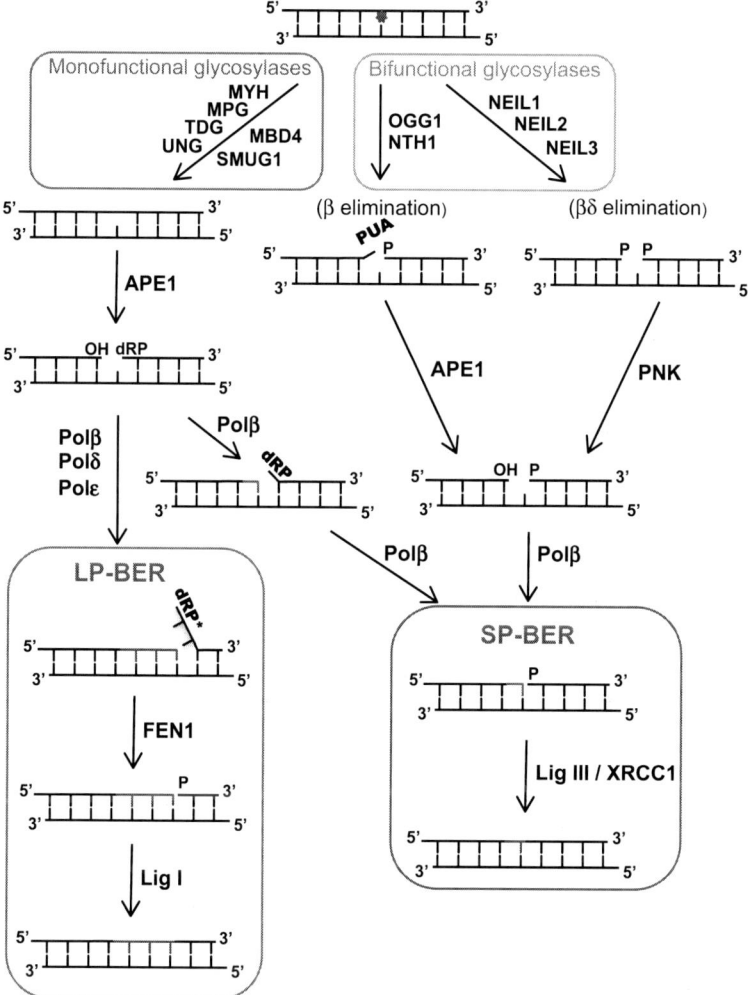

Figure 3.1.4 Base excision repair pathway. Monofunctional glycosylases and bifunctional glycosylases recognize and excise the DNA lesion, represented as a star. The AP site generated by monofunctional glycosylases is cleaved by APE1 and Polβ processes the resulting 5′ blocking end after filling in the gap with the proper nucleotide. LigIII, in complex with XRCC1, is then required to seal the nick in the DNA. Repair of oxidized DNA bases involves bifunctional glycosylases that cleave the DNA *via* β or βδ elimination. APE1 or PNK is required to clean 3′ blocking ends to allow Polβ to fill in the gap and LigIII to ligate the DNA. LP-BER is used when the 5′ dRP blocking end is not a substrate for the lyase activity of Polβ. Polβ, Polδ and Polε synthesize a fragment of 2–8 nucleotides whereas FEN1 removes the unpaired flap and LigI seals the DNA nick.

3.1.3.2 The DNA Glycosylases

The initiation of the BER pathway is carried out by the DNA *N*-glycosylases, a group of enzymes specialized in the recognition and the catalytic removal of a damaged base. The DNA glycosylases have developed a remarkable search mechanism that allows for the detection of a single damaged base in the context of thousands of normal bases. Although the crystal structures of many DNA glycosylases have been solved, the lesion detection mechanism still stands as an enigma for the DNA repair community.[158–160]

In mammals, eleven DNA glycosylases have been characterized so far, five of which are specific for oxidized bases (for reviews see refs. 142, 161 and 162). Although each of these enzymes recognizes and repairs a limited number of lesions, they all display a common excision mechanism involving the flipping out of the damaged base from the DNA helix followed by a nucleophilic attack at the C1′ of the target nucleotide promoting catalysis (for reviews see refs. 163 and 164). Interestingly, the glycosylases specialized in the detection and excision of oxidized lesions are exclusively bifunctional, and cleave AP sites *via* β- or β,δ-elimination reactions.

These enzymes are divided in two families based on structural and mechanistic criteria: the 'Helix-hairpin-Helix' (HhH) superfamily and the Fpg/Nei family (for reviews see refs. 163 and 165). The human enzymes, endonuclease III (NTH1) and 8-oxoguanine-DNA-glycosylase (OGG1) have been cloned and characterized, and belong to the HhH superfamily[166–173] named after the helix-hairpin-helix structural motif involved in DNA binding[174] that is present in all the members of the superfamily. The HhH enzymes, including NTH1 and OGG1, share several structural features as well as a common catalytic mechanism involving a lysine as the active site nucleophile.[175–178] The nucleophilic attack at the C1′ promotes the base excision that is promptly followed by a β-elimination reaction characteristic of this class of enzymes. Despite their similarities, NTH1 and OGG1 exhibit opposite substrate specificities. NTH1 recognizes and removes a large panel of oxidized pyrimidines, including thymine glycol (Tg), whereas OGG1 is responsible for the excision of 8-oxoguanine (8-oxoG) as well as other oxidized purines such as 8-oxoA and formamidopyrimidine (Fapy) lesions.[179–181] A third member of the HhH superfamily involved in oxidative damage processing is MUTYH. MUTYH, a monofunctional glycosylase, is an ortholog of *E. coli* MutY.[182–185] Both enzymes remove adenine misincorporated opposite 8-oxoG and FapyG and thus are important antimutators.

Three other oxidized base-specific glycosylases, endonuclease VIII-like 1 (NEIL1), NEIL2 and NEIL3 have been characterized in mammalian cells and are members of the Fpg/Nei family, named after the prototypical bacterial members Fpg (formamidopyrimidine DNA glycosylase or MutM) and Nei (endonuclease VIII).[101,186–190] Indeed, the three mammalian NEILs show a significant sequence homology with Fpg and Nei, suggesting that all three enzymes share a similar overall topology with the members of the Fpg/Nei superfamily. This assumption was confirmed by the crystal structure of human

NEIL1 showing that it is structurally related to Fpg and Nei.[191] As predicted from the primary sequences, the NEIL proteins harbor a 'helix-two turns-helix' (H2TH) motif, a hallmark of the Fpg/Nei superfamily known to interact with the DNA backbone. Interestingly, only NEIL2 and NEIL3 sequences feature the canonical zinc-binding residues required for the formation of a *C*-terminal zinc-finger motif—another Fpg/Nei signature motif involved in the DNA binding.[101,192] Although NEIL1 lacks the zinc ion coordination site at its *C*-terminus, the hNEIL1 structure revealed an anti-parallel β-hairpin that closely mimics the zinc-finger motif.[191] This zincless-finger structure also harbors a conserved arginine shown to be instrumental in glycosylase activity.[191,193] In addition to the structural similarity, the members of the Fpg/Nei superfamily exhibit a common catalytic mechanism involving the *N*-terminal residue (proline for NEIL1 and 2 and valine for NEIL3) as the active site nucleophile.[194,195] Like the other Fpg/Nei orthologs, the AP lyase activity of NEIL1 and 2 usually results in a β,δ-elimination product.[186–188]

Over the last decade, NEIL1 and NEIL2 have been extensively studied and it was at first established that these two enzymes exhibit marked preference for oxidized pyrimidines.[186–190] The best substrates for NEIL1 are Sp and Gh,[196,197] with the ring-opened purines (FapyG and FapyA)[198,199] and 8-oxoA[200] also being good substrates. Surprisingly, both NEIL1 and NEIL2 proteins show an unusual preference for lesions in bubble or forked DNA structures compared to double-stranded DNA,[197,201] suggesting that these enzymes may be involved in replication events or transcription. Although NEIL1, NEIL2 and NEIL3 were concurrently identified, the functionality of NEIL3 remained obscure until the recent expression, purification and characterization of the *Mus musculus* Neil3 by Liu and colleagues.[101] This study reports that MmuNeil3 exhibits a broad substrate specificity ranging from oxidized pyrimidines to ring-opened purines as well as the oxidation products of 8-oxoG (Sp and Gh), but with a distinct preference for single-stranded DNA and large bubble structures. In contrast to NEIL1 and NEIL2, NEIL3 utilizes an *N*-terminal valine as catalytic residue and primarily completes its lyase activity with a β-elimination reaction, although it still retains the ability to generate a β,δ-elimination product.[101]

Each mammalian DNA glycosylase exhibits a distinct substrate specificity, which is not restricted to one single lesion but rather encompasses a broad spectrum of lesions. Importantly, there are significant redundancies in the spectrum of lesions recognized by each of the glycosylases which constitute an important back-up strategy in case of glycosylase deficiency.

The vast majority of investigations on glycosylase activity have been carried out on naked DNA, while in mammalian cells, DNA repair needs to occur in the context of chromatin. This biological reality raises the question of how efficiently DNA glycosylases might access damaged DNA when tightly wrapped in nucleosomes. Using nucleosome substrates containing specific lesions, one study showed that lesion orientation had little impact on glycosylase access.[202] However, most studies demonstrated that the orientation of the lesion with respect to the histone octamer affects the ability of the glycosylase to access the lesion.[203-206] For example, lesions facing away from the histone octamer appear to be readily accessible, while inward facing lesions are much less accessible. Detailed analysis with NTH1 and NEIL1 are consistent with the hypothesis that access to an inwardly facing lesion can be accomplished by binding to DNA that is transiently unwound from the histone octamer.[204,206]

3.1.3.3   DNA Glycosylase Biology

Biochemical and structural studies have been instrumental in elucidating the mechanisms underpinning the BER pathway and its enzymes, although many aspects of BER biology in the context of the mammalian cell remain to be determined. Notably, the generation of BER deficient mouse models have provided insights into this problem. Nullizygous mice have been generated for four of the oxidized base-specific DNA glycosylases, Ogg1, Nth1, Neil1 and Neil3.[207-212] Despite the high toxicity of the unrepaired DNA lesions, these mutant mice, as well as the other mice deficient in a single DNA glycosylase, do not exhibit particular phenotypes as embryos or as young animals.[213] They are viable, fertile and are phenotypically comparable to the wild-type mice during their first months of life.[207-212] This absence of obvious phenotype strongly suggests that there exists compensatory mechanisms involving redundant repair activities. For example, biochemical studies showed that NEIL1 removes a wide range of DNA lesions, including Tg, Sp, Gh and formamidopyrimidines, and may therefore provide backup activity for NTH1 and OGG1, with the exception of 8-oxoG.[189] On the other hand, the majority of the glycosylase nullizygous mice showed an increase in base lesions in the genomic DNA of targeted organs as well as a significantly higher sensitivity to oxidative stress. For instance, an unusual accumulation of 8-oxoG lesions was observed exclusively in the hepatocytes of *Ogg1* null mice which led to a two- to three-fold increase of spontaneous mutations compared with wild-type mice.[207,208] Also FapyG levels were significantly elevated in the livers of $Nth^{-/-}$ and $Ogg^{-/-}$ mice and Fapy A levels were higher in the livers of $Nth^{-/-}$ mice.[214] A similar liver-specific accumulation of Tg lesions was reported in Nth1 deficient mice when treated with ionizing radiation, whereas *Neil1* null mice showed an increase of DNA lesions in the mitochondria in the absence of exogenous oxidative stress.[209,210] Interestingly, a number of *Neil1* knock-out mice, primarily males, developed symptoms of fatty liver disease and obesity associated with the human metabolic syndrome, which was attributed to the accumulation of unrepaired damage in the mitochondrial DNA.[212] However, this phenotype appears to have variable penetrance and the biological explanation of this phenotypic variability is currently unknown.

Unlike the mice lacking a single DNA glycosylase, double knock-out mice turn out to be tumor-prone, establishing a link between BER deficiency and tumorigenesis. In order to assess the toxicity of 8-oxoG accumulation in the cell, *Ogg1/Myh* double knock-out mice were generated.[215] Myh DNA glycosylase (MUTYH in humans) removes adenine mispaired with 8-oxoG during replication and participates with OGG1 in preventing G to T mutations. With the $Ogg1^{-/-}Myh^{-/-}$ mice, a significantly increased tumor incidence was observed compared to wild-type mice, with a higher frequency of lymphomas as well as lung and ovarian tumors.[215] Likewise, $Neil1^{-/-}Nth1^{-/-}$ mice display a high occurrence of lung and liver tumors compared to wild type or $Neil1^{-/-}$ and $Nth1^{-/-}$ null mice.[216] The latter result not only suggests that BER has a role in tumor prevention, but also clearly demonstrates the carcinogenicity of oxidative damage other than 8-oxoG.

Despite encouraging results obtained with knock-out mice and the obvious implication of unrepaired oxidative DNA damage in tumorigenesis,[217–221] the relationship between BER and carcinogenesis has been more difficult to establish with epidemiological studies in humans (for reviews see refs. 222 and 223). The major exception is MUTYH where individuals with biallalic mutations in this gene are predisposed to the familial adenomatous polyposis coli (APC) form of colon cancer with G–T mutations in the APC gene.[93,224–226]

Recent studies have focused on potential correlations between human polymorphic variants of BER proteins and cancer predisposition (for reviews see refs. 227–229). Associations between *hOGG1* and *hNEIL1* DNA glycosylase polymorphisms and specific types of cancer have been reported.[230–236] The most common single-nucleotide polymorphism (SNP) of *hOGG1* is the Ser326Cys variant.[227] Individuals homozygous for the Ser326Cys allele appeared to be at increased risk for lung cancer,[230,231] as well as prostate and nasopharyngeal carcinoma.[232,233] A recent study suggested that the carcinogenic character of the Ser326Cys polymorphism is probably not due to the moderate loss of activity of the resulting protein but rather its impaired ability to interact with other BER enzymes.[237] Significantly fewer SNPs of *NEIL1* have been characterized in humans compared to *OGG1*. Nonetheless, correlations between gastric cancer and inactive polymorphic variants of NEIL1,[234] as well as with genetic polymorphisms localized in the *NEIL1* promoter region,[238] have been found. Despite hundreds of clinical association studies involving BER SNPs, there are conflicting results regarding the involvement of defective BER genes in cancer risk and the combined data remain inconclusive (for reviews see refs. 222 and 223).

Over the last 30 years, the BER pathway has been extensively studied and characterized in the nucleus. However, oxidative damage is not limited to nuclear DNA (nDNA) and has been observed in mitochondrial DNA (mtDNA), which constitutes a substantial part of the total cellular DNA. Indeed, a mammalian cell harbors hundreds to thousands of mitochondria, each containing 2–10 copies of mtDNA encoding components of the electron transport chain. Moreover, mutations in mtDNA have been linked to aging,[239–241] hereditary diseases[242,243] and carcinogenesis.[244,245] Mitochondrial DNA is known to be associated with the inner membrane of the organelle and is therefore located in the vicinity of endogenous mitochondrial ROS generated by the nearby respiratory machinery. Coincidentally, it has been reported that the steady-state levels of oxidative damage are significantly higher in mitochondrial DNA than in nuclear DNA.[246,247] Although at first this discrepancy

was considered to be a consequence of the absence of DNA repair in mitochondria, it is now common knowledge that several DNA repair pathways,
including BER, actively take place in the context of mitochondria (for a review
see ref. 248). Several BER proteins have been identified in mammalian mitochondria, including three DNA glycosylases, Ogg1, Nth1 and Neil1. In
humans, nuclear and mitochondrial OGG1, αOGG1 and βOGG1, respectively,
result from alternative splicing of the human *OGG1* gene,[249,250] although no
apparent glycosylase activity has been yet detected for βOGG1 *in vitro*. In
contrast to the differential splicing described for OGG1, NTH1 contains a
putative mitochondrial targeting sequence (MTS)[249] and has been shown to
localize predominantly in mitochondria in mice.[251] Surprisingly, the human
homolog of NTH1 appears to be exclusively nuclear, in spite of the MTS.[249,251]
The presence of Neil1 has been reported in mouse liver mitochondria[214] and
Neil1 activity has been detected in mouse brain mitochondria.[252] The mitochondrial localization of Neil1 was confirmed by the unusual accumulation of
mitochondrial steady-state DNA damage and deletions observed in Neil1
knock-out mice.[212]

There also appears to be differential tissue specificities for several of the
glycosylases. Although a number of studies have reported that OGG1, NTH1
and NEIL1 were detected in all of the tissues, they nonetheless display distinct
expression patterns.[167,187,189,253] Whereas the expression levels of OGG1,
NTH1 and NEIL1 varied according to the tissue, both liver and brain constantly show high levels of the three glycosylases.[167,187,189,253] Interestingly,
NEIL2 exhibits strong tissue specificity with a high expression in skeletal
muscle and testis, contrary to NEIL1, which is weakly expressed in these
tissues.[188] In the same study, a moderate expression of NEIL2 was observed
in brain, heart and liver while the other organs tested display a very low level
of NEIL2.[188] In contrast to the ubiquitous expression of the glycosylases
described above, NEIL3 shows an expression restricted to testis and thymus.[189]
In mice Neil3 is found in neural stem cells but not in the mature brain.[254,255]
These observations suggest a specialized function for NEIL3.

The redundancy observed among DNA glycosylases specialized in oxidative
DNA damages attests to the important cellular role for these enzymes. As
discussed above, overlapping substrate specificities provide back-up functions
as suggested by the absence of a major phenotype in single glycosylase knock-
out mice. However, it has been postulated that NEIL1 and 2 are preferentially
involved in BER associated with genome replication and transcription,
respectively.[201] Indeed, the unusual preference of the NEIL proteins for single-
stranded substrates[101,197,201] links these enzymes to the repair of the single
stranded regions of the genomic DNA more likely to occur transiently during
replication and transcription. Moreover, NEIL1 is the only glycosylase that is
up-regulated during the S-phase of the cell cycle[187] and NEIL1 physically
interacts with a number of replication proteins,[256-258] which is consistent with
the repair of DNA prior to replication. Unlike NEIL1, NEIL2 is cell cycle
independent and its atypical preference for bubble substrates suggests an
association with the transcription machinery. A recent study has shown NEIL2

to associate with RNA polymerase II and to initiate transcription-dependent BER of 5-hydroxyuracil in the transcribed strand.[259] In contrast, the exclusive duplex DNA specificity of OGG1 and NTH1 as well as the constitutive expression of both enzymes has led investigators to postulate that OGG1 and NTH1 are housekeeping BER enzymes, working on both active and inactive genome sequences.[260]

3.1.4 The Downstream Enzymes

3.1.4.1 APE, Polβ, LigIIIα-XRCC1

Until recently, the BER pathway was thought to be a simple process relying on few enzymes. However, mammalian BER turns out to be much more complex than the core pathway described earlier. Indeed, BER promotes the repair of damaged DNA *via* two alternative sub-pathways: the short-patch BER (SP-BER) and long-patch BER (LP-BER).[261] In contrast to the SP-BER, which excises and repairs a single nucleotide, a fragment of 2–8 nucleotides is displaced and re-synthesized during LP-BER.[262] In addition, there is extensive evidence that BER enzymes interact with each other as well as non-BER proteins, promoting recruitment of downstream BER components and coordination of the global process (for a review see ref. 161).

In mammalian cells, oxidative DNA lesions are repaired by the BER system using two distinct sub-pathways. Although the first two steps of BER are invariably carried out by the DNA glycosylases and APE1, the resulting single-strand breaks (SSBs) can be processed by two different sets of BER enzymes specific for SP-BER or LP-BER. As described in Section 3.1.2.1, SP-BER affects only one nucleotide and requires Polβ and LigIII to synthesize the proper base and seal the nicked DNA, respectively. In contrast, the LP-BER first recruits Polβ, then the DNA polymerases δ/ε to fill a multi-nucleotide gap while displacing a fragment of 2–8 nucleotides.[262] The final step of DNA ligation is carried out by DNA ligase I. However, in order to allow DNA synthesis and ligation to occur in SP- and LP-BER, the 3′ and/or 5′ ends at the SSB intermediate need to be adequately processed. In both sub-pathways, the 3′ blocking ends generated by the DNA glycosylases β and βδ lyase activities are cleaned by APE1 exonuclease activity and PNK,[157] respectively (see Section 3.1.2.1). On the other hand, the removal of 5′ end adduct may influence the choice of the sub-pathway. During SP-BER, the 5′ dRP blocking end produced by APE1 endonuclease activity on an abasic site is removed by the 5′ dRP lyase activity of Polβ.[150] If the 5′ block is not a substrate for the lyase activity of Polβ, as occurs in many single-strand breaks, Polβ, in association with Polδ and ε, elongates the 3′ end by a few nucleotides while displacing a DNA strand corresponding to the newly synthesized fragment.[262,263] The Flap endonuclease 1 (FEN1) is then required to remove the resulting flap structure, including the 5′ blocking group, which prevents DNA ligation.[264,265] It appears that the choice of the sub-pathway in BER depends, to some extent, on the ability of

Polβ AP lyase activity to remove the 5′ blocking end. Recent studies have shown that high mobility group box 1 (HMGB1) regulates BER by inhibiting SP-BER and stimulating LP-BER (for a review see ref. 266).

Initially, mammalian BER enzymes were believed to work independently from one another but more recent data suggest that they actually interact with one another to increase repair efficiency. Indeed, many groups have established that transient BER protein complexes formed at the lesion site stimulate several BER enzymes as well as coordinating the different stages of the BER pathway. Multiple BER protein–protein interactions have been reported and summarized in reviews.[161,267] A number of studies suggest that, during SP-BER, the formation of BER complexes at the lesion site is a regulated process mediated by a protein known as X-ray cross complementing factor 1 (XRCC1).[268] XRCC1, with no known enzymatic activity, has been shown to interact with most of the BER proteins.[269] XRCC1, in complex with LigIIIα, is recruited to a DNA nick by poly(ADP-ribose) polymerases (PARP) 1 and 2, both specialized in the detection of SSBs.[270–272] Along with PARP 1 and 2, LigIIIα-XRCC1 complex acts as a scaffold protein upon which the repair complex is assembled[273] and this protein platform remains at the repair site until DNA ligation occurs.[267] The formation of the PARP1/2-LigIIIα-XRCC1 complex promotes the recruitment of Polβ to the damage site,[274] which completes the formation of the repair core complex. In addition, PNK and APE1 have been found to functionally interact with two proteins of the repair core complex, XRCC1 and Polβ.[275–277] As detailed earlier, the 5′dRP adduct is removed by Polβ, while APE1 and PNK remove the 3′ blocking adducts generated by bifunctional glycosylases. Therefore, the recruitment of APE1 or PNK to the SSB (BER intermediate) is mainly dictated by the glycosylase that initiates the BER pathway together with the chemical nature of the resulting 3′ end. However, XRCC1 coordination may include the initiation step of BER as suggested by the recruitment of XRCC1 to the lesion site by OGG1 and NEIL2.[268,269,278] Although biochemical studies have established that NTH1 and NEIL1 can interact with XRCC1,[269] recruitment of XRCC1 by these two enzymes has not been yet demonstrated.

In addition to the recruitment of BER proteins to the DNA, the repair complex modulates the activity of the interacting enzymes. XRCC1 has been shown to interact physically and functionally with hOGG1 and NTH1, stimulating their DNA glycosylase activity.[268,269] Similarly, APE1 enhances OGG1 and NTH1 DNA glycosylase activity while drastically reducing their AP lyase activity.[279,280] This specific interaction with APE1 accelerates the removal of the damaged base and favors the release of 5′ dRP ends (APE1 reaction product) at the SSB site. Likewise, evidence shows that the BER protein complex stimulates the APE1-independent BER pathway initiated by NEIL2 and possibly NEIL1.[269,277,278] Indeed, Das and colleagues demonstrated that NEIL2, PNK, XRCC1 and LigIIIα form a functional complex in human cells[278] and other groups showed that XRCC1 stimulates NEIL2 DNA glycosylase activity as well as the DNA kinase and DNA 3′phosphatase activities of PNK.[269,277] The ligation step of the SP-BER turns out to be completely dependent on the interaction between XRCC1 and LigIIIα, which not only stabilizes the enzyme but also restores the ligase activity of LigIIIα to an acceptable level.[281]

Although the modulation of the successive BER stages is dependent on protein–protein interactions involving different components of the repair complex, much of the evidence points to XRCC1 as the major player in the orchestration of the entire BER pathway. The BER system also undergoes an additional degree of regulation through post-translational protein modifications (for reviews see refs. 267, 282 and 283). Post-translational modifications of BER proteins include phosphorylation, acetylation, sumoylation, ubiquitylation and methylation.[267] Post-translational modifications, triggered by the cellular state or oxidative stress, may alter the function of BER proteins while influencing the BER protein-protein partnership, in order to adjust the DNA repair capacity to the cellular environment.[267,282,284–286]

The effect of nucleosome structures on base excision repair by the downstream proteins has also been examined *in vitro*. One group found that the entire BER reaction could occur on nucleosome substrates without disrupting them or altering the position of the histones,[202] while others found that Polβ was unable to access its substrate on nucleosomes.[203,287] Relevant to the final step in LP-BER, it was reported that DNA ligase I was about ten-fold less active on nucleosome substrates than on naked DNA[288] but FEN1, which often also acts in conjunction with DNA ligase I, exhibited higher endonuclease activity on nucleosomes than naked DNA.[289] Recent studies (Odell *et al.* under review) that examined the complete base excision repair on a nucleosome substrate as well as all of the nucleosome repair intermediates concluded that, like the DNA glycosylases, both APE and Polβ are inhibited to varying extents depending on the helical orientation of their substrates and both are able to act on their substrates without irreversibly altering the nucleosome or its translation position. However, the activity of ligase IIIα-XRCC1 on nicked nucleosome substrates was not affected by their helical orientation and the ligase IIIα-XRCC1 complex substantially disrupted the nucleosomes independently of any chromatin remodeling agents.

3.1.4.2 Biology of the Downstream Enzymes

In contrast to the DNA glycosylase knock-out mice discussed in Section 3.1.2.3, deletions of individual genes encoding downstream BER enzymes render the mice inviable. Mice nullizygous for *APE1*, *XRCC1* and *LigIII* result in early embryonic lethality,[290–293] emphasizing the importance of BER during embryogenesis. Likewise, knock-out mouse models of LP-BER specific enzymes, FEN1 and Lig I, died during embryogenesis.[294,295] Surprisingly, Polβ null mice survived embryonic development, only to die from lung failure immediately after birth.[296] In addition to a respiratory defect, Polβ null mice were significantly smaller than their wild-type littermates and histological examination revealed defective neurogenesis, suggesting that Polβ is important for neural development.[296] Because of the early lethality that affects these mice, it has not been possible to establish any relationship between a single gene deletion and an increase in mutagenesis and/or carcinogenesis. In order to circumvent the lethal phenotype associated with homozygous knock-out mice, heterozygous mice that target some BER intermediate stages have been generated. *APE1, FEN1* and *Polβ* heterozygous mice turned out to be viable and showed no abnormal phenotype compared with wild-type animals.[295,297] However, the deletion of one Ape1 allele led to a higher spontaneous mutation frequency in spleen and liver,[297] whereas the inactivation of one Fen1 allele appeared to predispose the mice to rapid cancer progression.[295] Predictably, Polβ heterozygous mice sustain higher levels of SSBs and chromosomal aberrations than wild-type mice.[298] In addition, Polβ $^{+/-}$ mice exhibit hypersensitivity to alkylating agents but not ROS,[298] as has been observed for Polβ$^{-/-}$ cells derived from embryonic fibroblasts.[299] Mosaic Polβ mice have been designed that were viable in spite of an elevated embryonic lethality, and unlike the heterozygous mice, Polβ mosaic animals had a significantly reduced size and weight compared with wild-type littermates.[300] This atypical phenotype is in keeping with the unusually small size of the Polβ knock-out embryos[296,300] and confirms the requirement of Polβ activity in development.

Although XRCC1 deficiency results in embryonic lethality, Tebbs and coworkers have been able to isolate XRCC1$^{-/-}$ embryonic fibroblasts from nullizygous early embryos.[292] Along with XRCC1 deficient Chinese hamster ovary cell lines, the XRCC1$^{-/-}$ cells derived from the mutant embryos are hypersensitive to a broad range of DNA damaging agents, except ROS.[292,301] The minimal hypersensitivity to oxidizing agents reported for XRCC1$^{-/-}$ and Polβ$^{-/-}$ cell lines suggests that oxidative DNA damage may also be repaired by XRCC1 and/or Polβ-independent BER sub-pathways. In contrast to DNA glycosylases, the BER downstream enzymes have unique specificity and activity that has proven to be critical for the proper functioning of BER, as suggested by the lethal phenotype associated with mice nullizygous for the BER downstream genes.

The severe phenotype observed for the knock-out mice, targeting APE1, Polβ, FEN1, XRCC1, LigIII or LigI, attests to the crucial importance of BER pathway in the maintenance of cellular integrity. In contrast to the DNA glycosylases, the proper functioning of intermediate and late stage BER proteins is an absolute requirement to prevent genomic instability. Interestingly, a large number of SNPs relative to the core BER proteins have been identified in humans, including APE1, Polβ and XRCC1. Several APE1 polymorphisms have been described in the literature, Asp148Glu being the most common substitution. Although the Aps148Glu variant was associated with hypersensitivity to ionizing radiation,[302] Hadi and colleagues have demonstrated that this single mutation had no impact on endonuclease and DNA binding activities.[303] However, the Asp148Glu variant might have altered interactions with other BER proteins and affect the overall efficiency of the APE1-dependent repair sub-pathway. On the other hand, association between the Asp148Glu polymorphism and lung/upper aerodigestive tract cancer risk has not been demonstrated.[304] A small-scale study has shown that approximately 30% of human tumors express variant forms of Polβ, half of which are single amino acid substitutions.[305] Functional characterizations of some Polβ mutants identified in human cancers have shown an altered catalytic efficiency that can potentially lead to genomic instability.[231,306,307] To date, more than sixty SNPs of XRCC1 gene have been identified in human population and abundantly studied; the vast majority of these investigations relate to Arg194Trp, Arg399Gln and Arg280His XRCC1 polymorphisms.[223,308] As observed for other BER SNPs, cancer association studies relative to XRCC1 SNPs have provided ambiguous data that could not establish a definitive connection between a particular SNP and an increase in cancer risk (for reviews see refs. 222 and 223). Despite the lack of reliable data linking a single BER SNP with cancer risk, it is likely that combinations of two or more BER SNPs may affect the frequency of cancer development, as suggested by the elevated risk of pancreatic cancer of individuals containing both APE1 Asp148Glu and XRCC1 Arg194Trp variants.[309]

In order to cope with the high level of oxidative DNA damage that threatens mitochondrial DNA, mammalian cells have developed an efficient mitochondrial BER system (mtBER). Although mitochondrial BER steps appear to be similar to nuclear BER, mtBER is an independent process that involves proteins encoded by nuclear genes.[310] The mtBER pathway consists of DNA glycosylases (see Section 3.1.2.2), APE, Polymerase γ (Polγ) and LigIII. Since a truncated version of APE was found in the mitochondria of bovine liver, it was suggested that mtAPE is an *N*-terminal truncated version of the full-length nuclear APE1; however, only full-length APE has been detected in various cell lines.[311] Thus, the identity of mtAPE remains unclear. The next step in the mtBER pathway is catalyzed by a single polymerase specific to the mitochondria, Polγ. Indeed, Polγ is involved in all replication, recombination and repair processes involving mtDNA. In addition to DNA synthesis activity, Polγ exhibits a 3′–5′ proofreading exonuclease activity, as well as a 5′ dRP lyase function used to process SSBs during BER (for a review see ref. 312). The final step of mtBER is catalyzed by a splice variant of LigIII, mtLigIII, which is encoded by the nuclear gene.[313] Unlike nuclear LigIII, mtLigIII does not interact with XRCC1, suggesting that the mitochondrial variant is independent of XRCC1.[314] Recent circumstantial evidence suggests that an additional BER sub-pathway, comparable to LP-BER, could take place in mitochondria[315–318] since FEN1 has been detected in human[317,318] and mouse[316,317] mitochondrial extracts.

3.1.5 Is there a Threshold for the Consequences of Oxidative DNA Damages?

There has always been a controversy around the possibility of a threshold for oxidative damage consequences.[319] This is primarily due to the fact that very high levels of cellular oxidative DNA damage are produced endogenously.[32] The argument has been that if you add a small increment of damage onto an already high level of damage, the consequences of doing so are going to be insignificant. For example, at the mean lethal dose of gamma rays to a mammalian cell (about 2 Gy), 2000 single strand breaks, 6000 base lesions and 80 double strand breaks are formed.[320] At this dose, on average, 37% of the population of cells survive and are thus able to repair this level of damage. The numbers of single-strand breaks and base lesions formed in this case are approximately equal to those produced endogenously by oxidative free radicals per human cell per day. Thus very low levels of ionizing radiation would produce incrementally small increases in the numbers of strand breaks and base lesions; even the level of double-strand breaks produced, which would be much lower, are presumably readily repaired.[320] The idea of a threshold dose is further supported by the fact that there are redundancies in the substrate specificities of the enzymes in the first step of the pathway, the DNA glycosylases. Moreover, there is a backup pathway for short-patch base excision repair, long-patch base excision repair,[261,321] and a backup for both of these, homology-directed repair.[137,138]

On the other hand, glycosylases seem to have functional specificities. For example, in humans, NTH1 and OGG1 appear to be responsible for the global recognition and removal of oxidative DNA base damage, while NEIL1 may be associated with repair just prior to the replication fork and NEIL2 with transcription.[201,256–259] There also appears to be tissue specificities for the DNA glycosylases. For example, NEIL3 is only found in the thymus[211] and in mice, and Neil3 is associated with embryonic development and is present in neural stem cells but not in the mature brain.[254,255] In addition, there have been a number of studies showing that the base excision repair system itself has to be carefully balanced[322–324] since during the repair of an oxidative DNA base damage, the subsequent intermediates in the pathway (abasic sites and single-strand breaks) are far more dangerous than the initial base damage. Also, the BER enzymes downstream from the DNA glycosylases are required for mammalian development because mice nullizygous for APE1, Pol β and LigIIIα are embryonic lethal.[290,293,325] There is also a second AP endonuclease present in human cells, APE2,[326,327] which may play a role in immune cell development[328–331] or possibly some other role in base excision repair.[332] Finally, even if the short patch base excision repair pathway is shuttled into long patch repair which can use replicative polymerases, APE1 is still required.

3.1.6   Conclusions

Clearly, an appropriate balance between ROS production and removal must be maintained because ROS are both cell signaling molecules and DNA damaging agents. Thus exogenous exposures that may alter this delicate balance are likely to be detrimental. Moreover, individual variations in BER and the damage response pathways in general as demonstrated by numerous polymorphic variants in the population can result in differential individual responses to genotoxic agents. Finally, humans are constantly confronted by oxidative stresses that are often beyond one's control, such as inflammation caused by infection. Therefore, setting a uniform threshold for exposure to exogenous agents that produce ROS is a potentially dangerous proposition.

Acknowledgments

Work in the authors' laboratories was supported by NIH grants R01 CA33657 and P01 CA098993 to S.S.W. and R01 CA052040 to S.D. The authors also wish to thank Dr Pierre Aller for providing Figure 3.1.3 and Debra Stern for preparing the manuscript and Figures 3.1.1 and 3.1.2.

References

1. B. Halliwell and J. M. C. Gutteridge, *Free Radicals in Biology and Medicine*, Oxford University Press, Oxford, 2002.
2. P. Jezek and L. Hlavata, Mitochondria in homeostasis of reactive oxygen species in cell, tissues, and organism, *Int. J. Biochem. Cell Biol.*, 2005, **37**, 2478–2503.
3. A. Boveris and E. Cadenas, Mitochondrial production of hydrogen peroxide regulation by nitric oxide and the role of ubisemiquinone, *IUBMB Life*, 2000, **50**, 245–250.
4. M. Balazy and S. Nigam, Aging, lipid modifications and phospholipases – New concepts, *Ageing Res. Rev..*, 2003, **2**, 209–291.
5. A. Boveris and E. Cadenas, *Cellular sources and steady-state levels of reactive oxygen species*, in *Oxygen, gene expression and cellular function*, ed. C. Mazzaro and L. Clerch, Marcel Dekker, New York, 1997, pp. 1–25.
6. R. A. Gottlieb, Cytochrome P450: major player in reperfusion injury, *Arch. Biochem.. Biophys.*, 2003, **420**, 262–267.
7. A. A. Caro and A. I. Cederbaum, Oxidative stress, toxicology, and pharmacology of CYP2E1, *Annu. Rev.. Pharmacol. Toxicol.*, 2004, **44**, 27–42.
8. M. Schrader and H. D. Fahimi, Mammalian peroxisomes and reactive oxygen species, *Histochem. Cell Biol.*, 2004, **122**, 383–393.
9. B. M. Babior, The NADPH oxidase of endothelial cells, *IUBMB Life*, 2000, **50**, 267–269.
10. B. M. Babior, J. D. Lambeth and W. Nauseef, The neutrophil NADPH oxidase, *Arch. Biochem.. Biophys.*, 2002, **397**, 342–344.
11. Y. Funato and H. Miki, Redox regulation of Wnt signalling *via* nucleoredoxin, *Free Radic.. Res.*, 2010, **44**, 379–388.
12. R. B. Hamanaka and N. S. Chandel, Mitochondrial reactive oxygen species regulate cellular signaling and dictate biological outcomes, *Trends Biochem.. Sci.*, 2010, **35**, 505–513.
13. N. Azad, A. Iyer, V. Vallyathan, L. Wang, V. Castranova, C. Stehlik and Y. Rojanasakul, Role of oxidative/nitrosative stress-mediated Bcl-2 regulation in apoptosis and malignant transformation, *Ann. N. Y. Acad. Sci.*, 2010, **1203**, 1–6.

14. S. Kar, S. Subbaram, P. M. Carrico and J. A. Melendez, Redox-control of matrix metalloproteinase-1: a critical link between free radicals, matrix remodeling and degenerative disease, *Respir. Physiol. Neurobiol.*, 2010, **174**, 299–306.

15. S. Pendyala and V. Natarajan, Redox regulation of Nox proteins, *Respir. Physiol. Neurobiol.*, 2010, **174**, 265–271.

16. M. J. Morgan and Z. G. Liu, Crosstalk of reactive oxygen species and NF-kappaB signaling, *Cell Res.*, 2011, **21**, 103–115.

17. H. Ohshima, M. Tatemichi and T. Sawa, Chemical basis of inflammation-induced carcinogenesis, *Arch. Biochem.. Biophys.*, 2003, **417**, 3–11.

18. S. Mena, A. Ortega and J. M. Estrela, Oxidative stress in environmental-induced carcinogenesis, *Mutat. Res.*, 2009, **674**, 36–44.

19. O. Handa, Y. Naito and T. Yoshikawa, *Helicobacter pylori*: a ROS-inducing bacterial species in the stomach, *Inflamm. Res.*, 2010, **59**, 997–1003.

20. F. Farinati, M. Piciocchi, E. Lavezzo, M. Bortolami and R. Cardin, Oxidative stress and inducible nitric oxide synthase induction in carcinogenesis, *Dig. Dis.*, 2010, **28**, 579–584.

21. C. von Sonntag. *The Chemical Basis of Radiation Biology*, Taylor and Francis, London, 1987.

22. M. Dizdaroglu, Oxidative damage to DNA in mammalian chromatin, *Mutat. Res.*, 1992, **275**, 331–342.

23. A. P. Breen and J. A. Murphy, Reactions of oxyl radicals with DNA, *Free Radic. Biol. Med.*, 1995, **18**, 1033–77.

24. M. Dizdaroglu, P. Jaruga, M. Birincioglu and H. Rodriguez, Free radical-induced damage to DNA: mechanisms and measurement, *Free Radic. Biol. Med.*, 2002, **32**, 1102–1115.

25. M. Dizdaroglu, E. Holwitt, M. P. Hagan and W. F. Blakely, Formation of cytosine glycol and 5,6-dihydroxycytosine in deoxyribonucleic acid on treatment with osmium tetroxide, *Biochem. J.*, 1986, **235**, 531–536.

26. M. Dizdaroglu, J. Laval and S. Boiteux, Substrate specificity of the *Escherichia coli* endonuclease III: excision of thymine- and cytosine-derived lesions in DNA produced by radiation-generated free radicals, *Biochemistry*, 1993, **32**, 12105–12111.

27. J. R. Wagner, Analysis of oxidative cytosine products in DNA exposed to ionizing radiation, *J. Chim. Phys.*, 1994, **91**, 1280–1286.

28. S. Steenken, Purine bases, nucleosides, and nucleotides: aqueous solution redox chemistry and transformation reactions of their radical cations and e⁻ and OH adducts, *Chem. Rev.*, 1989, **89**, 503–520.

29. M. Dizdaroglu, Mechanisms of free radical damage to DNA, in *DNA and Free Radicals: Techniques, Mechanisms and Applications*, ed. Ol. Aruoma and B. Halliwell, OICA International, St Lucia, 1998, pp. 3–26.

30. W. L. Neeley and J. M. Essigmann, Mechanisms of formation, genotoxicity, and mutation of guanine oxidation products, *Chem. Res. Toxicol.*, 2006, **19**, 491–505.

31. K. B. Lesiak and K. T. Wheeler, Formation of alpha-deoxyadenosine in polydeoxynucleotides exposed to ionizing radiation under anoxic conditions, *Radiat. Res.*, 1990, **121**, 328–337.

32. T. Lindahl and D. E. Barnes, Repair of endogenous DNA damage, *Cold Spring Harb. Symp. Quant. Biol.*, 2000, **65**, 127–133.

33. W. D. Henner, S. M. Grunberg and W. A. Haseltine, Sites and structure of gamma radiation-induced DNA strand breaks, *J. Biol. Chem.*, 1982, **257**, 11750–11754.

34. W. D. Henner, L. O. Rodriguez, S. M. Hecht and W. A. Haseltine, gamma Ray induced deoxyribonucleic acid strand breaks, 3′ glycolate termini, *J. Biol. Chem.*, 1983, **258**, 711–713.

35. P. C. Dedon, The chemical toxicology of 2-deoxyribose oxidation in DNA, *Chem. Res. Toxicol.*, 2008, **21**, 206–219.

36. K. Bebenek and T. A. Kunkel, Family growth: the eukaryotic DNA polymerase revolution, *Cell Mol. Life Sci.*, 2002, **59**, 54–57.

37. A. R. Lehmann, A. Niimi, T. Ogi, S. Brown, S. Sabbioneda, J. F. Wing, P. L. Kannouche and C. M. Green, Translesion synthesis: Y-family polymerases and the polymerase switch, *DNA Repair (Amst.)*, 2007, **6**, 891–899.

38. S. D. McCulloch and T. A. Kunkel, The fidelity of DNA synthesis by eukaryotic replicative and translesion synthesis polymerases, *Cell Res.*, 2008, **18**, 148–161.

39. L. S. Waters, B. K. Minesinger, M. E. Wiltrout, S. D'Souza, R. V. Woodruff and G. C. Walker, Eukaryotic translesion polymerases and their roles and regulation in DNA damage tolerance, *Microbiol. Mol. Biol. Rev.*, 2009, **73**, 134–154.

40. Z. Livneh, O. Ziv and S. Shachar, Multiple two-polymerase mechanisms in mammalian translesion DNA synthesis, *Cell Cycle*, 2010, **9**, 729–735.

41. A. P. Grollman and M. Moriya, Mutagenesis by 8-oxoguanine: an enemy within, *Trends Genet.*, 1993, **9**, 246–9.

42. S. S. Wallace, *Oxidative damage to DNA and its repair*, in *Oxidative Stress and the Molecular Biology of Antioxidant Defenses*, Cold Spring Harbor Laboratory Press, Cold Spring Harbor, 1997, pp. 49–90.

43. D. Wang, D. A. Kreutzer and J. M. Essigmann, Mutagenicity and repair of oxidative DNA damage: insights from studies using defined lesions, *Mutat. Res.*, 1998, **400**, 99–115.

44. S. S. Wallace, Biological consequences of free radical-damaged DNA bases, *Free Radic. Biol. Med.*, 2002, **33**, 1–14.

45. S. Bjelland and E. Seeberg, Mutagenicity, toxicity and repair of DNA base damage induced by oxidation, *Mutat. Res.*, 2003, **531**, 37–80.

46. B. Tudek, Imidazole ring-opened DNA purines and their biological significance, *J. Biochem. Mol. Biol.*, 2003, **36**, 12–19.

47. M. Dizdaroglu, G. Kirkali and P. Jaruga, Formamidopyrimidines in DNA: mechanisms of formation, repair, and biological effects, *Free Radic. Biol. Med.*, 2008, **45**, 1610–1621.

48. A. A. Purmal, Y. W. Kow and S. S. Wallace, Major oxidative products of cytosine, 5-hydroxycytosine and 5-hydroxyuracil, exhibit sequence context-dependent mispairing in vitro, *Nucleic Acids Res.*, 1994, **22**, 72–78.

49. A. A. Purmal, G. W. Lampman, J. P. Bond, Z. Hatahet and S. S. Wallace, Enzymatic processing of uracil glycol, a major oxidative product of DNA cytosine, *J. Biol. Chem.*, 1998, **273**, 10026–10035.

50. D. A. Kreutzer and J. M. Essigmann, Oxidized, deaminated cytosines are a source of C --> T transitions in vivo, *Proc. Natl. Acad. Sci. U. S. A.*, 1998, **95**, 3578–3582.

51. D. I. Feig, L. C. Sowers and L. A. Loeb, Reverse chemical mutagenesis: identification of the mutagenic lesions resulting from reactive oxygen species-mediated damage to DNA, *Proc. Natl. Acad. Sci. U. S. A.*, 1994, **91**, 6609–6613.

52. J. Y. Kao, I. Goljer, T. A. Phan and P. H. Bolton, Characterization of the effects of a thymine glycol residue on the structure, dynamics, and stability of duplex DNA by NMR, *J. Biol. Chem.*, 1993, **268**, 17787–17793.

53. J. Miller, K. Miaskiewicz and R. Osman, Structure-function studies of DNA damage using ab initio quantum mechanics and molecular dynamics simulation, *Ann. N. Y. Acad. Sci.*, 1994, **726**, 71–91.

54. K. Miaskiewicz, J. Miller, R. Ornstein and R. Osman, Molecular dynamics simulations of the effects of ring-saturated thymine lesions on DNA structure, *Biopolymers*, 1995, **35**, 113–124.

55. H. Ide, Y. W. Kow and S. S. Wallace, Thymine glycols and urea residues in M13 DNA constitute replicative blocks in vitro, *Nucleic Acids Res.*, 1985, **13**, 8035–8052.

56. P. Rouet and J. M. Essigmann, Possible role for thymine glycol in the selective inhibition of DNA synthesis on oxidized DNA templates, *Cancer Res.*, 1985, **45**, 6113–6118.

57. J. M. Clark and G. P. Beardsley, Thymine glycol lesions terminate chain elongation by DNA polymerase I *in vitro*, *Nucleic Acids Res.*, 1986, **14**, 737–749.

58. R. C. Hayes and J. E. LeClerc, Sequence dependence for bypass of thymine glycols in DNA by DNA polymerase I, *Nucleic Acids Res.*, 1986, **14**, 1045–1061.

59. J. M. McNulty, B. Jerkovic, P. H. Bolton and A. K. Basu, Replication inhibition and miscoding properties of DNA templates containing a site-specific cis-thymine glycol or urea residue, *Chem. Res. Toxicol.*, 1998, **11**, 666–673.

60. J. M. Clark and G. P. Beardsley, Template length, sequence context, and 3'-5' exonuclease activity modulate replicative bypass of thymine glycol lesions *in vitro*, *Biochemistry*, 1989, **28**, 775–779.

61. A. K. Basu, E. L. Loechler, S. A. Leadon and J. M. Essigmann, Genetic effects of thymine glycol: site-specific mutagenesis and molecular modeling studies, *Proc. Natl. Acad. Sci. U. S. A.*, 1989, **86**, 7677–76781.

62. J. M. Clark, N. Pattabiraman, W. Jarvis and G. P. Beardsley, Modeling and molecular mechanical studies of the cis-thymine glycol radiation damage lesion in DNA, *Biochemistry*, 1987, **26**, 5404–5409.
63. P. Aller, M. A. Rould, M. Hogg, S. S. Wallace and S. Doublié, A structural rationale for stalling of a replicative DNA polymerase at the most common oxidative thymine lesion, thymine glycol, *Proc. Natl. Acad. Sci. U. S. A.*, 2007, **104**, 814–818.
64. P. M. Achey and C. F. Wright, Inducible repair of thymine ring saturation damage in phi X174 DNA, *Radiat. Res.*, 1983, **93**, 609–612.
65. E. Moran and S. S. Wallace, The role of specific DNA base damages in the X-ray-induced inactivation of bacteriophage PM2, *Mutat. Res.*, 1985, **146**, 229–241.
66. R. C. Hayes, L. A. Petrullo, H. M. Huang, S. S. Wallace and J. E. LeClerc, Oxidative damage in DNA. Lack of mutagenicity by thymine glycol lesions, *J. Mol. Biol.*, 1988, **201**, 239–246.
67. M. F. Laspia and S. S. Wallace, Excision repair of thymine glycols, urea residues, and apurinic sites in *Escherichia coli*, *J. Bacteriol.*, 1988, **170**, 3359–3366.
68. M. F. Laspia and S. S. Wallace, SOS processing of unique oxidative DNA damages in *Escherichia coli*, *J. Mol. Biol.*, 1989, **207**, 53–60.
69. R. Kusumoto, C. Masutani, S. Iwai and F. Hanaoka, Translesion synthesis by human DNA polymerase eta across thymine glycol lesions, *Biochemistry*, 2002, **41**, 6090–6099.
70. P. L. Fischhaber, V. L. Gerlach, W. J. Feaver, Z. Hatahet, S. S. Wallace and E. C. Friedberg, Human DNA polymerase kappa bypasses and extends beyond thymine glycols during translesion synthesis *in vitro*, preferentially incorporating correct nucleotides, *J. Biol. Chem.*, 2002, **277**, 37604–37611.
71. R. E. Johnson, S. L. Yu, S. Prakash and L. Prakash, Yeast DNA polymerase zeta (zeta) is essential for error-free replication past thymine glycol, *Genes Dev.*, 2003, **17**, 77–87.
72. K. Takata, T. Shimizu, S. Iwai and R. D. Wood, Human DNA polymerase N (POLN) is a low fidelity enzyme capable of error-free bypass of 5S-thymine glycol, *J. Biol. Chem.*, 2006, **281**, 23445–23455.
73. T. J. Matray, K. J. Haxton and M. M. Greenberg, The effects of the ring fragmentation product of thymidine C5-hydrate on phosphodiesterases and klenow (exo-) fragment, *Nucleic Acids Res.*, 1995, **23**, 4642–4648.
74. M. M. Greenberg and T. J. Matray, Inhibition of klenow fragment (exo-) catalyzed DNA polymerization by (5R)-5,6-dihydro-5-hydroxythymidine and structural analogue 5,6-dihydro-5-methylthymidine, *Biochemistry*, 1997, **36**, 14071–14079.
75. A. M. Herrala and J. A. Vilpo, Template-primer activity of 5-(hydroxymethyl)uracil-containing DNA for prokaryotic and eukaryotic DNA and RNA polymerases, *Biochemistry*, 1989, **28**, 8274–8277.
76. D. D. Levy and G. W. Teebor, Site directed substitution of 5-hydroxymethyluracil for thymine in replicating phi X-174am3 DNA *via* synthesis

of 5-hydroxymethyl-2'-deoxyuridine-5'-triphosphate, *Nucleic Acids Res.*, 1991, **19**, 3337–3343.

77. Q. M. Zhang, H. Sugiyama, I. Miyabe, S. Matsuda, I. Saito and S. Yonei, Replication of DNA templates containing 5-formyluracil, a major oxidative lesion of thymine in DNA, *Nucleic Acids Res.*, 1997, **25**, 3969–3973.

78. A. Masaoka, H. Terato, M. Kobayashi, Y. Ohyama and H. Ide, Oxidation of thymine to 5-formyluracil in DNA promotes misincorporation of dGMP and subsequent elongation of a mismatched primer terminus by DNA polymerase, *J. Biol. Chem.*, 2001, **276**, 16501–16510.

79. I. Miyabe, Q. M. Zhang, H. Sugiyama, K. Kino and S. Yonei, Mutagenic effects of 5-formyluracil on a plasmid vector during replication in *Escherichia coli*, *Int J. Radiat. Biol.*, 2001, **77**, 53–58.

80. H. Kamiya, N. Murata-Kamiya, N. Karino, Y. Ueno, A. Matsuda and H. Kasai, Induction of T --> G and T --> A transversions by 5-formyluracil in mammalian cells, *Mutat. Res.*, 2002, **513**, 213–222.

81. D. Gasparutto, M. Ait-Abbas, M. Jaquinod, S. Boiteux and J. Cadet, Repair and coding properties of 5-hydroxy-5-methylhydantoin nucleosides inserted into DNA oligomers, *Chem. Res. Toxicol.*, 2000, **13**, 575–584.

82. J. O. Blaisdell, Z. Hatahet and S. S. Wallace, A novel role for *Escherichia coli* endonuclease VIII in prevention of spontaneous G-->T transversions, *J. Bacteriol.*, 1999, **181**, 6396–6402.

83. M. L. Wood, M. Dizdaroglu, E. Gajewski and J. M. Essigmann, Mechanistic studies of ionizing radiation and oxidative mutagenesis: genetic effects of a single 8-hydroxyguanine (7-hydro-8-oxoguanine) residue inserted at a unique site in a viral genome, *Biochemistry*, 1990, **29**, 7024–7032.

84. S. Shibutani, M. Takeshita and A. P. Grollman, Insertion of specific bases during DNA synthesis past the oxidation-damaged base 8-oxodG, *Nature*, 1991, **349**, 431–434.

85. M. Moriya, C. Ou, V. Bodepudi, F. Johnson, M. Takeshita and A. P. Grollman, Site-specific mutagenesis using a gapped duplex vector: a study of translesion synthesis past 8-oxodeoxyguanosine in *E. coli*, *Mutat. Res.*, 1991, **254**, 281–288.

86. K. C. Cheng, D. S. Cahill, H. Kasai and S. Nishimura, L. A. Loeb, 8-Hydroxyguanine, an abundant form of oxidative DNA damage, causes G----T and A----C substitutions, *J. Biol. Chem.*, 1992, **267**, 166–172.

87. L. G. Brieba, B. F. Eichman, R. J. Kokoska, S. Doublié, T. A. Kunkel and T. Ellenberger, Structural basis for the dual coding potential of 8-oxoguanosine by a high-fidelity DNA polymerase, *EMBO J.*, 2004, **23**, 3452–3461.

88. G. W. Hsu, M. Ober, T. Carell and L. S. Beese, Error-prone replication of oxidatively damaged DNA by a high-fidelity DNA polymerase, *Nature*, 2004, **431**, 217–221.

89. J. C. Klein, M. J. Bleeker, C. P. Saris, H. C. Roelen, H. F. Brugghe, H. van den Elst, G. A. van der Marel, J. H. van Boom, J. G. Westra,

E. Kriek and A. J. M. Berns, Repair and replication of plasmids with site-specific 8-oxodG and 8-AAFdG residues in normal and repair-deficient human cells, *Nucleic Acids Res.*, 1992, **20**, 4437–4443.

90. M. Moriya, Single-stranded shuttle phagemid for mutagenesis studies in mammalian cells: 8-oxoguanine in DNA induces targeted G.C--> T.A transversions in simian kidney cells, *Proc. Natl. Acad. Sci. U. S. A.*, 1993, **90**, 1122–1126.

91. J. Tchou, V. Bodepudi, S. Shibutani, I. Antoshechkin, J. Miller, A. P. Grollman and F. Johnson, Substrate specificity of Fpg protein. Recognition and cleavage of oxidatively damaged DNA, *J. Biol. Chem.*, 1994, **269**, 15318–15324.

92. C. J. Wiederholt, M. O. Delaney, M. A. Pope, S. S. David and M. M. Greenberg, Repair of DNA containing Fapy.dG and its beta-C-nucleoside analogue by formamidopyrimidine DNA glycosylase and MutY, *Biochemistry*, 2003, **42**, 9755–9760.

93. N. H. Chmiel, A. L. Livingston and S. S. David, Insight into the functional consequences of inherited variants of the hMYH adenine glycosylase associated with colorectal cancer: complementation assays with hMYH variants and pre-steady-state kinetics of the corresponding mutated *E.coli* enzymes, *J. Mol. Biol.*, 2003, **327**, 431–443.

94. S. Boiteux and J. Laval, Imidazole open ring 7-methylguanine: an inhibitor of DNA synthesis, *Biochem. Biophys Res. Commun.*, 1983, **110**, 552–558.

95. T. R. O'Connor and S. Boiteux, J. Laval, Ring-opened 7-methylguanine residues in DNA are a block to in vitro DNA synthesis, *Nucleic Acids Res.*, 1988, **16**, 5879–5894.

96. B. Tudek and S. Boiteux, J. Laval, Biological properties of imidazole ring-opened N7-methylguanine in M13mp18 phage DNA, *Nucleic Acids Res.*, 1992, **20**, 3079–3084.

97. K. Asagoshi, H. Terato, Y. Ohyama and H. Ide, Effects of a guanine-derived formamidopyrimidine lesion on DNA replication: translesion DNA synthesis, nucleotide insertion, and extension kinetics, *J. Biol. Chem.*, 2002, **277**, 14589–14597.

98. C. J. Wiederholt and M. M. Greenberg, Fapy.dG instructs Klenow exo(-) to misincorporate deoxyadenosine, *J. Am. Chem. Soc.*, 2002, **124**, 7278–7279.

99. M. A. Kalam, K. Haraguchi, S. Chandani, E. L. Loechler, M. Moriya, M. M. Greenberg and A. K. Basu, Genetic effects of oxidative DNA damages: comparative mutagenesis of the imidazole ring-opened formamidopyrimidines (Fapy lesions) and 8-oxo-purines in simian kidney cells, *Nucleic Acids Res.*, 2006, **34**, 2305–2315.

100. J. N. Patro, C. J. Wiederholt, Y. L. Jiang, J. C. Delaney, J. M. Essigmann and M. M. Greenberg, Studies on the replication of the ring opened formamidopyrimidine, Fapy.dG in *Escherichia coli*, *Biochemistry*, 2007, **46**, 10202–10212.

101. M. Liu, V. Bandaru, J. P. Bond, P. Jaruga, X. Zhao, P. P. Christov, C. J. Burrows, C. J. Rizzo, M. Dizdaroglu and S. S. Wallace, The mouse

ortholog of NEIL3 is a functional DNA glycosylase in vitro and in vivo, *Proc. Natl. Acad. Sci. U. S. A.*, 2010, **107**, 4925–4930.

102. W. Guschlbauer, A. M. Duplaa, A. Guy, R. Teoule and G. V. Fazakerley, Structure and in vitro replication of DNA templates containing 7,8-dihydro-8-oxoadenine, *Nucleic Acids Res.*, 1991, **19**, 1753–1758.

103. M. L. Wood, A. Esteve, M. L. Morningstar, G. M. Kuziemko and J. M. Essigmann, Genetic effects of oxidative DNA damage: comparative mutagenesis of 7,8-dihydro-8-oxoguanine and 7,8-dihydro-8-oxoadenine in *Escherichia coli*, *Nucleic Acids Res.*, 1992, **20**, 6023–6032.

104. S. Shibutani, V. Bodepudi, F. Johnson and A. P. Grollman, Translesional synthesis on DNA templates containing 8-oxo-7,8-dihydrodeoxy-adenosine, *Biochemistry*, 1993, **32**, 4615–4621.

105. H. Kamiya, H. Miura, N. Murata-Kamiya, H. Ishikawa, T. Sakaguchi, H. Inoue, T. Sasaki, C. Masutani, F. Hanaoka, S. Nishimura and E. Ohtsuka, 8-Hydroxyadenine (7,8-dihydro-8-oxoadenine) induces mis-incorporation in *in vitro* DNA synthesis and mutations in NIH 3T3 cells, *Nucleic Acids Res.*, 1995, **23**, 2893–2899.

106. H. Kamiya, N. Murata-Kamiya, S. Koizume, H. Inoue, S. Nishimura and E. Ohtsuka, 8-Hydroxyguanine (7,8-dihydro-8-oxoguanine) in hot spots of the c-Ha-ras gene: effects of sequence contexts on mutation spectra, *Carcinogenesis*, 1995, **16**, 883–889.

107. M. A. Graziewicz, T. H. Zastawny, R. Olinski, E. Speina, J. Siedlecki and B. Tudek, Fapyadenine is a moderately efficient chain terminator for prokaryotic DNA polymerases, *Free Radic. Biol. Med.*, 2000, **28**, 75–83.

108. M. O. Delaney, C. J. Wiederholt and M. M. Greenberg, Fapy.dA induces nucleotide misincorporation translesionally by a DNA polymerase, *Angew. Chem., Int. Ed. Engl.*, 2002, **41**, 771–773.

109. H. Kamiya, T. Ueda, T. Ohgi, A. Matsukage and H. Kasai, Misincorporation of dAMP opposite 2-hydroxyadenine, an oxidative form of adenine, *Nucleic Acids Res.*, 1995, **23**, 761–766.

110. H. Kamiya and H. Kasai, Effect of sequence contexts on misincorpora-tion of nucleotides opposite 2-hydroxyadenine, *FEBS Lett.*, 1996, **391**, 113–116.

111. H. Kamiya and H. Kasai, Substitution and deletion mutations induced by 2-hydroxyadenine in *Escherichia coli*: effects of sequence contexts in leading and lagging strands, *Nucleic Acids Res.*, 1997, **25**, 304–311.

112. H. Kamiya and H. Kasai, Mutations induced by 2-hydroxyadenine on a shuttle vector during leading and lagging strand syntheses in mammalian cells, *Biochemistry*, 1997, **36**, 11125–11130.

113. H. Ide, T. Yamaoka and Y. Kimura, Replication of DNA templates containing the alpha-anomer of deoxyadenosine, a major adenine lesion produced by hydroxyl radicals, *Biochemistry*, 1994, **33**, 7127–7133.

114. H. Shimizu, R. Yagi, Y. Kimura, K. Makino, H. Terato, Y. Ohyama and H. Ide, Replication bypass and mutagenic effect of alpha-deoxyadenosine site-specifically incorporated into single-stranded vectors, *Nucleic Acids Res.*, 1997, **25**, 597–603.

115. V. Duarte, J. G. Muller and C. J. Burrows, Insertion of dGMP and dAMP during in vitro DNA synthesis opposite an oxidized form of 7,8-dihydro-8-oxoguanine, *Nucleic Acids Res.*, 1999, **27**, 496–502.

116. O. Kornyushyna, A. M. Berges, J. G. Muller and C. J. Burrows, In vitro nucleotide misinsertion opposite the oxidized guanosine lesions spiroiminodihydantoin and guanidinohydantoin and DNA synthesis past the lesions using *Escherichia coli* DNA polymerase I (Klenow fragment), *Biochemistry*, 2002, **41**, 15304–15314.

117. P. Aller, Y. Ye, S. S. Wallace, C. J. Burrows and S. Doublié, Crystal structure of a replicative DNA polymerase bound to the oxidized guanine lesion guanidinohydantoin, *Biochemistry*, 2010, **49**, 2502–2509.

118. K. D. Sugden and B. D. Martin, Guanine and 7,8-dihydro-8-oxoguanine-specific oxidation in DNA by chromium(V), *Environ. Health Perspect.*, 2002, **110**(Suppl 5), 725–728.

119. P. T. Henderson, J. C. Delaney, J. G. Muller, W. L. Neeley, S. R. Tannenbaum, C. J. Burrows and J. M. Essigmann, The hydantoin lesions formed from oxidation of 7,8-dihydro-8-oxoguanine are potent sources of replication errors in vivo, *Biochemistry*, 2003, **42**, 9257–9262.

120. P. T. Henderson, J. C. Delaney, F. Gu, S. R. Tannenbaum and J. M. Essigmann, Oxidation of 7,8-dihydro-8-oxoguanine affords lesions that are potent sources of replication errors *in vivo*, *Biochemistry*, 2002, **41**, 914–921.

121. S. Boiteux and J. Laval, Coding properties of poly(deoxycytidylic acid) templates containing uracil or apyrimidinic sites: in vitro modulation of mutagenesis by deoxyribonucleic acid repair enzymes, *Biochemistry*, 1982, **21**, 6746–6751.

122. T. A. Kunkel, R. M. Schaaper and L. A. Loeb, Depurination-induced infidelity of deoxyribonucleic acid synthesis with purified deoxyribonucleic acid replication proteins *in vitro*, *Biochemistry*, 1983, **22**, 2378–2384.

123. D. Sagher and B. Strauss, Insertion of nucleotides opposite apurinic/apyrimidinic sites in deoxyribonucleic acid during in vitro synthesis: uniqueness of adenine nucleotides, *Biochemistry*, 1983, **22**, 4518–4526.

124. B. S. Strauss, Translesion DNA synthesis: polymerase response to altered nucleotides, *Cancer Surv.*, 1985, **4**, 493–516.

125. D. Hevroni and Z. Livneh, Bypass and termination at apurinic sites during replication of single-stranded DNA in vitro: a model for apurinic site mutagenesis, *Proc. Natl. Acad. Sci. U. S. A.*, 1988, **85**, 5046–5050.

126. M. Hogg, S. S. Wallace and S. Doublié, Crystallographic snapshots of a replicative DNA polymerase encountering an abasic site, *EMBO J.*, 2004, **23**, 1483–1493.

127. H. Ling, F. Boudsocq, R. Woodgate and W. Yang, Snapshots of replication through an abasic lesion, structural basis for base substitutions and frameshifts, *Mol. Cell*, 2004, **13**, 751–762.

128. M. Hogg, S. S. Wallace and S. Doublié, Bumps in the road: how replicative DNA polymerases see DNA damage, *Curr. Opin. Struct. Biol.*, 2005, **15**, 86–93.

129. K. E. Zahn, S. S. Wallace and S. Doublié, DNA polymerases provide a canon of strategies for translesion synthesis past oxidatively generated lesions, *Curr. Opin. Struct. Biol.*, 2011, **21**, 358–369.
130. R. M. Schaaper and L. A. Loeb, Depurination causes mutations in SOS-induced cells, *Proc. Natl. Acad. Sci. U. S. A.*, 1981, **78**, 1773–1777.
131. R. M. Schaaper, T. A. Kunkel and L. A. Loeb, Infidelity of DNA synthesis associated with bypass of apurinic sites, *Proc. Natl. Acad. Sci. U. S. A.*, 1983, **80**, 487–491.
132. J. Evans, M. Maccabee, Z. Hatahet, J. Courcelle, R. Bockrath, H. Ide and S. Wallace, Thymine ring saturation and fragmentation products: lesion bypass, misinsertion and implications for mutagenesis, *Mutat. Res.*, 1993, **299**, 147–156.
133. T. A. Kunkel, Mutational specificity of depurination, *Proc. Natl. Acad. Sci. U. S. A.*, 1984, **81**, 1494–1498.
134. C. W. Lawrence, A. Borden, S. K. Banerjee and J. E. LeClerc, Mutation frequency and spectrum resulting from a single abasic site in a single-stranded vector, *Nucleic Acids Res.*, 1990, **18**, 2153–2157.
135. J. B. Neto, A. Gentil, R. E. Cabral and A. Sarasin, Mutation spectrum of heat-induced abasic sites on a single-stranded shuttle vector replicated in mammalian cells, *J. Biol. Chem.*, 1992, **267**, 19718–19723.
136. Y. W. Kow, G. Faundez, R. J. Melamede and S. S. Wallace, Processing of model single-strand breaks in phi X-174 RF transfecting DNA by *Escherichia coli*, *Radiat. Res.*, 1991, **126**, 357–366.
137. M. Shrivastav, L. P. De Haro and J. A. Nickoloff, Regulation of DNA double-strand break repair pathway choice, *Cell Res.*, 2008, **18**, 134–147.
138. E. M. Kass and M. Jasin, Collaboration and competition between DNA double-strand break repair pathways, *FEBS Lett.*, 2010, **584**, 3703–3708.
139. H. E. Krokan, H. Nilsen, F. Skorpen, M. Otterlei and G. Slupphaug, Base excision repair of DNA in mammalian cells, *FEBS Lett.*, 2000, **476**, 73–77.
140. S. Mitra, T. Izumi, I. Boldogh, K. K. Bhakat, J. W. Hill and T. K. Hazra, Choreography of oxidative damage repair in mammalian genomes, *Free Radic. Biol. Med.*, 2002, **33**, 15–28.
141. T. Izumi, L. R. Wiederhold, G. Roy, R. Roy, A. Jaiswal, K. K. Bhakat, S. Mitra and T. K. Hazra, Mammalian DNA base excision repair proteins: their interactions and role in repair of oxidative DNA damage, *Toxicology*, 2003, **193**, 43–65.
142. J. C. Fromme and G. L. Verdine, Base excision repair, *Adv. Protein Chem.*, 2004, **69**, 1–41.
143. D. E. Barnes and T. Lindahl, Repair and genetic consequences of endogenous DNA base damage in mammalian cells, *Annu. Rev. Genet.*, 2004, **38**, 445–476.
144. T. Lindahl, An N-glycosidase from *Escherichia coli* that releases free uracil from DNA containing deaminated cytosine residues, *Proc. Natl. Acad. Sci. U. S. A.*, 1974, **71**, 3649–3653.

145. S. Mitra, T. K. Hazra, R. Roy, S. Ikeda, T. Biswas, J. Lock, I. Boldogh and T. Izumi, Complexities of DNA base excision repair in mammalian cells, *Mol. Cell*, 1997, **7**, 305–312.

146. S. S. Wallace, Enzymatic processing of radiation-induced free radical damage in DNA, *Radiat. Res.*, 1998, **150**, S60–79.

147. L. Gros, M. K. Saparbaev and J. Laval, Enzymology of the repair of free radicals-induced DNA damage, *Oncogene*, 2002, **21**, 8905–8925.

148. D. O. Zharkov, Base excision DNA repair, *Cell Mol. Life Sci.*, 2008, **65**, 1544–1565.

149. P. W. Doetsch and R. P. Cunningham, The enzymology of apurinic/apyrimidinic endonucleases, *Mutat. Res.*, 1990, **236**, 173–201.

150. Y. Matsumoto and K. Kim, Excision of deoxyribose phosphate residues by DNA polymerase beta during DNA repair, *Science*, 1995, **269**, 699–702.

151. R. W. Sobol, J. K. Horton, R. Kuhn, H. Gu, R. K. Singhal, R. Prasad, K. Rajewsky and S. H. Wilson, Requirement of mammalian DNA polymerase-beta in base-excision repair, *Nature*, 1996, **379**, 183–186.

152. R. Prasad, W. A. Beard, P. R. Strauss and S. H. Wilson, Human DNA polymerase beta deoxyribose phosphate lyase. Substrate specificity and catalytic mechanism, *J. Biol. Chem.*, 1998, **273**, 15263–15270.

153. R. Prasad, D. D. Shock, W. A. Beard and S. H. Wilson, Substrate channeling in mammalian base excision repair pathways: passing the baton, *J. Biol. Chem.*, 2010, **285**, 40479–40488.

154. A. E. Tomkinson and Z. B. Mackey, Structure and function of mammalian DNA ligases, *Mutat. Res.*, 1998, **407**, 1–9.

155. V. Bailly and W. G. Verly, *Escherichia coli* endonuclease III is not an endonuclease but a beta-elimination catalyst, *Biochem. J.*, 1987, **242**, 565–572.

156. V. Bailly, M. Derydt and W. G. Verly, Delta-elimination in the repair of AP (apurinic/apyrimidinic) sites in DNA, *Biochem. J.*, 1989, **261**, 707–713.

157. L. Wiederhold, J. B. Leppard, P. Kedar, F. Karimi-Busheri, A. Rasouli-Nia, M. Weinfeld, A. E. Tomkinson, T. Izumi, R. Prasad, S. H. Wilson, S. Mitra and T. K. Hazra, AP endonuclease-independent DNA base excision repair in human cells, *Mol. Cell*, 2004, **15**, 209–220.

158. A. Banerjee, W. L. Santos and G. L. Verdine, Structure of a DNA glycosylase searching for lesions, *Science*, 2006, **311**, 1153–1157.

159. R. H. Porecha and J. T. Stivers, Uracil DNA glycosylase uses DNA hopping and short-range sliding to trap extrahelical uracils, *Proc. Natl. Acad. Sci. U. S. A.*, 2008, **105**, 10791–10796.

160. Y. Qi, M. C. Spong, K. Nam, M. Karplus and G. L. Verdine, Entrapment and structure of an extrahelical guanine attempting to enter the active site of a bacterial DNA glycosylase, MutM, *J. Biol. Chem.*, 2010, **285**, 1468–1478.

161. M. L. Hegde, T. K. Hazra and S. Mitra, Early steps in the DNA base excision/single-strand interruption repair pathway in mammalian cells, *Cell Res.*, 2008, **18**, 27–47.

162. I. R. Grin and D. O. Zharkov, Eukaryotic endonuclease VIII-Like proteins: New components of the base excision DNA repair system, *Biochemistry (Moscow)*, 2011, **76**, 80–93.

163. A. K. McCullough, M. L. Dodson and R. S. Lloyd, Initiation of base excision repair: glycosylase mechanisms and structures, *Annu. Rev. Biochem.*, 1999, **68**, 255–285.

164. J. T. Stivers and Y. L. Jiang, A mechanistic perspective on the chemistry of DNA repair glycosylases, *Chem. Rev.*, 2003, **103**, 2729–2759.

165. J. L. Huffman, O. Sundheim and J. A. Tainer, DNA base damage recognition and removal: new twists and grooves, *Mutat. Res.*, 2005, **577**, 55–76.

166. H. Aburatani, Y. Hippo, T. Ishida, R. Takashima, C. Matsuba, T. Kodama, M. Takao, A. Yasui, K. Yamamoto and M. Asano, Cloning and characterization of mammalian 8-hydroxyguanine-specific DNA glycosylase/apurinic, apyrimidinic lyase, a functional mutM homologue, *Cancer Res.*, 1997, **57**, 2151–2156.

167. J. P. Radicella, C. Dherin, C. Desmaze, M. S. Fox and S. Boiteux, Cloning and characterization of hOGG1, a human homolog of the OGG1 gene of *Saccharomyces cerevisiae*, *Proc. Natl. Acad. Sci. U. S. A.*, 1997, **94**, 8010–8015.

168. R. Aspinwall, D. G. Rothwell, T. Roldan-Arjona, C. Anselmino, C. J. Ward, J. P. Cheadle, J. R. Sampson, T. Lindahl, P. C. Harris and I. D. Hickson, Cloning and characterization of a functional human homolog of *Escherichia coli* endonuclease III, *Proc. Natl. Acad. Sci. U. S. A.*, 1997, **94**, 109–114.

169. T. P. Hilbert, W. Chaung, R. J. Boorstein, R. P. Cunningham and G. W. Teebor, Cloning and expression of the cDNA encoding the human homologue of the DNA repair enzyme, *Escherichia coli* endonuclease III, *J. Biol. Chem.*, 1997, **272**, 6733–6740.

170. T. A. Rosenquist, D. O. Zharkov and A. P. Grollman, Cloning and characterization of a mammalian 8-oxoguanine DNA glycosylase, *Proc. Natl. Acad. Sci. U. S. A.*, 1997, **94**, 7429–7434.

171. T. Roldan-Arjona, Y. F. Wei, K. C. Carter, A. Klungland, C. Anselmino, R. P. Wang, M. Augustus and T. Lindahl, Molecular cloning and functional expression of a human cDNA encoding the antimutator enzyme 8-hydroxyguanine-DNA glycosylase, *Proc. Natl. Acad. Sci. U. S. A.*, 1997, **94**, 8016–8020.

172. K. Arai, K. Morishita, K. Shinmura, T. Kohno, S. R. Kim, T. Nohmi, M. Taniwaki, S. Ohwada and J. Yokota, Cloning of a human homolog of the yeast OGG1 gene that is involved in the repair of oxidative DNA damage, *Oncogene*, 1997, **14**, 2857–2861.

173. A. H. Sarker, S. Ikeda, H. Nakano, H. Terato, H. Ide, K. Imai, K. Akiyama, K. Tsutsui, Z. Bo, K. Kubo, K. Yamamoto, A. Yasui, M. C. Yoshida and S. Seki, Cloning and characterization of a mouse homologue (mNthl1) of *Escherichia coli* endonuclease III, *J. Mol. Biol.*, 1998, **282**, 761–774.

174. M. M. Thayer, H. Ahern, D. Xing, R. P. Cunningham and J. A. Tainer, Novel DNA binding motifs in the DNA repair enzyme endonuclease III crystal structure, *EMBO J.*, 1995, **14**, 4108–4120.

175. H. M. Nash, R. Lu, W. S. Lane and G. L. Verdine, The critical active-site amine of the human 8-oxoguanine DNA glycosylase, hOgg1: direct identification, ablation and chemical reconstitution, *Chem. Biol.*, 1997, **4**, 693–702.

176. X. Liu and R. Roy, Mutation at active site lysine 212 to arginine uncouples the glycosylase activity from the lyase activity of human endonuclease III, *Biochemistry*, 2001, **40**, 13617–13622.

177. L. Eide, L. Luna, E. C. Gustad, P. T. Henderson, J. M. Essigmann, B. Demple and E. Seeberg, Human endonuclease III acts preferentially on DNA damage opposite guanine residues in DNA, *Biochemistry*, 2001, **40**, 6653–6659.

178. S. Ikeda, T. Biswas, R. Roy, T. Izumi, I. Boldogh, A. Kurosky, A. H. Sarker, S. Seki and S. Mitra, Purification and characterization of human NTH1, a homolog of *Escherichia coli* endonuclease III. Direct identification of Lys-212 as the active nucleophilic residue, *J. Biol. Chem.*, 1998, **273**, 21585–21593.

179. D. O. Zharkov, T. A. Rosenquist, S. E. Gerchman and A. P. Grollman, Substrate specificity and reaction mechanism of murine 8-oxoguanine-DNA glycosylase, *J. Biol. Chem.*, 2000, **275**, 28607–28617.

180. K. Asagoshi, T. Yamada, Y. Okada, H. Terato, Y. Ohyama, S. Seki and H. Ide, Recognition of formamidopyrimidine by *Escherichia coli* and mammalian thymine glycol glycosylases. Distinctive paired base effects and biological and mechanistic implications, *J. Biol. Chem.*, 2000, **275**, 24781–24786.

181. K. Asagoshi, T. Yamada, H. Terato, Y. Ohyama, Y. Monden, T. Arai, S. Nishimura, H. Aburatani, T. Lindahl and H. Ide, Distinct repair activities of human 7,8-dihydro-8-oxoguanine DNA glycosylase and formamidopyrimidine DNA glycosylase for formamidopyrimidine and 7,8-dihydro-8-oxoguanine, *J. Biol. Chem.*, 2000, **275**, 4956–4964.

182. J. P. McGoldrick, Y. C. Yeh, M. Solomon, J. M. Essigmann and A. L. Lu, Characterization of a mammalian homolog of the *Escherichia coli* MutY mismatch repair protein, *Mol. Cell Biol.*, 1995, **15**, 989–996.

183. M. M. Slupska, C. Baikalov, W. M. Luther, J. H. Chiang, Y. F. Wei and J. H. Miller, Cloning and sequencing a human homolog (hMYH) of the *Escherichia coli* mutY gene whose function is required for the repair of oxidative DNA damage, *J. Bacteriol*, 1996, **178**, 3885–3892.

184. M. M. Slupska, W. M. Luther, J. H. Chiang, H. Yang and J. H. Miller, Functional expression of hMYH, a human homolog of the *Escherichia coli* MutY protein, *J. Bacteriol.*, 1999, **181**, 6210–6213.

185. M. A. Pope and S. S. David, DNA damage recognition and repair by the murine MutY homologue, *DNA Repair (Amst.)*, 2005, **4**, 91–102.

186. V. Bandaru, S. Sunkara, S. S. Wallace and J. P. Bond, A novel human DNA glycosylase that removes oxidative DNA damage and is

homologous to *Escherichia coli* endonuclease VIII, *DNA Repair (Amst.)*, 2002, **1**, 517–529.

187. T. K. Hazra, T. Izumi, I. Boldogh, B. Imhoff, Y. W. Kow, P. Jaruga, M. Dizdaroglu and S. Mitra, Identification and characterization of a human DNA glycosylase for repair of modified bases in oxidatively damaged DNA, *Proc. Natl. Acad. Sci. U. S. A.*, 2002, **99**, 3523–3528.

188. T. K. Hazra, Y. W. Kow, Z. Hatahet, B. Imhoff, I. Boldogh, S. K. Mokkapati, S. Mitra and T. Izumi, Identification and characterization of a novel human DNA glycosylase for repair of cytosine-derived lesions, *J. Biol. Chem.*, 2002, **277**, 30417–30420.

189. I. Morland, V. Rolseth, L. Luna, T. Rognes, M. Bjoras and E. Seeberg, Human DNA glycosylases of the bacterial Fpg/MutM superfamily: an alternative pathway for the repair of 8-oxoguanine and other oxidation products in DNA, *Nucleic Acids Res.*, 2002, **30**, 4926–4936.

190. M. Takao, S. Kanno, K. Kobayashi, Q. M. Zhang, S. Yonei, G. T. van der Horst and A. Yasui, A back-up glycosylase in Nth1 knock-out mice is a functional Nei (endonuclease VIII) homologue, *J. Biol. Chem.*, 2002, **277**, 42205–42213.

191. S. Doublié, V. Bandaru, J. P. Bond and S. S. Wallace, The crystal structure of human endonuclease VIII-like 1 (NEIL1) reveals a zincless finger motif required for glycosylase activity, *Proc. Natl. Acad. Sci. U. S. A.*, 2004, **101**, 10284–10289.

192. A. Das, L. Rajagopalan, V. S. Mathura, S. J. Rigby, S. Mitra and T. K. Hazra, Identification of a zinc finger domain in the human NEIL2 (Nei-like-2) protein, *J. Biol. Chem.*, 2004, **279**, 47132–47138.

193. D. O. Zharkov, G. Golan, R. Gilboa, A. S. Fernandes, S. E. Gerchman, J. H. Kycia, R. A. Rieger, A. P. Grollman and G. Shoham, Structural analysis of an *Escherichia coli* endonuclease VIII covalent reaction intermediate, *EMBO J.*, 2002, **21**, 789–800.

194. D. O. Zharkov, R. A. Rieger, C. R. Iden and A. P. Grollman, NH2-terminal proline acts as a nucleophile in the glycosylase/AP-lyase reaction catalyzed by *Escherichia coli* formamidopyrimidine-DNA glycosylase (Fpg) protein, *J. Biol. Chem.*, 1997, **272**, 5335–5341.

195. S. Burgess, P. Jaruga, M. L. Dodson, M. Dizdaroglu and R. S. Lloyd, Determination of active site residues in *Escherichia coli* endonuclease VIII, *J. Biol. Chem.*, 2002, **277**, 2938–2944.

196. N. Krishnamurthy, X. Zhao, C. J. Burrows and S. S. David, Superior removal of hydantoin lesions relative to other oxidized bases by the human DNA glycosylase hNEIL1, *Biochemistry*, 2008, **47**, 7137–7146.

197. X. Zhao, N. Krishnamurthy, C. J. Burrows and S. S. David, Mutation *versus* repair: NEIL1 removal of hydantoin lesions in single-stranded, bulge, bubble, and duplex DNA contexts, *Biochemistry*, 2010, **49**, 1658–1666.

198. A. Katafuchi, T. Nakano, A. Masaoka, H. Terato, S. Iwai, F. Hanaoka and H. Ide, Differential specificity of human and *Escherichia coli*

endonuclease III and VIII homologues for oxidative base lesions, *J. Biol. Chem.*, 2004, **279**, 14464–14471.

199. P. Jaruga, M. Birincioglu, T. A. Rosenquist and M. Dizdaroglu, Mouse NEIL1 protein is specific for excision of 2,6-diamino-4-hydroxy-5-formamidopyrimidine and 4,6-diamino-5-formamidopyrimidine from oxidatively damaged DNA, *Biochemistry*, 2004, **43**, 15909–15914.

200. I. R. Grin, G. L. Dianov and D. O. Zharkov, The role of mammalian NEIL1 protein in the repair of 8-oxo-7,8-dihydroadenine in DNA, *FEBS Lett.*, 2010, **584**, 1553–1557.

201. H. Dou, S. Mitra and T. K. Hazra, Repair of oxidized bases in DNA bubble structures by human DNA glycosylases NEIL1 and NEIL2, *J. Biol. Chem.*, 2003, **278**, 49679–49684.

202. H. Nilsen, T. Lindahl and A. Verreault, DNA base excision repair of uracil residues in reconstituted nucleosome core particles, *EMBO J.*, 2002, **21**, 5943–5952.

203. B. C. Beard, S. H. Wilson and M. J. Smerdon, Suppressed catalytic activity of base excision repair enzymes on rotationally positioned uracil in nucleosomes, *Proc. Natl. Acad. Sci. U. S. A.*, 2003, **100**, 7465–7470.

204. A. Prasad, S. S. Wallace and D. S. Pederson, Initiation of base excision repair of oxidative lesions in nucleosomes by the human, bifunctional DNA glycosylase NTH1, *Mol. Cell Biol.*, 2007, **27**, 8442–8453.

205. J. M. Hinz, Y. Rodriguez and M. J. Smerdon, Rotational dynamics of DNA on the nucleosome surface markedly impact accessibility to a DNA repair enzyme, *Proc. Natl. Acad. Sci. U. S. A.*, 2010, **107**, 4646–4651.

206. I. D. Odell, K. Newick, N. H. Heintz, S. S. Wallace and D. S. Pederson, Non-specific DNA binding interferes with the efficient excision of oxidative lesions from chromatin by the human DNA glycosylase, NEIL1, *DNA Repair (Amst.)*, 2010, **9**, 134–143.

207. A. Klungland, I. Rosewell, S. Hollenbach, E. Larsen, G. Daly, B. Epe, E. Seeberg, T. Lindahl and D. E. Barnes, Accumulation of premutagenic DNA lesions in mice defective in removal of oxidative base damage, *Proc. Natl. Acad. Sci. U. S. A.*, 1999, **96**, 13300–13305.

208. O. Minowa, T. Arai, M. Hirano, Y. Monden, S. Nakai, M. Fukuda, M. Itoh, H. Takano, Y. Hippou, H. Aburatani, K. Masumura, T. Nohmi, S. Nishimura and T. Noda, Mmh/Ogg1 gene inactivation results in accumulation of 8-hydroxyguanine in mice, *Proc. Natl. Acad. Sci. U. S. A.*, 2000, **97**, 4156–4161.

209. M. Osterod, S. Hollenbach, J. G. Hengstler, D. E. Barnes, T. Lindahl and B. Epe, Age-related and tissue-specific accumulation of oxidative DNA base damage in 7,8-dihydro-8-oxoguanine-DNA glycosylase (Ogg1) deficient mice, *Carcinogenesis*, 2001, **22**, 1459–1463.

210. M. T. Ocampo, W. Chaung, D. R. Marenstein, M. K. Chan, A. Altamirano, A. K. Basu, R. J. Boorstein, R. P. Cunningham and G. W. Teebor, Targeted deletion of mNth1 reveals a novel DNA repair enzyme activity, *Mol. Cell. Biol.*, 2002, **22**, 6111–6121.

211. K. Torisu, D. Tsuchimoto, Y. Ohnishi and Y. Nakabeppu, Hemato-poietic tissue-specific expression of mouse Neil3 for endonuclease VIII-like protein, *J. Biochem.*, 2005, **138**, 763–772.

212. V. Vartanian, B. Lowell, I. G. Minko, T. G. Wood, J. D. Ceci, S. George, S. W. Ballinger, C. L. Corless, A. K. McCullough and R. S. Lloyd, The metabolic syndrome resulting from a knockout of the NEIL1 DNA glycosylase, *Proc. Natl. Acad. Sci. U. S. A.*, 2006, **103**, 1864–1869.

213. E. C. Friedberg and L. B. Meira, Database of mouse strains carrying targeted mutations in genes affecting biological responses to DNA damage Version 7, *DNA Repair (Amst.)*, 2006, **5**, 189–209.

214. J. Hu, N. C. de Souza-Pinto, K. Haraguchi, B. A. Hogue, P. Jaruga, M. M. Greenberg, M. Dizdaroglu and V. A. Bohr, Repair of for-mamidopyrimidines in DNA involves different glycosylases: role of the OGG1, NTH1, and NEIL1 enzymes, *J. Biol. Chem.*, 2005, **280**, 40544–40551.

215. Y. Xie, H. Yang, C. Cunanan, K. Okamoto, D. Shibata, J. Pan, D. E. Barnes, T. Lindahl, M. McIlhatton, R. Fishel and J. H. Miller, Defi-ciencies in mouse Myh and Ogg1 result in tumor predisposition and G to T mutations in codon 12 of the K-ras oncogene in lung tumors, *Cancer Res.*, 2004, **64**, 3096–3102.

216. M. K. Chan, M. T. Ocampo-Hafalla, V. Vartanian, P. Jaruga, G. Kirkali, K. L. Koenig, S. Brown, R. S. Lloyd, M. Dizdaroglu and G. W. Teebor, Targeted deletion of the genes encoding NTH1 and NEIL1 DNA N-glycosylases reveals the existence of novel carcinogenic oxidative damage to DNA, *DNA Repair (Amst.)*, 2009, **8**, 786–794.

217. M. S. Cooke, J. Lunec and M. D. Evans, Progress in the analysis of urinary oxidative DNA damage, *Free Radic. Biol. Med.*, 2002, **33**, 1601–1614.

218. C. Jungst, B. Cheng, R. Gehrke, V. Schmitz, H. D. Nischalke, J. Ramakers, P. Schramel, P. Schirmacher, T. Sauerbruch and W. H. Caselmann, Oxidative damage is increased in human liver tissue adjacent to hepatocellular carcinoma, *Hepatology*, 2004, **39**, 1663–1672.

219. A. R. Trzeciak, S. G. Nyaga, P. Jaruga, A. Lohani, M. Dizdaroglu and M. K. Evans, Cellular repair of oxidatively induced DNA base lesions is defective in prostate cancer cell lines, PC-3 and DU-145, *Carcinogenesis*, 2004, **25**, 1359–1370.

220. M. S. Cooke, R. Rozalski, R. Dove, D. Gackowski, A. Siomek, M. D. Evans and R. Olinski, Evidence for attenuated cellular 8-oxo-7,8-dihydro-2′-deoxyguanosine removal in cancer patients, *Biol. Chem.*, 2006, **387**, 393–400.

221. P. Karihtala and Y. Soini, Reactive oxygen species and antioxidant mechanisms in human tissues and their relation to malignancies, *APMIS*, 2007, **115**, 81–103.

222. G. Frosina, Commentary: DNA base excision repair defects in human pathologies, *Free Radic. Res.*, 2004, **38**, 1037–1054.

223. R. J. Hung, J. Hall, P. Brennan and P. Boffetta, Genetic polymorphisms in the base excision repair pathway and cancer risk: a HuGE review, *Am. J. Epidemiol*, 2005, **162**, 925–942.

224. S. Jones, P. Emmerson, J. Maynard, J. M. Best, S. Jordan, G. T. Williams, J. R. Sampson and J. P. Cheadle, Biallelic germline mutations in MYH predispose to multiple colorectal adenoma and somatic G:C-->T:A mutations, *Hum. Mol. Genet.*, 2002, **11**, 2961–2967.

225. S. Jones, S. Lambert, G. T. Williams, J. M. Best, J. R. Sampson and J. P. Cheadle, Increased frequency of the k-ras G12C mutation in MYH polyposis colorectal adenomas, *Br. J. Cancer*, 2004, **90**, 1591–1593.

226. M. A. Pope, N. H. Chmiel and S. S. David, Insight into the functional consequences of hMYH variants associated with colorectal cancer: distinct differences in the adenine glycosylase activity and the response to AP endonucleases of Y150C and G365D murine MYH, *DNA Repair (Amst.)*, 2005, **4**, 315–325.

227. T. Nohmi, S. R. Kim and M. Yamada, Modulation of oxidative mutagenesis and carcinogenesis by polymorphic forms of human DNA repair enzymes, *Mutat. Res.*, 2005, **591**, 60–73.

228. K. K. Chan, Q. M. Zhang and G. L. Dianov, Base excision repair fidelity in normal and cancer cells, *Mutagenesis*, 2006, **21**, 173–178.

229. A. A. Nemec, S. S. Wallace and J. B. Sweasy, Variant base excision repair proteins: contributors to genomic instability, *Semin. Cancer Biol.*, 2010, **20**, 320–328.

230. H. Sugimura, T. Kohno, K. Wakai, K. Nagura, K. Genka, H. Igarashi, B. J. Morris, S. Baba, Y. Ohno, C. Gao, Z. Li, J. Wang, T. Takezaki, K. Tajima, T. Varga, T. Sawaguchi, J. K. Lum, J. J. Martinson, S. Tsugane, T. Iwamasa, K. Shinmura and J. Yokota, hOGG1 Ser326Cys polymorphism and lung cancer susceptibility, *Cancer Epidemiol. Biomarkers Prev.*, 1999, **8**, 669–674.

231. E. L. Goode, C. M. Ulrich and J. D. Potter, Polymorphisms in DNA repair genes and associations with cancer risk, *Cancer Epidemiol. Biomarkers Prev.*, 2002, **11**, 1513–1530.

232. L. Chen, A. Elahi, J. Pow-Sang, P. Lazarus and J. Park, Association between polymorphism of human oxoguanine glycosylase 1 and risk of prostate cancer, *J. Urol.*, 2003, **170**, 2471–2474.

233. E. Y. Cho, A. Hildesheim, C. J. Chen, M. M. Hsu, I. H. Chen, B. F. Mittl, P. H. Levine, M. Y. Liu, J. Y. Chen, L. A. Brinton, Y. J. Cheng and C. S. Yang, Nasopharyngeal carcinoma and genetic polymorphisms of DNA repair enzymes XRCC1 and hOGG1, *Cancer Epidemiol. Biomarkers Prev.*, 2003, **12**, 1100–1104.

234. K. Shinmura, H. Tao, M. Goto, H. Igarashi, T. Taniguchi, M. Maekawa, T. Takezaki and H. Sugimura, Inactivating mutations of the human base excision repair gene NEIL1 in gastric cancer, *Carcinogenesis*, 2004, **25**, 2311–2317.

235. P. Broderick, T. Bagratuni, J. Vijayakrishnan, S. Lubbe, I. Chandler and R. S. Houlston, Evaluation of NTHL1, NEIL1, NEIL2, MPG, TDG,

UNG and SMUG1 genes in familial colorectal cancer predisposition, *BMC Cancer*, 2006, **6**, 243.

236. L. M. Roy, P. Jaruga, T. G. Wood, A. K. McCullough, M. Dizdaroglu and R. S. Lloyd, Human polymorphic variants of the NEIL1 DNA glycosylase, *J. Biol. Chem.*, 2007, **282**, 15790–15798.

237. V. S. Sidorenko, A. P. Grollman, P. Jaruga, M. Dizdaroglu and D. O. Zharkov, Substrate specificity and excision kinetics of natural polymorphic variants and phosphomimetic mutants of human 8-oxoguanine-DNA glycosylase, *FEBS J.*, 2009, **276**, 5149–5162.

238. M. Goto, K. Shinmura, H. Tao, S. Tsugane and H. Sugimura, Three novel NEIL1 promoter polymorphisms in gastric cancer patients, *World J. Gastrointest. Oncol.*, 2010, **2**, 117–120.

239. G. A. Cortopassi, D. Shibata, N. W. Soong and N. Arnheim, A pattern of accumulation of a somatic deletion of mitochondrial DNA in aging human tissues, *Proc. Natl. Acad. Sci. U. S. A.*, 1992, **89**, 7370–7374.

240. D. C. Wallace, A mitochondrial paradigm for degenerative diseases and ageing, *Novartis Found. Symp*, 2001, **235**, 247–263, discussion 263–236.

241. K. Tonska, A. Solyga and E. Bartnik, Mitochondria and aging: innocent bystanders or guilty parties?, *J. Appl. Genet.*, 2009, **50**, 55–62.

242. D. C. Wallace, Mitochondrial DNA mutations in diseases of energy metabolism, *J. Bioenerg. Biomembr.*, 1994, **26**, 241–250.

243. E. A. Schon, Mitochondrial genetics and disease, *Trends Biochem. Sci.*, 2000, **25**, 555–560.

244. K. Polyak, Y. Li, H. Zhu, C. Lengauer, J. K. Willson, S. D. Markowitz, M. A. Trush, K. W. Kinzler and B. Vogelstein, Somatic mutations of the mitochondrial genome in human colorectal tumours, *Nat. Genet.*, 1998, **20**, 291–293.

245. M. S. Fliss, H. Usadel, O. L. Caballero, L. Wu, M. R. Buta, S. M. Eleff, J. Jen and D. Sidransky, Facile detection of mitochondrial DNA mutations in tumors and bodily fluids, *Science*, 2000, **287**, 2017–2019.

246. D. L. Croteau and V. A. Bohr, Repair of oxidative damage to nuclear and mitochondrial DNA in mammalian cells, *J. Biol. Chem.*, 1997, **272**, 25409–25412.

247. M. L. Hamilton, Z. Guo, C. D. Fuller, H. Van Remmen, W. F. Ward, S. N. Austad, D. A. Troyer, I. Thompson and A. Richardson, A reliable assessment of 8-oxo-2-deoxyguanosine levels in nuclear and mitochondrial DNA using the sodium iodide method to isolate DNA, *Nucleic Acids Res.*, 2001, **29**, 2117–2126.

248. P. Liu and B. Demple, DNA repair in mammalian mitochondria: Much more than we thought?, *Environ. Mol. Mutagen.*, 2010, **51**, 417–426.

249. M. Takao, H. Aburatani, K. Kobayashi and A. Yasui, Mitochondrial targeting of human DNA glycosylases for repair of oxidative DNA damage, *Nucleic Acids Res.*, 1998, **26**, 2917–2922.

250. K. Nishioka, T. Ohtsubo, H. Oda, T. Fujiwara, D. Kang, K. Sugimachi and Y. Nakabeppu, Expression and differential intracellular localization

of two major forms of human 8-oxoguanine DNA glycosylase encoded by alternatively spliced OGG1 mRNAs, *Mol. Biol. Cell*, 1999, **10**, 1637–1652.

251. S. Ikeda, T. Kohmoto, R. Tabata and Y. Seki, Differential intracellular localization of the human and mouse endonuclease III homologs and analysis of the sorting signals, *DNA Repair (Amst.)*, 2002, **1**, 847–854.

252. R. Gredilla, C. Garm, R. Holm, V. A. Bohr and T. Stevnsner, Differential age-related changes in mitochondrial DNA repair activities in mouse brain regions, *Neurobiol. Aging*, 2010, **31**, 993–1002.

253. K. Imai, A. H. Sarker, K. Akiyama, S. Ikeda, M. Yao, K. Tsutsui, T. Shohmori and S. Seki, Genomic structure and sequence of a human homologue (NTHL1/NTH1) of *Escherichia coli* endonuclease III with those of the adjacent parts of TSC2 and SLC9A3R2 genes, *Gene*, 1998, **222**, 287–295.

254. V. Rolseth, E. Runden-Pran, L. Luna, C. McMurray, M. Bjoras and O. P. Ottersen, Widespread distribution of DNA glycosylases removing oxidative DNA lesions in human and rodent brains, *DNA Repair (Amst.)*, 2008, **7**, 1578–1588.

255. G. A. Hildrestrand, C. G. Neurauter, D. B. Diep, C. G. Castellanos, S. Krauss, M. Bjoras and L. Luna, Expression patterns of Neil3 during embryonic brain development and neoplasia, *BMC Neurosci.*, 2009, **10**, 45.

256. M. L. Hegde, C. A. Theriot, A. Das, P. M. Hegde, Z. Guo, R. K. Gary, T. K. Hazra, B. Shen and S. Mitra, Physical and functional interaction between human oxidized base-specific DNA glycosylase NEIL1 and flap endonuclease 1, *J. Biol. Chem.*, 2008, **283**, 27028–27037.

257. H. Dou, C. A. Theriot, A. Das, M. L. Hegde, Y. Matsumoto, I. Boldogh, T. K. Hazra, K. K. Bhakat and S. Mitra, Interaction of the human DNA glycosylase NEIL1 with proliferating cell nuclear antigen. The potential for replication-associated repair of oxidized bases in mammalian genomes, *J. Biol. Chem.*, 2008, **283**, 3130–3140.

258. C. A. Theriot, M. L. Hegde, T. K. Hazra and S. Mitra, RPA physically interacts with the human DNA glycosylase NEIL1 to regulate excision of oxidative DNA base damage in primer-template structures, *DNA Repair (Amst.)*, 2010, **9**, 643–652.

259. D. Banerjee, S. M. Mandal, A. Das, M. L. Hegde, S. Das, K. K. Bhakat, I. Boldogh, P. S. Sarkar, S. Mitra and T. K. Hazra, Preferential repair of oxidized base damage in the transcribed genes of mammalian cells, *J. Biol. Chem.*, 2011, **286**, 6006–6016.

260. T. K. Hazra, A. Das, S. Das, S. Choudhury, Y. W. Kow and R. Roy, Oxidative DNA damage repair in mammalian cells: a new perspective, *DNA Repair (Amst.)*, 2007, **6**, 470–480.

261. G. Frosina, P. Fortini, O. Rossi, F. Carrozzino, G. Raspaglio, L. S. Cox, D. P. Lane, A. Abbondandolo and E. Dogliotti, Two pathways for base excision repair in mammalian cells, *J. Biol. Chem.*, 1996, **271**, 9573–9578.

262. P. Fortini, B. Pascucci, E. Parlanti, M. D'Errico, V. Simonelli and E. Dogliotti, The base excision repair: mechanisms and its relevance for cancer susceptibility, *Biochimie*, 2003, **85**, 1053–1071.

263. R. Prasad, G. L. Dianov, V. A. Bohr and S. H. Wilson, FEN1 stimulation of DNA polymerase beta mediates an excision step in mammalian long patch base excision repair, *J. Biol. Chem.*, 2000, **275**, 4460–4466.

264. A. Klungland and T. Lindahl, Second pathway for completion of human DNA base excision-repair: reconstitution with purified proteins and requirement for DNase IV (FEN1), *EMBO J.*, 1997, **16**, 3341–3348.

265. K. Kim, S. Biade and Y. Matsumoto, Involvement of flap endonuclease 1 in base excision DNA repair, *J. Biol. Chem.*, 1998, **273**, 8842–8848.

266. Y. Liu, R. Prasad and S. H. Wilson, HMGB1: roles in base excision repair and related function, *Biochim. Biophys. Acta*, 2010, **1799**, 119–130.

267. K. H. Almeida and R. W. Sobol, A unified view of base excision repair: lesion-dependent protein complexes regulated by post-translational modification, *DNA Repair (Amst.)*, 2007, **6**, 695–711.

268. S. Marsin, A. E. Vidal, M. Sossou, J. Menissier-de Murcia, F. Le Page, S. Boiteux, G. de Murcia and J. P. Radicella, Role of XRCC1 in the coordination and stimulation of oxidative DNA damage repair initiated by the DNA glycosylase hOGG1, *J. Biol. Chem.*, 2003, **278**, 44068–44074.

269. A. Campalans, S. Marsin, Y. Nakabeppu, R. O'Connor T, S. Boiteux and J. P. Radicella, XRCC1 interactions with multiple DNA glycosylases: a model for its recruitment to base excision repair, *DNA Repair (Amst.)*, 2005, **4**, 826–835.

270. K. W. Caldecott, S. Aoufouchi, P. Johnson and S. Shall, XRCC1 polypeptide interacts with DNA polymerase beta and possibly poly (ADP-ribose) polymerase, and DNA ligase III is a novel molecular 'nick-sensor' *in vitro*, *Nucleic Acids Res.*, 1996, **24**, 4387–4394.

271. V. Schreiber, J. C. Ame, P. Dolle, I. Schultz, B. Rinaldi, V. Fraulob, J. Menissier-de Murcia and G. de Murcia, Poly(ADP-ribose) polymerase-2 (PARP-2) is required for efficient base excision DNA repair in association with PARP-1 and XRCC1, *J. Biol. Chem.*, 2002, **277**, 23028–23036.

272. J. B. Leppard, Z. Dong, Z. B. Mackey and A. E. Tomkinson, Physical and functional interaction between DNA ligase IIIalpha and poly(ADP-Ribose) polymerase 1 in DNA single-strand break repair, *Mol. Cell Biol.*, 2003, **23**, 5919–5927.

273. M. Masson, C. Niedergang, V. Schreiber, S. Muller and J. Menissier-de, Murcia and G. de Murcia, XRCC1 is specifically associated with poly-(ADP-ribose) polymerase and negatively regulates its activity following DNA damage, *Mol. Cell. Biol.*, 1998, **18**, 3563–3571.

274. Y. Kubota, R. A. Nash, A. Klungland, P. Schar, D. E. Barnes and T. Lindahl, Reconstitution of DNA base excision-repair with purified human proteins: interaction between DNA polymerase beta and the XRCC1 protein, *EMBO J.*, 1996, **15**, 6662–6670.

275. R. A. Bennett, D. M. Wilson, 3rd, D. Wong and B. Demple, Interaction of human apurinic endonuclease and DNA polymerase beta in the

base excision repair pathway, *Proc. Natl. Acad. Sci. U. S. A.*, 1997, **94**, 7166–7169.

276. A. E. Vidal, S. Boiteux, I. D. Hickson and J. P. Radicella, XRCC1 coordinates the initial and late stages of DNA abasic site repair through protein-protein interactions, *EMBO J.*, 2001, **20**, 6530–6539.

277. C. J. Whitehouse, R. M. Taylor, A. Thistlethwaite, H. Zhang, F. Karimi-Busheri, D. D. Lasko, M. Weinfeld and K. W. Caldecott, XRCC1 stimulates human polynucleotide kinase activity at damaged DNA termini and accelerates DNA single-strand break repair, *Cell*, 2001, **104**, 107–117.

278. A. Das, L. Wiederhold, J. B. Leppard, P. Kedar, R. Prasad, H. Wang, I. Boldogh, F. Karimi-Busheri, M. Weinfeld, A. E. Tomkinson, S. H. Wilson, S. Mitra and T. K. Hazra, NEIL2-initiated, APE-independent repair of oxidized bases in DNA: Evidence for a repair complex in human cells, *DNA Repair (Amst.)*, 2006, **5**, 1439–1448.

279. A. E. Vidal, I. D. Hickson, S. Boiteux and J. P. Radicella, Mechanism of stimulation of the DNA glycosylase activity of hOGG1 by the major human AP endonuclease: bypass of the AP lyase activity step, *Nucleic Acids Res.*, 2001, **29**, 1285–1292.

280. D. R. Marenstein, M. K. Chan, A. Altamirano, A. K. Basu, R. J. Boorstein, R. P. Cunningham and G. W. Teebor, Substrate specificity of human endonuclease III (hNTH1). Effect of human APE1 on hNTH1 activity, *J. Biol. Chem.*, 2003, **278**, 9005–9012.

281. K. W. Caldecott, C. K. McKeown, J. D. Tucker, S. Ljungquist and L. H. Thompson, An interaction between the mammalian DNA repair protein XRCC1 and DNA ligase III, *Mol. Cell. Biol.*, 1994, **14**, 68–76.

282. C. S. Busso, M. W. Lake and T. Izumi, Posttranslational modification of mammalian AP endonuclease (APE1), *Cell Mol. Life Sci.*, 2010, **67**, 3609–3620.

283. J. Fan and D. M. Wilson, 3rd, Protein-protein interactions and posttranslational modifications in mammalian base excision repair, *Free Radic. Biol. Med.*, 2005, **38**, 1121–1138.

284. F. Dantzer, L. Luna, M. Bjoras and E. Seeberg, Human OGG1 undergoes serine phosphorylation and associates with the nuclear matrix and mitotic chromatin *in vivo*, *Nucleic Acids Res.*, 2002, **30**, 2349–2357.

285. K. K. Bhakat, T. K. Hazra and S. Mitra, Acetylation of the human DNA glycosylase NEIL2 and inhibition of its activity, *Nucleic Acids Res.*, 2004, **32**, 3033–3039.

286. J. Hu, S. Z. Imam, K. Hashiguchi, N. C. de Souza-Pinto and V. A. Bohr, Phosphorylation of human oxoguanine DNA glycosylase (alpha-OGG1) modulates its function, *Nucleic Acids Res.*, 2005, **33**, 3271–3282.

287. S. Nakanishi, R. Prasad, S. H. Wilson and M. Smerdon, Different structural states in oligonucleosomes are required for early *versus* late steps of base excision repair, *Nucleic Acids Res.*, 2007, **35**, 4313–4321.

288. D. R. Chafin, J. M. Vitolo, L. A. Henricksen, R. A. Bambara and J. J. Hayes, Human DNA ligase I efficiently seals nicks in nucleosomes, *EMBO J.*, 2000, **19**, 5492–5501.

289. C. F. Huggins, D. R. Chafin, S. Aoyagi, L. A. Henricksen, R. A. Bambara and J. J. Hayes, Flap endonuclease 1 efficiently cleaves base excision repair and DNA replication intermediates assembled into nucleosomes, *Mol. Cell*, 2002, **10**, 1201–1011.

290. H. Gu, J. D. Marth, P. C. Orban, H. Mossmann and K. Rajewsky, Deletion of a DNA polymerase beta gene segment in T cells using cell type-specific gene targeting, *Science*, 1994, **265**, 103–106.

291. D. L. Ludwig, M. A. MacInnes, Y. Takiguchi, P. E. Purtymun, M. Henrie, M. Flannery, J. Meneses, R. A. Pedersen and D. J. Chen, A murine AP-endonuclease gene-targeted deficiency with post-implantation embryonic progression and ionizing radiation sensitivity, *Mutat. Res.*, 1998, **409**, 17–29.

292. R. S. Tebbs, M. L. Flannery, J. J. Meneses, A. Hartmann, J. D. Tucker, L. H. Thompson, J. E. Cleaver and R. A. Pedersen, Requirement for the Xrcc1 DNA base excision repair gene during early mouse development, *Dev. Biol.*, 1999, **208**, 513–529.

293. N. Puebla-Osorio, D. B. Lacey, F. W. Alt and C. Zhu, Early embryonic lethality due to targeted inactivation of DNA ligase III, *Mol. Cell. Biol.*, 2006, **26**, 3935–3941.

294. D. Bentley, J. Selfridge, J. K. Millar, K. Samuel, N. Hole, J. D. Ansell and D. W. Melton, DNA ligase I is required for fetal liver erythropoiesis but is not essential for mammalian cell viability, *Nat. Genet.*, 1996, **13**, 489–491.

295. M. Kucherlapati, K. Yang, M. Kuraguchi, J. Zhao, M. Lia, J. Heyer, M. F. Kane, K. Fan, R. Russell, A. M. Brown, B. Kneitz, W. Edelmann, R. D. Kolodner, M. Lipkin and R. Kucherlapati, Haploinsufficiency of Flap endonuclease (Fen1) leads to rapid tumor progression, *Proc. Natl. Acad. Sci. U. S. A.*, 2002, **99**, 9924–9929.

296. N. Sugo, Y. Aratani, Y. Nagashima, Y. Kubota and H. Koyama, Neonatal lethality with abnormal neurogenesis in mice deficient in DNA polymerase beta, *EMBO J.*, 2000, **19**, 1397–1404.

297. J. Huamani, C. A. McMahan, D. C. Herbert, R. Reddick, J. R. McCarrey, M. I. MacInnes, D. J. Chen and C. A. Walter, Spontaneous mutagenesis is enhanced in Apex heterozygous mice, *Mol. Cell. Biol.*, 2004, **24**, 8145–8153.

298. D. C. Cabelof, Z. Guo, J. J. Raffoul, R. W. Sobol, S. H. Wilson, A. Richardson and A. R. Heydari, Base excision repair deficiency caused by polymerase beta haploinsufficiency: accelerated DNA damage and increased mutational response to carcinogens, *Cancer Res.*, 2003, **63**, 5799–5807.

299. J. K. Horton, D. F. Joyce-Gray, B. F. Pachkowski, J. A. Swenberg and S. H. Wilson, Hypersensitivity of DNA polymerase beta null mouse fibroblasts reflects accumulation of cytotoxic repair intermediates from site-specific alkyl DNA lesions, *DNA Repair (Amst.)*, 2003, **2**, 27–48.

300. U. A. Betz, C. A. Vosshenrich, K. Rajewsky and W. Muller, Bypass of lethality with mosaic mice generated by Cre-loxP-mediated recombination, *Curr. Biol.*, 1996, **6**, 1307–1316.

301. J. K. Horton, M. Watson, D. F. Stefanick, D. T. Shaughnessy, J. A. Taylor and S. H. Wilson, XRCC1 and DNA polymerase beta in cellular protection against cytotoxic DNA single-strand breaks, *Cell Res.*, 2008, **18**, 48–63.

302. J. J. Hu, T. R. Smith, M. S. Miller, H. W. Mohrenweiser, A. Golden and L. D. Case, Amino acid substitution variants of APE1 and XRCC1 genes associated with ionizing radiation sensitivity, *Carcinogenesis*, 2001, **22**, 917–922.

303. M. Z. Hadi, M. A. Coleman, K. Fidelis, H. W. Mohrenweiser and D. M. Wilson, 3rd, Functional characterization of Ape1 variants identified in the human population, *Nucleic Acids Res.*, 2000, **28**, 3871–3879.

304. D. Gu, M. Wang, Z. Zhang and J. Chen, The DNA repair gene APE1 T1349G polymorphism and cancer risk: a meta-analysis of 27 case-control studies, *Mutagenesis*, 2009, **24**, 507–512.

305. D. Starcevic, S. Dalal and J. B. Sweasy, Is there a link between DNA polymerase beta and cancer?, *Cell Cycle*, 2004, **3**, 998–1001.

306. T. Xi, I. M. Jones and H. W. Mohrenweiser, Many amino acid substitution variants identified in DNA repair genes during human population screenings are predicted to impact protein function, *Genomics*, 2004, **83**, 970–979.

307. S. Dalal, A. Chikova, J. Jaeger and J. B. Sweasy, The Leu22Pro tumor-associated variant of DNA polymerase beta is dRP lyase deficient, *Nucleic Acids Res.*, 2008, **36**, 411–422.

308. Z. Hu, H. Ma, D. Lu, J. Zhou, Y. Chen, L. Xu, J. Zhu, X. Huo, J. Qian, Q. Wei and H. Shen, A promoter polymorphism (-77T>C) of DNA repair gene XRCC1 is associated with risk of lung cancer in relation to tobacco smoking, *Pharmacogenet. Genomics*, 2005, **15**, 457–463.

309. L. Jiao, M. L. Bondy, M. M. Hassan, R. A. Wolff, D. B. Evans, J. L. Abbruzzese and D. Li, Selected polymorphisms of DNA repair genes and risk of pancreatic cancer, *Cancer Detect. Prev.*, 2006, **30**, 284–291.

310. V. A. Bohr, Repair of oxidative DNA damage in nuclear and mitochondrial DNA, and some changes with aging in mammalian cells, *Free Radic. Biol. Med.*, 2002, **32**, 804–812.

311. R. Chattopadhyay, L. Wiederhold, B. Szczesny, I. Boldogh, T. K. Hazra, T. Izumi and S. Mitra, Identification and characterization of mitochondrial abasic (AP)-endonuclease in mammalian cells, *Nucleic Acids Res.*, 2006, **34**, 2067–2076.

312. M. A. Graziewicz, M. J. Longley and W. C. Copeland, DNA polymerase gamma in mitochondrial DNA replication and repair, *Chem. Rev.*, 2006, **106**, 383–405.

313. U. Lakshmipathy and C. Campbell, The human DNA ligase III gene encodes nuclear and mitochondrial proteins, *Mol. Cell. Biol.*, 1999, **19**, 3869–3876.

314. U. Lakshmipathy and C. Campbell, Mitochondrial DNA ligase III function is independent of Xrcc1, *Nucleic Acids Res.*, 2000, **28**, 3880–3886.

315. M. Akbari, T. Visnes, H. E. Krokan and M. Otterlei, Mitochondrial base excision repair of uracil and AP sites takes place by single-nucleotide

insertion and long-patch DNA synthesis, *DNA Repair (Amst.)*, 2008, **7**, 605–616.

316. P. Liu, L. Qian, J. S. Sung, N. C. de Souza-Pinto, L. Zheng, D. F. Bogenhagen, V. A. Bohr, D. M. Wilson, 3rd, B. Shen and B. Demple, Removal of oxidative DNA damage *via* FEN1-dependent long-patch base excision repair in human cell mitochondria, *Mol. Cell. Biol.*, 2008, **28**, 4975–4987.

317. B. Szczesny, A. W. Tann, M. J. Longley, W. C. Copeland and S. Mitra, Long patch base excision repair in mammalian mitochondrial genomes, *J. Biol. Chem.*, 2008, **283**, 26349–26356.

318. L. Kalifa, G. Beutner, N. Phadnis, S. S. Sheu and E. A. Sia, Evidence for a role of FEN1 in maintaining mitochondrial DNA integrity, *DNA Repair (Amst.)*, 2009, **8**, 1242–1249.

319. G. J. Jenkins, Z. Zair, G. E. Johnson and S. H. Doak, Genotoxic thresholds, DNA repair, and susceptibility in human populations, *Toxicology*, 2010, **278**, 305–310.

320. J. F. Ward, DNA damage as the cause of ionizing radiation-induced gene activation, *Radiat. Res.*, 1994, **138**, S85–S88.

321. A. B. Robertson, A. Klungland, T. Rognes and I. Leiros, DNA repair in mammalian cells: Base excision repair: the long and short of it, *Cell Mol. Life Sci.*, 2009, **66**, 981–993.

322. L. J. Hofseth, M. A. Khan, M. Ambrose, O. Nikolayeva, M. Xu-Welliver, M. Kartalou, S. P. Hussain, R. B. Roth, X. Zhou, L. E. Mechanic, I. Zurer, V. Rotter, L. D. Samson and C. C. Harris, The adaptive imbalance in base excision-repair enzymes generates microsatellite instability in chronic inflammation, *J. Clin. Invest.*, 2003, **112**, 1887–1894.

323. R. W. Sobol, M. Kartalou, K. H. Almeida, D. F. Joyce, B. P. Engelward, J. K. Horton, R. Prasad, L. D. Samson and S. H. Wilson, Base excision repair intermediates induce p53-independent cytotoxic and genotoxic responses, *J. Biol. Chem.*, 2003, **278**, 39951–39999.

324. I. Rusyn, R. C. Fry, T. J. Begley, J. Klapacz, J. P. Svensson, M. Ambrose and L. D. Samson, Transcriptional networks in *S. cerevisiae* linked to an accumulation of base excision repair intermediates, *PLoS One*, 2007, **2**, e1252.

325. S. Xanthoudakis, R. J. Smeyne, J. D. Wallace and T. Curran, The redox/ DNA repair protein, Ref-1, is essential for early embryonic development in mice, *Proc. Natl. Acad. Sci. U. S. A.*, 1996, **93**, 8919–8923.

326. M. Z. Hadi and D. M. Wilson, 3rd, Second human protein with homology to the *Escherichia coli* abasic endonuclease exonuclease III, *Environ. Mol. Mutagen.*, 2000, **36**, 312–324.

327. M. Z. Hadi, K. Ginalski, L. H. Nguyen and D. M. Wilson, 3rd, Determinants in nuclease specificity of Ape1 and Ape2, human homologues of *Escherichia coli* exonuclease III, *J. Mol. Biol.*, 2002, **316**, 853–866.

328. Y. Ide, D. Tsuchimoto, Y. Tominaga, M. Nakashima, T. Watanabe, K. Sakumi, M. Ohno and Y. Nakabeppu, Growth retardation and dyslymphopoiesis accompanied by G2/M arrest in APEX2-null mice, *Blood*, 2004, **104**, 4097–4103.

329. J. E. Guikema, E. K. Linehan, D. Tsuchimoto, Y. Nakabeppu, P. R. Strauss, J. Stavnezer and C. E. Schrader, APE1- and APE2-dependent DNA breaks in immunoglobulin class switch recombination, *J. Exp. Med.*, 2007, **204**, 3017–3026.
330. C. E. Schrader, J. E. Guikema, X. Wu and J. Stavnezer, The roles of APE1, APE2, DNA polymerase beta and mismatch repair in creating S region DNA breaks during antibody class switch, *Philos. Trans. R. Soc. Lond. B, Biol. Sci.*, 2009, **364**, 645–652.
331. J. E. Guikema, R. M. Gerstein, E. K. Linehan, E. K. Cloherty, E. Evan-Browning, D. Tsuchimoto, Y. Nakabeppu and C. E. Schrader, Apurinic/Apyrimidinic endonuclease 2 is necessary for normal B cell development and recovery of lymphoid progenitors after chemotherapeutic challenge, *J. Immunol.*, 2011, **186**, 1943–1950.
332. P. Burkovics, I. Hajdu, V. Szukacsov, I. Unk and L. Haracska, Role of PCNA-dependent stimulation of 3′-phosphodiesterase and 3′-5′ exonuclease activities of human Ape2 in repair of oxidative DNA damage, *Nucleic Acids Res.*, 2009, **37**, 4247–4255.

CHAPTER 3.2

The Plasticity of DNA Damage Response during Cell Differentiation: Pathways and Consequences

PAOLA FORTINI, CHIARA FERRETTI AND
EUGENIA DOGLIOTTI*

Department of Environment and Primary Prevention, Istituto Superiore di
Sanità, Rome, Italy
*Email: eugenia.dogliotti@iss.it

3.2.1 Introduction

Embryonic stem cells (ESC) are rapidly proliferating, self-renewing and pluripotent cells. Upon differentiation they give rise to all type of cells in organisms, including germ cells. Since these cells are critical for proper embryo formation they are expected to have stringent control of their genome integrity. Tissue-specific adult stem cells (ASC) are quiescent cells already differentiation-committed that are functionally defined by their ability to self-perpetuate (self-renewal) and, when activated, accomplish their differentiation programme. They are found in many highly regenerative organs (e.g. blood, skin, digestive tract) and also in non-renewing organs such as muscle where they allow repair after tissue damage. During maturation, the propagation of ASC with un-repaired lesions should be avoided but once differentiation is complete the

Issues in Toxicology No. 13
The Cellular Response to the Genotoxic Insult: The Question of Threshold for
Genotoxic Carcinogens
Edited by Helmut Greim and Richard J. Albertini

control of genome integrity may become dispensable although tissue integrity and function should be maintained. In this chapter the information available on the plasticity of the mechanisms of DNA repair, damage signalling and cell death active in different stages of the differentiation programme are reviewed.

3.2.2   DNA Damage Control

Given the cost of mutation propagation in stem cells (SC), these cells have evolved multiple systems to limit DNA damage accumulation and respond effectively to it. The first line of defence is aimed at hampering DNA damage production. Stem cells from a variety of tissues present high levels of ATP-binding cassette (ABC) transporters that protect cells against toxic substances by pumping them across cell membranes.[1] Stem cells rely mostly on glycolysis for energy supply and their mitochondria are rather immature with low O_2 consumption and thus low levels of endogenous DNA damage.[2] The maintenance of an hypoxic environment is very important to preserve the SC function as shown by the effects of altered reactive oxygen species (ROS) regulation on SC renewal and differentiation (see Section 3.2.3).

3.2.3   DNA Repair

To ensure genome integrity, cells have evolved several mechanisms of DNA repair: homologous recombination (HR), non-homologous end joining (NHEJ), nucleotide excision repair (NER), base excision repair (BER) and mismatch repair (MMR). Depending on cell type, differentiation stage and age, cells may have different need to withstand genotoxic stress and therefore might have different priorities in the use of the various DNA repair pathways.

3.2.3.1   Homologous Recombination and Non-homologous End Joining

Double-strand breaks (DSB) are the most lethal form of DNA damage, so an accurate mechanism to repair DSB is required to maintain genome integrity. Mammalian cells present two major pathways for repair of DNA DSB: homologous recombination (HR) and non-homologous end joining (NHEJ) (reviewed in ref. 3).

Homologous recombination is a high-fidelity mechanism which uses an undamaged template containing hundreds of base pairs of sequence homology to replace an adjacent damaged one. Because sister chromatids are available as templates upon DNA replication, homologous recombination is active mainly in S/G2 phases of the cell cycle. In mammals, the MRN complex (consisting of Mre11-Rad50-Nbs1) together with the DNA endonuclease CtIP carries out the initial processing of the DNA ends. Following re-section, the exposed single-strand DNA (ssDNA) is coated by replication protein A (RPA), which recruits the Rad52 epistasis group of proteins to enable Rad51 filament formation. The breast cancer susceptibility genes BRCA1 and BRCA2 are also involved in facilitating orderly HR. The DNA ends can then be repaired using different mechanisms.

NHEJ requires little or no sequence homology for efficient repair and can be error-prone. This mechanism is active during all cell cycle phases and is the predominant pathway for DSB repair in mammalian cells. NHEJ facilitates direct modification and ligation of the two broken DNA ends which involve Ku heterodimers (Ku70/XRCC6 and Ku80/XRCC5), DNA-PK catalytic subunit (DNA-PKcs), DNA ligase IV (LigIV), XRCC4 and Cernunnos.

Defects in DSB repair lead to premature aging, neurodegeneration and increased cancer susceptibility. Mutations in the HR proteins BRCA1/2 that cause increased breast and ovarian cancers in women have been linked to the accumulation of genetically unstable mammary stem cells[4] and mutations in model mice defective in NHEJ components have been shown to display hematopoietic stem cells (HSC) self-renewal defects.[5,6]

DSB repair is expected to be important in ESC since they have higher basal level of γ-H2AX staining (a DSB marker) when compared to mouse embryonic fibroblasts (MEF). Furthermore, ESC are suggested to repair DSB more quickly than mouse 3T3 cells after exposure to ionizing radiation (IR).[7] Consistent with the high efficiency of DNA repair in stem cells, several DSB repair genes, as BRCA1 and XRCC5, are up-regulated in human ESC[8] and Rad51 protein level is 20-fold higher in ESC than in MEF.[7]

Because of the requirement for high-fidelity DSB repair in stem cells, it is conceivable to suppose that ESC utilize HR rather than NHEJ as the main mechanism of repair. A comparative analysis of the DSB repair capacity of ESC and differentiated cells in mouse shows that HR is the major mechanism to repair DSB in ESC with a minimal role played by NHEJ. In contrast, when ESC are induced to differentiate, NHEJ becomes the predominant pathway and HR is reduced. The protein level of DNA Lig IV seems to be rate-limiting for NHEJ. Indeed, the abundance of Lig IV is low in ESC and increased in MEF concomitantly with NHEJ activity.[9]

In addition, human ESC (hESC) use HR extensively to repair DSB and its efficiency decreases when hESC are induced to differentiate in neural progenitors and in astrocytes.[10] Despite the preferential use of HR, hESC repair DSB induced by I-Sce endonuclease utilizing NHEJ *via* a pathway that is independent of DNA-PK.[10] NHEJ is also the repair mode for hESC exposed to radiation during the late G2 phase of the cell cycle,[11] but in this case DNA-PK is required and is responsible for the high level of misrejoining.

An elegant study analyses the selective requirement for HR or NHEJ during nervous system development.[12] By using mice carrying a germ line disruption of XRCC2 (HR defective) and LIG4 (NHEJ defective), the two pathways for recombination are found to be spatiotemporally distinct: HR inactivation is crucial from the early steps of embryogenesis leading to abundant apoptosis, whereas the disruption of NHEJ has deleterious consequences only at later developmental stages (not before E12). Since the late stages of the embryogenesis are characterized by massive differentiation, these results imply that the HR pathway has an essential protective role against DSB-induced cytotoxicity in proliferating cells becoming dispensable in post-mitotic cells where NHEJ is the pathway of election. The key role of the NHEJ in differentiated long-lived cells was also shown in an *in vitro* murine adipogenesis cell system.[13] A faster DSB repair kinetics in adipocytes compared to their proliferating precursors is found after exposure to a radiomimetic chemical or ionizing radiation. The increased ability of adipocytes to repair DSB is mainly ascribed to the up-regulation of DNA-PK expression and activity.

Interestingly, a recent study shows that the induction of NHEJ after H_2O_2 treatment (working concentration close to concentration in normal brain) in post-mitotic neurons requires G_0 exit mediated by pRb phosphorylation directed by cyclin C. Abrogation of G_0–G_1 transition by silencing of cyclin C compromises NHEJ activation, while forcing G_1 entry determines NHEJ induction even in absence of DNA damage.[14] Hence, the re-entering into the cell cycle to activate NHEJ can explain the expression of cell cycle markers observed in damaged neurons of normal brain.[15]

In the hematopoietic system, the induction of NHEJ in stem cells can be detrimental for genome stability. Hematopoietic development starts from largely quiescent, slowly cycling HSC that, in response to environmental stimuli, are able to give rise to a series of proliferating committed progenitors and mature cells. In contrast to human HSC (hHSC) that undergo apoptosis upon IR exposure, murine HSC (mHSC) survive and repair DNA damage by error-prone NHEJ that can lead to accumulation of gene aberrations. Differently, myeloid progenitors from mouse mainly die after IR treatment and surviving cells use high-fidelity HR (see also Section 3.2.4).[16,17]

3.2.3.2 Nucleotide Excision Repair

The removal of ultraviolet (UV) light-induced photoproducts, bulky chemical adducts and intra-strand DNA cross-links depends on nucleotide excision repair (NER). This mechanism consists of three sub-pathways: namely global genome repair (GGR), transcription-coupled repair (TCR) and domain-associated repair (DAR). While GGR operates on the entire genome, TCR and DAR participate in the repair of active genes. In particular, TCR provides efficient repair of the transcribed strand only and DAR repairs both strands of active genes and is detectable only when GGR is attenuated.[18]

At mechanistic level it is known that, during GGR, the double helix distortion caused by DNA damage is sensed by the heterotrimer XPC/HR23B/Centrin2, with the contribution of the DDB heterodimer for some lesions. The next step is the opening of a denaturation bubble around the lesion by the transcription factor TFIIH and later the damaged strand is nicked by XPG on the $3'$ side of the lesion and by the heterodimer ERCC1/XPF on the $5'$ side. Finally, after the displacement of an oligonucleotide encompassing the lesion, the undamaged strand is used as a template to fill the gap. The mechanism of TCR is very similar, differing only in the recognition step that involves RNA polymerase II (RNAPII) rather than xeroderma pigmentosum, complementation group C (XPC) and damage specific DNA binding protein 1 (DDB) complexes. Cockayne Syndrome (CS) proteins associate with UV-stalled RNAPII and are required for efficient TCR by recruiting NER factors and chromatin remodelers.[19] The importance of functional NER is illustrated by genetic disorders like Cockayne syndrome (CS) and trichothiodistrophy (TTD) defective in TC-NER, and Xeroderma pigmentosum (XP) characterized by GGR defects or total NER deficiencies. Interestingly, mice models of TTD[6] and NER (*i.e.* ERCC1 defective[20]) present decreased hematopoietic SC function with reduced self-renewal potential and increased apoptosis indicating that a defect in NER affects also the SC compartments.

ESC and differentiated cells have different capacities for repairing UV light-induced DNA damage. A comparative study on UV-C irradiated cells treated with or without T4-endonuclease V (an enzyme that generates single-strand breaks at pyrimidine dimers) suggests that hESC present faster repair of T4-endonuclease V-sensitive sites compared to fibroblast cell lines.[8]

Pioneering studies carried out by Hanawalt's team in several human terminally differentiated tissues (*i.e.* striated muscle, macrophages and neurons) show that NER is generally attenuated during cell differentiation. More precisely, in terminally differentiated cells GGR is lower than the undifferentiated counterpart whereas, within active genes, not only the transcribed, but also the non-transcribed strand are efficiently repaired.[18,21] The attenuation of GGR could be ascribed to incomplete phosphorylation of ubiquitin-activating E1 enzyme Ube1 upon differentiation that could lead to a reduction in the ubiquitination of TFIIH, decreasing its activity in GGR.[22]

This phenomenon also occurs in human B lymphocytes in which NER at global level is down-regulated.[23] B lymphocytes are G_0-arrested cells that retain the ability to re-enter the cell cycle upon a proper stimulus. Like terminally differentiated cells, in these quiescent B cells, a decrease in phosphorylation of Ube1 and GGR impairment is observed, while transcribed genes are efficiently repaired by TCR and DAR. Interestingly, an increase in phosphorylation of Ube1 and NER recovery is detectable upon proliferation. It is important to note that NER impairment during quiescence could lead to DNA damage accumulation at global genome level that would increase the likelihood of mutation fixation upon proliferative stimuli promoting B cell malignancies or abnormal immune functions.[23]

The different contribution of NER sub-pathways in protection during differentiation is also detected by analysing the UV sensitivity of stem cells and differentiated cells.[24] UV sensitivity of MEF depends on the capacity of perform TC-NER, as shown by marked UV-sensitivity of $Xpa^{-/-}$ and $Csb^{-/-}$ MEF compared with GGR defective MEF. Conversely, GGR is the main determinant of UV sensitivity in ESC as shown by severe UV sensitivity of $Xpa^{-/-}$ and $Xpc^{-/-}$ ESC compared with $Csb^{-/-}$ TC-NER defective ESC.

Together these studies suggest that NER down-regulation is a common feature in G_0-arrested cells, either quiescent or terminally differentiated. It is reasonable that non-dividing cells, which only have to maintain genomic integrity of transcribed genes and preserve tissue specificity, do not need a severe surveillance of the entire genome but an efficient repair of active genes.

In contrast, ESC cannot rely on only TCR or DAR because the accumulation of damage and mutations in non-transcribed genes would be detrimental for a proper development and lead to genetic instability.

3.2.3.3 Base Excision Repair

Structurally non-distorting lesions in DNA, such as alkylated, oxidised bases and abasic site, are mainly processed by BER. The first step of BER implies the removal of the damaged base by specific DNA glycosylases, which give rise to abasic sites rapidly converted into single-strand breaks (SSB) by an apurinic/apyrimidinic (AP) endonuclease. After that, the polymerization step involves either the short patch BER (SP-BER) or the long patch BER (LP-BER) pathway that differ for the repair-patch length and the specific players involved. DNA polymerase β and DNA ligase III are responsible for the filling-in and the sealing step of the SP-BER, respectively, whereas in LP-BER, these reactions are catalysed by the PCNA-dependent DNA polymerases δ/ε and/or DNA polymerase β and DNA ligase I.[25]

BER is the repair mechanism of election for oxidative DNA damage that is also produced spontaneously as a consequence of aerobic metabolism. While stem cells are characterized by low metabolism, some post-mitotic tissues such as muscle and brain present high metabolism and therefore are particularly susceptible to ROS-induced damage (see Section 3.2.2). The BER capacity in skeletal muscle was investigated by Narciso and coworkers using an *in vitro* murine system which recapitulates the *in vivo* differentiation process.[26] Muscle satellite cells (myoblasts) isolated from mouse thigh are ASC that are able to proliferate and differentiate, in an appropriate culture medium, in multi-nucleated myotubes. A comparative analysis of BER efficiency in myoblasts *versus* myotubes showed a clear impairment of BER in post-mitotic muscle cells. Both the SP-BER and LP-BER sub-pathways were affected although the LP-BER, which shares several partners with DNA replication, was more severely compromised. At molecular level the BER impairment was ascribed to the nearly complete lack of DNA ligase I and to the strong down-regulation of XRCC1, a scaffold protein known to be essential for DNA ligase III stabilization. Consistently, XRCC1 is a transcriptional target for FoxM1 and E2F1 which activate several cell cycle genes and are both down-regulated upon cell cycle exit in myotubes.[27] In contrast, down-regulation of XRCC1 is not observed in post-mitotic neural cells where this enzyme has a protective role against the cytotoxic effects induced by oxidative DNA damage.[28] Moreover, XRCC1 seems to be important during neurogenesis since its inactivation leads to loss of cerebellar interneurons and abnormal hippocampal functions.[29]

Since the mitochondrial genome is a susceptible target for oxidative damage, its repair should play an important role particularly in post-mitotic tissues with high metabolism. It has been recently discovered that repair in mitochondria occurs not only by SP-BER but also by LP-BER. The LP-BER is involved in the processing of oxidative lesions[30] and requires the nuclease activity of FEN1 to repair damage both in nuclear and mitochondrial genome.[31,32] FEN1 is not only involved in LP-BER, but it is also essential for DNA replication. FEN1 is strongly down-regulated in terminally differentiated cells. How this might impact on the efficiency of mitochondrial BER in differentiated tissues under normal and pathological conditions should be investigated.

A comparison of BER efficiency between stem and differentiated cells shows that murine ESC have a higher BER activity compared with MEF as assessed by using an *in vitro* SP-BER assay and human ESC have a faster repair than fibroblast cell lines after oxidative stress. BER proteins such as Ape1, DNA Ligase III, PCNA, UNG2 and XRCC1, are expressed in murine ESC at higher level than MEF and this is also true for αOGG1 and Ape1 in hESC after H_2O_2 treatment.[8,33] It is of note that in post-mitotic skeletal muscle cells, UNG2 that is involved in the processing of post-replicative lesions is down-regulated (Fortini, P. and Dogliotti E., unpublished data) suggesting that stem cells need to up-regulate DNA repair to counteract DNA damage that also arises during replication.

If in general DNA repair is down-regulated during differentiation, a notable exception is the defective BER observed in monocytes and its enhancement during dendritic cell maturation.[34] The BER defect in monocytes may account for their selective killing during tumour therapy with alkylating agents.

3.2.3.4   Mismatch Repair

Mismatches in DNA can occur not only in proliferating cells but also in non-dividing cells because of base deamination or during repair attempts by error-prone DNA polymerases. Few studies have compared the MMR efficiency of stem cells and differentiated cells. MMR has been reported to be active in differentiated neurons[35] and as efficient as in undifferentiated cells.[36,37] In contrast, one report indicates that MMR proteins, namely MSH2 and MSH6, are highly expressed in mouse ESC and MMR is significantly more active in ESC than in MEF.[33] Mice lacking MSH2 present defective HSC activity with microsatellite instability detected in their progeny.[38]

3.2.4   DNA Damage Signalling

DNA damage induces an evolutionary conserved signalling pathway defined as DNA damage response (DDR) that is a network of interacting pathways coordinating repair and cell cycle progression. The DDR involves damage sensors such as the MRN complex and the kinases ATM, ATR and DNA-PK, signal transducers such as 53BP1and BRCA-1, and effectors such as the checkpoint kinases Chk1 and Chk2 that all together decide the cell fate by operating repair, cell cycle arrest, senescence or cell death. Several human diseases have been described that present defects in the DDR. Patients with mutations in the kinase ATM present blood vessels abnormalities, cerebellar degeneration, immunodeficiency and increased cancer risk. Many of these defects can be linked to defects in SC function as shown by increased ROS levels and the decreased number and function of HSC in a mouse model of ATM deficiency.[39,40] In this regard the recent finding that ATM protects cells from ROS accumulation by regulating the pentose phosphate pathway is noteworthy.[41]

3.2.4.1 DNA Damage Response in Embryo Stem Cells

Before summarizing the outcome of several studies it is important to remember that there are species-specific differences between mouse and human stem cells regarding DDR and that any extrapolation from experiments carried out in mouse SC to human SC should therefore be made with caution.

The pluripotent ESC are rapidly dividing and present a cell cycle that is shorter than somatic cells mainly due to a shortened G1 phase and facilitated transition from G1 to S.[42] The rapid progression through the cell cycle may expose ESC to increased risk of replication errors. It is well-established that upon DNA damage mouse ESC lack a G1 checkpoint and undergo a p53-independent apoptosis as the preferential route to clear highly damaged cells.[43] The lack of a functional G1 checkpoint is partly due to sequestration of p53 in the cytoplasm, thereby interfering with the induction of p21 and the sequestration of Chk2 at centrosomes that compromises the ATM-Chk2 pathway.[44] However, the checkpoint signalling cascade is not fully inactive since irradiated human ESC activate ATM-dependent checkpoint signalling cascade, including phosphorylation of p53, and arrest in the G2/M stage of the cell cycle but not in the G1 phase. This arrest is temporary and might indicate an attempt to repair by HR when the sister chromatid is present.[45,46]

3.2.4.2 DNA Damage Response in Adult Stem Cells

ASC from most tissues are largely quiescent and this state has been postulated as a further level of preservation of genomic integrity. However, DNA damage checkpoints and DNA repair pathways are cell-cycle dependent and this may lead to accumulation of DNA damage and ultimately to aging. Even more threatening is the fact that, when ASC damaged cells enter the cell cycle, DNA damage will be primarily repaired by NHEJ which is error-prone (see Section 3.2.3). The outcome of DDR in these cells seems to be species-specific and dependent on the developmental stage. An example is provided by the hematopoietic system where upon irradiation, foetal hHSC respond mostly with apoptosis and over cell elimination[47] whereas mHSC either proliferating or quiescent efficiently repair and survive.[17] In quiescent mHSC, survival is guaranteed by enhanced expression of prosurvival genes such as bcl-2 which inhibits p53-dependent cell death; but since these cells use NHEJ to repair damage, they may suffer of genetic instability. Conversely, proliferating HSC use the high-fidelity HR pathway, thus maintaining genomic integrity. Similar to HSC, in the skin epidermis the multipotent hair follicle bulge SC (BSC) are more resistant to DNA damage than the other cells of the epidermis.[48] The mechanisms underlying this phenomenon are also similar, with higher levels of the antiapoptotic protein Bcl-2 and accelerated DNA repair (by NHEJ) as hallmarks of DDR in these cells. Interestingly, upon irradiation, p53 is also activated in these cells but for a short time and a drastic decrease occurs by 24 h after irradiation, likely contributing to the resistance of BSC to apoptosis. The epidermis offers also an example of cell-type specificity within the same hair follicle niche. Melanocyte SC (MSC) respond very differently to DNA damage compared with BSC. In particular, although both cell types survive after damage, MSC counteract the load of damage by premature differentiation without involving p53. Conversely, deficiency of ATM sensitizes MSC to ectopic differentiation, demonstrating that this kinase protects MSC from their premature differentiation.[49]

3.2.4.3 DNA Damage Response in Post-mitotic Cells

If accumulation of damage occurs in post-mitotic cells, it can also contribute to disease progression or lead to loss of tissue homeostasis particularly when such cells provide a support function for tissue stem cells. Conditions such as neuro-degeneration and cardiomiopathy may arise as a consequence of damage in terminally differentiated cells.

The muscle and nervous system have been broadly interrogated for changes in DDR along the process of terminal differentiation also because of the availability of well-characterized *in vitro* cell systems amenable to fine mole-cular/biochemical characterization. Twenty years ago the first evidence was published that DNA damage can block cell differentiation.[50] More recently, a DNA damage-activated differentiation checkpoint was identified in skeletal muscle progenitors that coordinates repair and the expression of differentiation-specific genes during cell cycle arrest.[51] Upon exposure to genotoxic stress, this checkpoint holds the transcription of differentiation genes while DNA is repaired. The transient phosphorylation of MyoD by DNA damage activated c-abl that inhibits MyoD-dependent transcription of muscle genes is likely to be the key event in the coordination of repair and differentiation. The mechanism of inhibition of muscle gene transcription depends on the cell cycle phase at which myoblasts arrest in response to damage.[52] This is in striking contrast with what occurs in stem cells exposed to genotoxic agents that differentiate due to the p53-mediated repression of the stem cell marker Nanog (see Section 3.2.5.1).[53] A number of proteins implicated in DDR are down-regulated during terminal differentiation. For instance, myotubes are deprived of ATR and Chk1 but the ATM-Chk2 branch is fully functional upon radiation damage[54] as well as after exposure to a variety of DNA break-inducing agents[54](Fortini *et al.*, submitted). Ataxia telangiectasia mutated (ATM) mediates the response to DNA breaks as in proliferating cells but p53, which is down-regulated in myotubes, is not activated thus preventing cell death. A few types of damage (*e.g.* doxorubicin) are able to activate p53 and then restore apoptotic pathways (see Section 3.2.4). ATM also mediates the response to DNA breaks in human neuron-like cells.[55,56] Following genotoxic stress, post-mitotic neurons attempt to re-enter the cell cycle contributing to neuronal cell death. This has been also documented in a wide variety of neurodegenerative diseases including ataxia telangiectasia (AT). This mechanism seems to be ATM-dependent.[57] The Chk1/Cdc25A axis has been suggested to participate in the activation of cell cycle-mediated neuronal death.[58]

3.2.5 Cell Death

Whenever unprocessed DNA lesions persist after activation of DDR and DNA repair, proliferating cells can either halt the cell cycle progression in G1 (cellular senescence) or trigger programmed cell death pathways such as apoptosis (self-killing) and autophagy (self-eating). Quiescent/post-mitotic cells cannot adopt the replicative senescence or the canonical cell cycle arrest-dependent apoptosis as strategies to get rid of damaged cells but they are provided of alternative strategies such as autophagy.

3.2.5.1 Apoptosis

Upon DNA damage, ESC are unable to arrest the cell cycle (see Section 3.2.3.1) and readily undergo apoptosis by a p53-independent apoptotic pathway. Consistently, ESC are hypersensitive to the cytotoxic effect of a variety of DNA damaging agents.[59] Besides apoptosis that efficiently clear highly damaged cells, ESC can engage an alternative response to damage. Mouse ESC (mESC) activate p53 that then binds to the promoter of Nanog, a gene required for ESC self-renewal, leading to its repression. The repression of Nanog stimulates mESC differentiation thus promoting the removal of the damaged cells from the replicating stem cell pool (reviewed in ref. 60).

In striking contrast, terminally differentiated cells are resistant to DNA damage-induced cell death. Post-mitotic skeletal muscle cells (myotubes) are radioresistant compared with the corresponding ASC (proliferating myoblasts) and show reduced sensitivity to DNA single strand break-induced agents[54] (Fortini *et al.* submitted). The radioresistance has been molecularly characterized and shown to be due to the uncoupling between ATM activation and p53 phosphorylation,[54] which is required for apoptosis induction. This crosstalk was restored after exposure to doxorubicin, indicating that an apoptotic pathway is still active in post-mitotic muscle cells although the potential contribution of other cell death pathways (necrosis, autophagy) has not been investigated.

Like myotubes, neurons are also resistant to a variety of stressors. The execution of cell death by apoptosis in mature neurons has been well-characterized.[61] It seems to require an additional step compared with proliferative cells, *i.e.* the removal of the block due to the direct binding between caspases and the X-linked inhibitor of apoptosis protein (XIAP). After DNA damage, the XIAP mediated inhibition can be overcome *via* a p53-dependent induction of Apaf-1. Consistently, p53 deficient neurons are refractory to relieve the XIAP 'brake' and are therefore resistant to DNA damage. A similar process seems to account for the resistance of myotubes to cytosolic cytochrome c microinjection.[62] The restricted apoptosis could be physiologically essential for long-term tissue maintenance, but whether this is general phenomenon in post-mitotic tissues awaits to be clarified.

As previously mentioned post-mitotic neurons in G_0 phase are able to re-enter in the cell cycle after repairable DNA damage induction (see Sections 3.2.2 and 3.2.3). Recent studies clarified that this process is cyclin C dependent and leads to the activation of NHEJ and apoptosis. The restoration of G_0–G_1

and G_1-S progression is critical for efficient DNA repair and apoptotic signalling, respectively.[14] Whether the concurrent reduced DNA repair capacity and restricted apoptosis typical of post-mitotic cells could lead to potentially detrimental outcomes such as transcriptional mutagenesis awaits to be clarified.

3.2.5.2   Autophagy

Autophagy is an evolutionarily conserved mechanism which takes place at basal levels in all eukaryotic cells. Long-lived macromolecules and cytoplasmic components, such as misfolded proteins and dysfunctional organelles, are sequestered in the autophagosome, a double-membrane vesicle able to fuse with the lysosome. The content of the autophagolysosomes is then hydrolysed, leading to cell energy regeneration.

A set of autophagy-related (ATG) genes required for autophagy have been identified. Emerging evidence indicates that autophagy has an essential role in both cell differentiation and in the preservation of the homeostasis of cells, tissues and organisms (for a review see refs. 63 and 64). As an example, primary MEF induced for adipocyte differentiation activate a massive autophagic response and, if autophagy is defective (ATG5 is missing), adipogenesis is aberrant and significantly reduced.[65] Consistently, ATG5-defective mouse embryos, which cannot survive to the neonatal starvation period, show an irregular development of the white fat tissue at birth thus corroborating the *in vitro* evidence.[66] SIRT-1, a nicotinamide adenine dinucleotide (NAD+)-dependent deacetylase protein with a pleiotropic role in the maintenance of homeostasis following genotoxic stress, is one of the major player involved in the crosstalk between autophagy and cell differentiation.[66–68]

If the analysis of a variety of systemic and tissue-specific knockout models of ATG genes has led to the clear notion that autophagy plays an important role in mammalian development and differentiation (reviewed in ref. 64), its role in the response to DNA damage is less clear. However, several reports indicate that autophagy can be induced by DNA damage and it can be seen as an integral part of the DDR. The first molecular link between DDR and autophagy was the discovery that PARP-1 is involved in autophagy.[69] Under mild genotoxic stimuli, PARP1 behaves as a survival factor whereas in the presence of massive DNA damage PARP1 becomes a mediator of cell death.[70] The PARP1 activation in the presence of severe DNA damage can lead to cell removal by apoptotis whereas the overactivation of PARP1, due to extensive DNA breakage, causes cell energy depletion which blocks the apoptotic machinery and culminates in cell leakage and necrosis. More recently, it has been shown that PARP1 is activated in MEF after exposure to doxorubicin and that PARP1 deficient or inhibited cells are partially protected against the cytotoxic effect induced by this compound. This reduced cell sensitivity is due to the lack of the PARP1-dependent energy collapse and the consequent activation of mTOR, the negative regulator of autophagy. Nevertheless, in a PARP1 proficient background, the pharmacological inhibition of autophagy by 3-methyladenine confers resistance to the doxorubicin-dependent cell death.[69] These observations strongly suggest that there is a fine balance between DNA damage/repair and energy crisis regulated by the PARP-1 activation level.

To sustain the idea that autophagy is part of DDR, a crosstalk between autophagy and a new cytoplasmic function of ATM in response to oxidative stress has been recently described.[71,72] Under ROS overproduction, ATM activates TSC2 *via* LKB1/AMPK metabolic pathway repressing mTORC1 and thus inducing autophagy. It should be stressed that this novel pathway can be activated in the presence of an inhibitor of ATM nuclear factor, indicating that this mechanism is independent of the canonical role of ATM in DNA damage signalling (see Section 3.23). Consistently, the ATM-mediated repression of mTORC1 is p53 independent.

A further link between DDR and autophagy emerges from a recent study in yeast.[73] Upon inhibition of histone deacetylase (HDAC) activity, resulting in hyperacetylation of proteins, lack of DDR and DSB repair was observed. This was due to decreased association of Sae2 and Exo1 (both involved in DNA end re-section during HR) with DSB ends and decreased levels of Sae2 protein. Interestingly, inhibition of mTOR activity and thus triggering of autophagy also resulted in decreased levels of Sae2. The challenging model proposed by the authors is that the severely damaged DNA and associated machinery are removed from the nucleus *via* an autophagic process regulated by the acetylation status of key repair proteins. This may be a mechanism for keeping DNA repair enzymes away from cellular DNA that is not damaged, preventing accidental 'repair' of replicating DNA.

In view of the key role of autophagy in the control of post-mitotic tissue integrity, a better definition of the role of autophagy in the control of genome integrity in different stages of the differentiation program is strongly recommended.

3.2.6   Conclusions

DNA damage response (including DNA repair) changes during differentiation, thus setting the tolerance to DNA damage to different levels. The risk for mutation propagation remains very high when the DDR does not work properly in SC. The current genotoxicity tests are based on highly proliferating somatic cells that are equipped with efficient systems to repair DNA damage or to get rid of highly damaged cells. The message we learn from the plasticity of DDR during differentiation is that ESC are equipped with metabolic and DDR systems that minimize insults to genomic integrity but that ASC are at risk of damage accumulation and mutation propagation once they are stimulated to enter the cell cycle because of the exclusive use of NHEJ (Figure 3.2.1).

The discussion on the validity of the linear, non-threshold dose–response curve for cancer induction by DNA reactive genotoxic chemicals, which is currently the default assumption of several regulatory agencies, should take into account the emerging knowledge that a number of human cancers develop from a small subset of cells with self-renewal properties functionally resembling tissue-specific stem cells.[74] As summarized in this chapter, the sensitivity of these cells to DNA damage accumulation and mutation induction is very

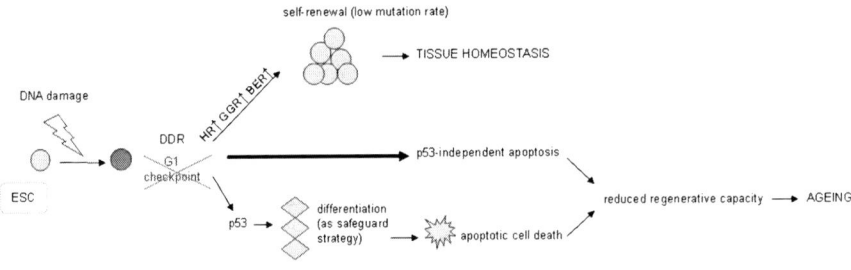

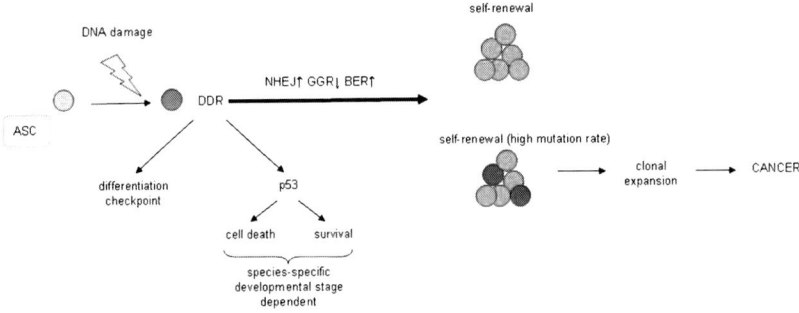

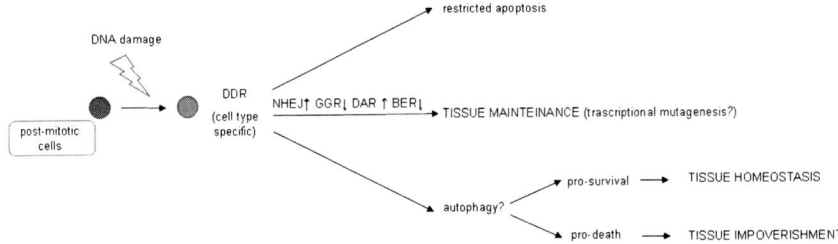

Figure 3.2.1 DNA damage response: from embryonic stem cells to post-mitotic cells. The differences in the impact of DNA damage and activated pathways in embryonic stem cells (ESC), adult stem cells (ASC) and post-mitotic cells are illustrated. Black circles represent damaged cells; bold arrows indicate the main pathways adopted to handle DNA damage; the outcome is highlighted in capital letters.

different compared to that of somatic proliferating cells (Table 3.2.1) and different SC compartments are provided with mechanisms that either limit or accelerate DNA damage accrual. An additional relevant question is how deregulation of these mechanisms leads to cancer and how and whether the DDR of cancer stem cells affects the response to chemo- and radio-therapy.

Table 3.2.1 Differences in physiological status and response to DNA damage between stem cells and post-mitotic/somatic cells.

Process	ESC	ASC	Post-mitotic cells	Somatic cells
Proliferation	YES	Quiescent	Irreversible exit from cell cycle	YES
Metabolic activity	Low/glycolysis	Low	High	High
ROS steady-state level	Low	Low	High	High
Mutation rate	Low	High	Transcriptional mutagenesis?	Low
Turnover	High (elimination of damaged cells)	Persistence of damaged cells	Restricted apoptosis	High (elimination of damaged cells)
ABC transporter activity	High			Low
DNA repair	Efficient	Efficient/less efficient	Down-regulated	Efficient
DSBR (double strand break repair)	HR	NHEJ (when stimulated to replicate)	NHEJ	HR
NER	Efficient GGR	Down-regulated	Down-regulated but DAR is active	Efficient GGR and TCR
BER	Efficient	Efficient*	Down-regulated	Efficient
DDR	Lack of G1 checkpoint	Species-specific developmental stage dependent	Not fully active (lack of ATR-Chk1 branch)	Active

*As measured in proliferating ASC.

References

1. M. Huls, F. G. Russel and R. Masereeuw, The role of ATP binding cassette transporters in tissue defense and organ regeneration, *J. Pharmacol. Exp. Ther.*, 2009, **328**, 3–9.
2. S. Varum, A. S. Rodrigues, M. B. Moura, O. Momcilovic, C. A. Easley 4th, J. Ramalho-Santos, B. Van Houten and G. Schatten, Energy metabolism in human pluripotent stem cells and their differentiated counterparts, *PLoS One*, 2011, **6**(6), e20914.
3. K. A. Bernstein and R. Rothstein, At loose ends: resecting a double-strand break, *Cell*, 2009, **137**, 807–810.
4. P. Liu, L. Qian, J. S. Sung, N. C. de Souza-Pinto, L. Zheng, D. F. Bogenhagen, V. Bohr, D. M. Wilson 3rd, B. Shen and B. Demple, Removal of oxidative DNA damage *via* FEN1-dependent long-patch base excision repair in human cell mitochondria, *Mol. Cell. Biol.*, 2008, **28**, 4975–4987.
5. J. Kenyon and S. L. Gerson, The role of DNA damage repair in aging of adult stem cells, *Nucleic Acids Res.*, 2007, **35**, 7557–7565.
6. D. J. Rossi, D. Bryder, J. Seita, A. Nussenzweig, J. Hoeijmakers and I. L. Weissman, Deficiencies in DNA damage repair limit the function of haematopoietic stem cells with age, *Nature*, 2007, **447**(7145), 725–729.
7. E. D. Tichy and P. J. Stambrook, DNA repair in murine embryonic stem cells and differentiated cells, *Exp. Cell. Res*, 2008, **314**, 1929–1936.
8. S. Maynard, A. M. Swistikowa, J. W. Lee, J. Liu, S.-T. Liu, A. Da Cruz, M. Rao, N. de Souza-Pinto, X. Zeng and V. A. Bohr, Human embryonic stem cells have enhanced repair of multiple forms of DNA damage, *Stem Cells*, 2008, **26**, 2266–2274.
9. E. D. Tichy, R. Pillai, L. Deng, L. Liang, J. Tischfield, S. J. Schwemberger, G. F. Babcock and P. J. Stambrook, Mouse embryonic stem cells, but not somatic cells, predominantly use homologous recombination to repair double-strand DNA breaks, *Stem Cells Dev.*, 2010, **19**, 1699–16711.
10. B. R. Adams, A. J. Hawkings, L. F. Povirk and K. Valerie, ATM-independent, high-fidelity nonhomologous end joining predominates in human embryonic stem cells, *Aging*, 2010, **2**, 582–596.
11. A. N. Bogomazova, M. A. Lagarkova, L. W. Tskhovrebova, M. V. Shutova and S. L. Kiselev, Error-prone nonhomologous end joining repair operates in human pluripotent stem cells during late G2, *Aging*, 2011, **3**(6), 1–13.
12. K. E. Orii, Y. Lee, N. Kondo and P. J. McKinnon, Selective utilization of non homologous end-joining and homologous recombination DNA repair pathways during nervous system development, *Proc. Natl. Acad. Sci. U. S. A.*, 2006, **103**(26), 10017–10022.
13. A. Meulle, B. Salles, D. Daviaud, P. Valet and C. Muller, Positive regulation of DNA double strand break repair activity during differentiation of long life span cells: the example of adipogenesis, *PLoS One*, 2008, **3**(10), e3345.
14. A. Tomashevski, D. R. Webster, P. Grammas, M. Gorospe and I. I. Kruman, Cyclin C-dependent cell cycle entry is required for activation of

nonhomologous end joining DNA repair in postmitotic neurons, *Cell Death Differ.*, 2010, **17**(7), 1189–1198.

15. S. Schmetsdorf, E. Arnold, M. Holzerm, T. Arendt and U. Gärtner, A putative role for cell cycle-related proteins in microtubule-based neuroplasticity, *Eur. J. Neurosci.*, 2009, **29**, 1096–1107.

16. K. Naka and A. Hirao, Maintenance of genomic integrity in hematopoietic stem cells, *Int. J. Hematol.*, 2011, **93**, 434–439.

17. M. Mohrin, E. Bourke, D. Alexander, M. R. Warr, K. Barry-Holson, M. M. Le Beau, C. G. Morrison and E. Passegué, Hematopoietic stem cell quiescence promotes error-prone DNA repair and mutagenesis, *Cell Stem Cell*, 2010, **7**, 174–185.

18. T. Nouspikel, DNA repair in mammalian cells: so DNA repair really is that important?, *Cell. Mol. Life Sci.*, 2009, **66**, 965–967.

19. M. Fousteri and L. H. Mullenders, Transcription-coupled nucleotide excision repair in mammalian cells: molecular mechanisms and biological effects, *Cell Res.*, 2008, **18**(1), 73–84.

20. J. M. Prasher, A. S. Lalai, C. Heijmans-Antonissen, R. E. Ploemacher, J. H. Hoeijmakers, I. P. Touw and L. J. Niedernhofer, Reduced hemato-poietic reserves in DNA interstrand crosslink repair-deficient Ercc1-/-mice, *EMBO J.*, 2005, **24**, 861–871.

21. T. Nouspikel, DNA repair in differentiated cells: some new answers to old questions, *Neuroscience*, 2007, **145**, 1213–1221.

22. T. Nouspikel and P. C. Hanawalt, Impaired nucleotide excision repair upon macrophage differentiation is corrected by E1 ubiquitin-activating enzyme, *Proc. Natl. Acad. Sci. U. S. A.*, 2006, **103**(44), 16188–16193.

23. N. Hyka-Nouspikel, K. Lemonidis, W. T. Lu and T. Nouspikel, Circu-lating human B lymphocytes are deficient in nucleotide excision repair and accumulate mutations upon proliferation, *Blood*, 2011, **117**(23), 6277–6286.

24. H. de Waard, E. Sonneveld, J. de Wit, R. Esveldt-van Lange, J. H. Hoeijmakers, H. Vrieling and G. T. J. van der Horst, Cell-type-specific consequences of nucleotide excision repair deficiencies: embryonic stem cells *versus* fibroblasts, *DNA Repair (Amst.)*, 2008, **7**(10), 1659–1669.

25. P. Fortini and E. Dogliotti, Base damage and single-strand break repair: mechanisms and functional significance of short- and long-patch repair subpathways, *DNA Repair (Amst.)*, 2007, **6**(4), 398–409.

26. L. Narciso, P. Fortini, D. Pajalunga, A. Franchitto, P. Liu, P. Degan, M. Frechet, B. Demple, M. Crescenzi and E. Dogliotti, Terminally differ-entiated muscle cells are defective in base excision DNA repair and hypersensitive to oxygen injury, *Proc. Natl. Acad. Sci. U. S. A.*, 2007, **104**(43), 17010–17015.

27. P. Fortini and E. Dogliotti, Mechanisms of dealing with DNA damage in terminally differentiated cells, *Mutat. Res.*, 2010, **685**(1–2), 38–44.

28. A. Kulkarni, D. R. McNeill, M. Gleichmann, M. P. Mattson and D. M. Wilson III, XRCC1 protects against the lethality of induced oxidative DNA damage in nondividing neural cells, *Nucleic Acids Res.*, 2008, **36**(15), 5111–5121.

29. Y. Lee, S. Katyal, Y. Li, S. F. El-Khamisy, H. R. Russell, K. W. Caldecott and P. J. McKinnon, Genesis of cerebellar interneurons and the prevention of neural DNA damage requires XRCC1, *Nat. Neurosci.*, 2009, **12**(8), 973–980.

30. J. S. Sung, M. S. DeMott and B. Demple, Long-patch base excision DNA repair of 2-deoxyribonolactone prevents the formation of DNA-protein cross-links with DNA polymerase beta, *J. Biol. Chem.*, 2005, **280**(47), 39095–39103.

31. P. Liu, L. Qian, J. S. Sung, N. C. de Souza-Pinto, L. Zheng, D. F. Bogenhagen, V. A. Bohr, D. M. Wilson 3rd, B. Shen and B. Demple, Removal of oxidative DNA damage *via* FEN1-dependent long-patch base excision repair in human cell mitochondria, *Mol. Cell. Biol.*, 2008, **28**(16), 4975–4987.

32. B. Szczesny, A. W. Tann, M. J. Longley, W. C. Copeland and S. Mitra, Long patch base excision repair in mammalian mitochondrial genomes, *J. Biol. Chem.*, 2008, **283**(39), 26349–26356.

33. E. D. Tichy, L. Liang, L. Deng, J. Tischfield, S. J. Schwemberger, G. F. Babcock and P. J. Stambrook, Mismatch and base excision repair proficiency in murine embryonic stem cells, *DNA repair (Amst.)*, 2011, **10**(4), 445–451.

34. M. Briegert and B. Kaina, Human monocytes, but not dendritic cells derived from them, are defective in base excision repair and hypersensitive to methylating agents, *Cancer Res.*, 2007, **67**(1), 26–31.

35. M. Belloni, D. Uberti, C. Rizzini, J. Jiricny and M. Memo, Induction of two DNA mismatch repair proteins, MSH2 and MSH6, in differentiated human neuroblastoma SH-SY5Y cells exposed to doxorubicin, *J. Neurochem.*, 1999, **72**(3), 974–979.

36. P. David, E. Efrati, G. Tocco, S. W. Krauss and M. F. Goodman, DNA replication and postreplication mismatch repair in cell-free extracts from cultured human neuroblastoma and fibroblast cells, *J. Neurosci.*, 1997, **17**(22), 8711–8720.

37. G. B. Panigrahi, R. Lau, S. E. Montgomery, M. R. Leonard and C. E. Pearson, Slipped (CTG)*(CAG) repeats can be correctly repaired, escape repair or undergo errorprone repair, *Nat. Struct. Mol. Biol.*, 2005, **12**(8), 654–662.

38. J. S. Reese, L. Liu and S. L. Gerson, Repopulating defect of mismatch repair-deficient hematopoietic stem cells, *Blood*, 2003, **102**(5), 1626–1633.

39. K. Ito, A. Hirao, F. Arai, S. Matsuoka, K. Takubo, I. Hamaguchi, K. Nomiyama, K. Hosokawa, K. Sakurada, N. Nakagata, Y. Ikeda, T. W. Mak and T. Suda, Regulation of oxidative stress by ATM is required for self-renewal of haematopoietic stem cells, *Nature*, 2004, **431**(7011), 997–1002.

40. K. Ito, A. Hirao, F. Arai, K. Takubo, S. Matsuoka, K. Miyamoto, M. Ohmura, K. Naka, K. Hosokawa, Y. Ikeda and T. Suda, Reactive oxygen species act through p38 MAPK to limit the lifespan of hematopoietic stem cells, *Nat Med.*, 2006, **12**(4), 446–451.

41. C. Cosentino, D. Grieco and V. Costanzo, ATM activates the pentose phosphate pathway promoting anti-oxidant defence and DNA repair, *EMBO J.*, 2011, **30**(3), 546–555.

42. K. A. Becker, P. N. Ghule, J. A. Therrien, J. B. Lian, J. L. Stein, A. J. van Wijnen and G. S. Stein, Self-renewal of human embryonic stem cells is supported by a shortened G1 cell cycle phase, *J. Cell Physiol.*, 2006, **209**(3), 883–893.

43. M. I. Aladjem, B. T. Spike, L. W. Rodewald, T. J. Hope, M. Klemm, R. Jaenisch and G. M. Wahl, ES cells do not activate p53-dependent stress responses and undergo p53-independent apoptosis in response to DNA damage, *Curr. Biol.*, 1998, **8**(3), 145–155.

44. Y. Hong and P. J. Stambrook, Restoration of an absent G1 arrest and protection from apoptosis in embryonic stem cells after ionizing radiation, *Proc. Natl. Acad. Sci. U. S. A.*, 2004, **101**(40), 14443–14448.

45. O. Momcilović, S. Choi, S. Varum, C. Bakkenist, G. Schatten and C. Navara, Ionizing radiation induces ataxia telangiectasia mutated-dependent checkpoint signaling and G(2) but not G(1) cell cycle arrest in pluripotent human embryonic stem cells, *Stem Cells*, 2009, **27**(8), 1822–1835.

46. O. Momcilović, L. Knobloch, J. Fornsaglio, S. Varum, C. Easley and G. Schatten, DNA damage responses in human induced pluripotent stem cells and embryonic stem cells, *PLoS One*, 2010, **5**(10), e13410.

47. M. Milyavsky, O. I. Gan, M. Trottier, M. Komosa, O. Tabach, F. Notta, E. Lechman, K. G. Hermans, K. Eppert, Z. Konovalova, O. Ornatsky, E. Domany, M. S. Meyn and J. E. Dick, A distinctive DNA damage response in human hematopoietic stem cells reveals an apoptosis-independent role for p53 in self-renewal, *Cell Stem Cell*, 2010, **7**(2), 186–197.

48. P. A. Sotiropoulou, A. Candi, G. Mascré, S. De Clercq, K. K. Youssef, G. Lapouge, E. Dahl, C. Semeraro, G. Denecker, J. C. Marine and C. Blanpain, Bcl-2 and accelerated DNA repair mediates resistance of hair follicle bulge stem cells to DNA-damage-induced cell death, *Nat Cell Biol.*, 2010, **12**(6), 572–582.

49. K. Inomata, T. Aoto, N. T. Binh, N. Okamoto, S. Tanimura, T. Wakayama, S. Iseki, E. Hara, T. Masunaga, H. Shimizu and E. K. Nishimura, Genotoxic stress abrogates renewal of melanocyte stem cells by triggering their differentiation, *Cell.*, 2009, **137**(6), 1088–1099.

50. H. Ito, S. C. Miller, M. E. Billingham, H. Akimoto, S. V. Torti, R. Wade, R. Gahlmann, G. Lyons, L. Kedes and F. M. Torti, Doxorubicin selectively inhibits muscle gene expression in cardiac muscle cells *in vivo* and *in vitro*, *Proc. Natl. Acad. Sci. U. S. A.*, 1990, **87**(11), 4275–4279.

51. P. L. Puri, K. Bhakta, L. D. Wood, A. Costanzo, J. Zhu and J. Y. Wang, A myogenic differentiation checkpoint activated by genotoxic stress, *Nat. Genet.*, 2002, **32**(4), 585–593.

52. M. Simonatto, L. Giordani, F. Marullo, G. C. Minetti, P. L. Puri and L. Latella, Coordination of cell cycle, DNA repair and muscle gene expression in myoblasts exposed to genotoxic stress, *Cell Cycle*, 2011, **10**(14), 2355–2363.

53. T. Lin, C. Chao, S. Saito, S. J. Mazur, M. E. Murphy, E. Appella and Y. Xu, p53 induces differentiation of mouse embryonic stem cells by suppressing Nanog expression, *Nat. Cell Biol.*, 2005, **7**(2), 165–171.
54. L. Latella, J. Lukas, C. Simone, P. L. Puri and J. Bartek, Differentiation-induced radioresistance in muscle cells, *Mol. Cell. Biol.*, 2004, **24**(14), 6350–6361.
55. S. Biton, I. DarI, L. Mittelman, Y. Pereg, A. Barzilai and Y. Shiloh, Nuclear ataxia-telangiectasia mutated (ATM) mediates the cellular response to DNA double strand breaks in human neuron-like cells, *J. Biol. Chem.*, 2006, **281**(25), 17482–17491.
56. S. Biton, M. Gropp, P. Itsykson, Y. Pereg, L. Mittelman, K. Johe, B. Reubinoff and Y. Shiloh, ATM-mediated response to DNA double strand breaks in human neurons derived from stem cells, *DNA Repair (Amst.)*, 2007, **6**(1), 128–134.
57. I. I. Kruman, R. P. Wersto, F. Cardozo-Pelaez, L. Smilenov, S. L. Chan, F. J. Chrest, R. Emokpae, Jr., M. Gorospe and M. P Mattson, Cell cycle activation linked to neuronal cell death initiated by DNA damage, *Neuron.*, 2004, **41**(4), 549–561.
58. Y. Zhang, D. Qu, E. J. Morris, M. J. O'Hare, S. M. Callaghan, R. S. Slack, H. M. Geller and D. S. Park, The Chk1/Cdc25A pathway as activators of the cell cycle in neuronal death induced by camptothecin, *J. Neurosci.*, 2006, **26**(34), 8819–8828.
59. W. P. Roos, M. Christmann, S. T. Fraser and B. Kaina, Mouse embryonic stem cells are hypersensitive to apoptosis triggered by the DNA damage O(6)-methylguanine due to high E2F1 regulated mismatch repair, *Cell Death Differ.*, 2007, **14**(8), 1422–1432.
60. M. H. Sherman, C. H. Bassing and M. A. Teitell, Regulation of cell differentiation by the DNA damage response, *Trends Cell Biol.*, 2011, **21**(5), 312–319.
61. A. E. Vaughn and M. Deshmukh, Essential postmitochondrial function of p53 uncovered in DNA damage-induced apoptosis in neurons, *Cell Death Differ.*, 2007, **14**(5), 973–981.
62. M. I. Smith, Y. Y. Huang and M. Deshmukh, Skeletal muscle differentiation evokes endogenous XIAP to restrict the apoptotic pathway, *PLoS One*, 2009, **4**(3), e5097.
63. N. Mizushima and B. Levine, Autophagy in mammalian development and differentiation, *Nat. Cell Biol.*, 2010, **12**(9), 823–830.
64. G. Mariño, F. Madeo and G. Kroemer, Autophagy for tissue homeostasis and neuroprotection, *Curr. Opin. Cell Biol.*, 2011, **23**(2), 198–206.
65. R. Baerga, Y. Zhang, P. H. Chen, S. Goldman and S. Jin, Targeted deletion of autophagy-related 5 (atg5) impairs adipogenesis in a cellular model and in mice, *Autophagy*, 2009, **5**(8), 1118–1130.
66. M. Fulco, R. L. Schiltz, S. Iezzi, M. T. King, P. Zhao, Y. Kashiwaya, E. Hoffman, R. L. Veech and V. Sartorelli, Sir2 regulates skeletal muscle differentiation as a potential sensor of the redox state, *Mol. Cell*, 2003, **12**(1), 51–62.

67. F. Picard, M. Kurtev, N. Chung, A. Topark-Ngarm, T. Senawong, R. Machado De Oliveira, M. Leid, M. W. McBurney and L. Guarente, Sirt1 promotes fat mobilization in white adipocytes by repressing PPAR-gamma, *Nature*, 2004, **429**(6993), 771–776.
68. E. Aymard, V. Barruche, T. Naves, S. Bordes, B. Closs, M. Verdier and M. H. Ratinaud, Autophagy in human keratinocytes: an early step of the differentiation?, *Exp. Dermatol.*, 2011, **20**(3), 263–268.
69. J. A. Muñoz-Gámez, J. M. Rodríguez-Vargas, R. Quiles-Pérez, R. Aguilar-Quesada, D. Martín-Oliva, G. de Murcia, J. Menissier de Murcia, A. Almendros, M. Ruiz de Almodóvar and F. J. Oliver, PARP-1 is involved in autophagy induced by DNA damage, *Autophagy*, 2009, **5**(1), 61–74.
70. H. C. Ha and S. H. Snyder, Poly(ADP-ribose) polymerase is a mediator of necrotic cell death by ATP depletion, *Proc. Natl. Acad. Sci. U. S. A.*, 1999, **96**(24), 13978–13982.
71. A. Alexander, S. L. Cai, J. Kim, A. Nanez, M. Sahin, K. H. MacLean, K. Inoki, K. L. Guan, J. Shen, M. D. Person, D. Kusewitt, G. B. Mills, M. B. Kastan and C. L. Walker, ATM signals to TSC2 in the cytoplasm to regulate mTORC1 in response to ROS, *Proc. Natl. Acad. Sci. U. S. A.*, 2010, **107**(9), 4153–4158.
72. A. Alexander, J. Kim and C. L. Walker, ATM engages the TSC2/mTORC1 signaling node to regulate autophagy, *Autophagy*, 2010, **6**(5), 672–673.
73. T. Robert, F. Vanoli, I. Chiolo, G. Shubassi, K. A. Bernstein, R. Rothstein, O. A. Botrugno, D. Parazzoli, A. Oldani, S. Minucci and M. Foiani, HDACs link the DNA damage response, processing of double-strand breaks and autophagy, *Nature*, 2011, **471**(7336), 74–79.
74. M. F. Clarke and M. Fuller, Stem cells and cancer: two faces of eve, 2006, *Cell*, **124**(6), 1111–1115.

Tumour Suppressor Protein-mediated Regulation of Base Excision Repair in Response to DNA Damage

GRIGORY L. DIANOV,* GIULIA ORLANDO AND JASON L. PARSONS

Gray Institute for Radiation Oncology and Biology, Department of Oncology, University of Oxford, Old Road Campus Research Building, Roosevelt Drive, Oxford OX3 7DQ, UK
*Email: grigory.dianov@oncology.ox.ac.uk

3.3.1 Base Excision Repair

Base excision repair (BER) is initiated by damage-specific DNA glycosylases that identify and release the corrupted base by hydrolysis of the N-glycosylic bond linking the DNA base to the sugar phosphate backbone. In mammalian cells, the arising abasic site (AP-site) is further processed by AP-endonuclease 1 (APE1) that cleaves the phosphodiester bond 5' to the AP-site, generating a DNA single-strand break (SSB) with a 5'-sugar phosphate. This SSB is then repaired by a DNA repair complex that includes DNA polymerase β (Pol β), XRCC1 and DNA ligase IIIα (Lig III).[1,2] Pol β possesses AP lyase activity that removes the 5'-sugar phosphate and also, functioning as a DNA

Issues in Toxicology No. 13
The Cellular Response to the Genotoxic Insult: The Question of Threshold for Genotoxic Carcinogens
Edited by Helmut Greim and Richard J. Albertini
© The Royal Society of Chemistry 2012
Published by the Royal Society of Chemistry, www.rsc.org

polymerase, adds one nucleotide to the 3′-end of the arising single-nucleotide gap. Finally, XRCC1-Lig III complex seals the DNA ends, therefore accomplishing DNA repair (reviewed in[3]). This pathway is commonly referred to as the short-patch BER pathway, through which cells are accomplishing the majority of repair (Figure 3.3.1, left branch). However, if the 5′-sugar phosphate is resistant to cleavage by Pol β, then a switch to Pol δ/ε occurs that adds 2–8 more nucleotides into the repair gap, therefore generating a flap structure that is removed by flap endonuclease-1 (FEN-1) in a PCNA-dependent manner. DNA ligase I then seals the remaining nick in the DNA backbone and this process is commonly referred to as long-patch BER[4,5] (Figure 3.3.1 right branch).

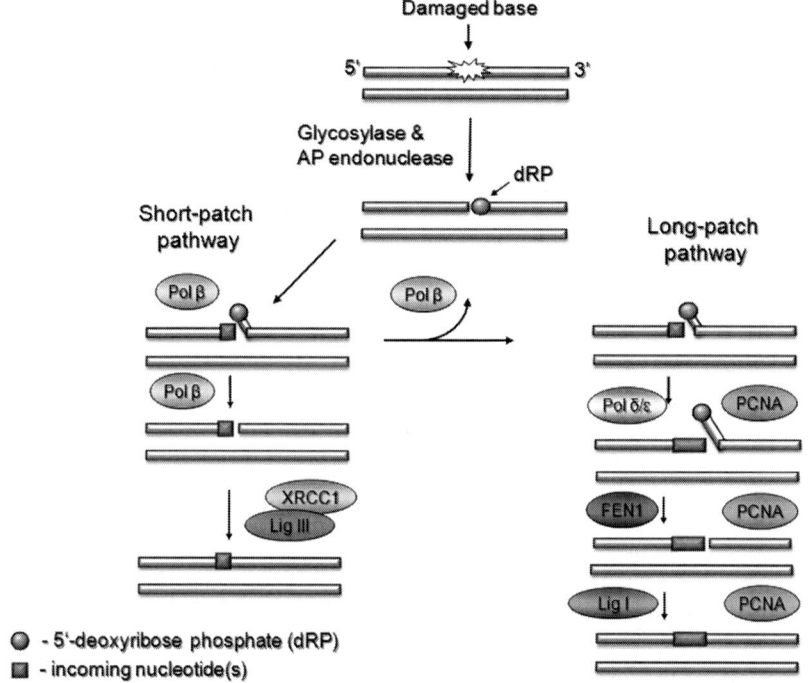

Figure 3.3.1 Base excision repair (BER) pathways. BER is initiated by a damage-specific DNA glycosylase and APE1 that excise the damaged base and incise the arising abasic site, respectively, generating a DNA single-strand break with a 5′-sugar phosphate. In the short-patch pathway (**left branch**), Pol β removes the 5′-sugar phosphate and adds one nucleotide into the single-nucleotide gap. Finally XRCC1-Lig III complex seals the remaining DNA ends. However, if the 5′-sugar phosphate is resistant to cleavage by Pol β, then the long-patch pathway proceeds that requires Pol δ/ε to add 2–8 more nucleotides into the repair gap, generating a flap structure, which is then removed by FEN-1 in a PCNA-dependent manner, prior to sealing of the remaining DNA ends by Lig I.

3.3.2   BER Regulation

To support error-free transcription and replication, BER proteins should be present in adequate amounts to be able to repair DNA promptly. Indeed, mutations affecting the amounts or enzymatic activities of these proteins increase genome instability and reduce cell viability in response to DNA damage.[6-15] However, the amount of BER enzymes should be tightly controlled since, when overproduced, BER enzymes may affect other DNA transactions and can also lead to genome instability and cancer.[16,17]

The amount of DNA lesions in human cells originates from the chemical instability of the DNA molecule itself, but also depends on cellular metabolism and exposure to exogenous mutagens.[18] A combination of these factors leads to variations in DNA damage levels and BER should be responsive to the changing environment. Indeed, as it has been recently demonstrated by our laboratory, the amount of BER enzymes present within a cell at any time (the steady-state level of BER enzymes) is tightly regulated and is linked to the amount of DNA lesions.[19] This is achieved by controlling the cytoplasmic pool of the three major BER enzymes (XRCC1, Lig III and Pol β) through targeted proteasomal degradation. Proteins targeted for degradation are marked with a chain of ubiquitin (small 76 amino acid protein) molecules, in which ubiquitin is linked *via* its *C*-terminus onto the ε-amino group of a lysine residue in the substrate protein that is catalysed by a cascade of ubiquitin activating enzymes.

Firstly, ubiquitin (Ub) is activated through an ATP-dependent reaction by a ubiquitin activating enzyme (E1) to form an E1-Ub thioester. Secondly, the activated ubiquitin is delivered to a ubiquitin conjugating enzyme (E2) and finally, a complex is formed between the E2-Ub thioester, the target protein and a ubiquitin ligase (E3) that conjugates ubiquitin to the protein. Since ubiquitin itself contains seven lysines, these can serve as targets sites for ubiquitin chain formation and this subsequent polyubiquitylation of proteins are recognised by the 26S proteasome that unfolds the protein, removes the polyubiquitin chains and degrades the protein (for review see ref. 20).

We have previously demonstrated that if the levels of BER proteins exceed the level of DNA lesions, then the excessive BER enzymes are ubiquitylated and thus labelled for proteasomal degradation.[19] For example, in human cells, Pol β is sequentially ubiquitylated by the E3 ubiquitin ligases Mule and CHIP, which leads to its proteasomal degradation and subsequently down-regulates its steady-state level.[21] Interestingly, the E3 ligase activity of Mule is regulated by the ARF protein (Alternate Reading Frame protein also known as p14 in humans and p19 in mice),[22,23] which is a well-known tumour suppressor protein whose release is modulated in response to DNA damage.[24] Release of ARF and the consequent inhibition of Mule activity leads to an accumulation of Pol β and an increased rate of DNA repair (Figure 3.3.2). In this scenario, ARF is acting as a signalling molecule transmitting information about the DNA damage and coordinating the cellular responses. Furthermore, activation of DNA repair by ARF is accompanied by inhibition of the Mdm2 ubiquitin ligase that subsequently causes activation of p53-dependent cell cycle arrest to

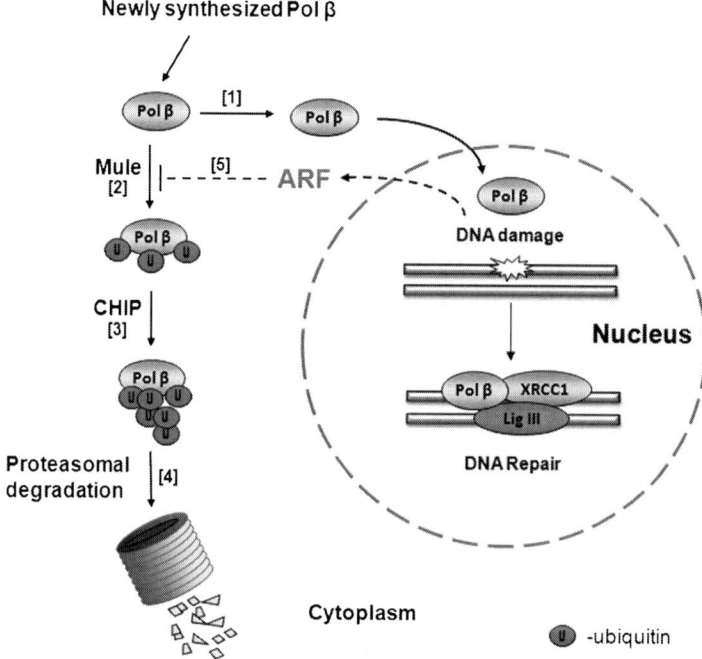

Figure 3.3.2   Regulation of BER in response to DNA damage. Newly synthesised Pol
β is either directly transferred to the nucleus [1] or is monoubiquitylated
by Mule [2], which is then a target for CHIP-mediated polyubiquitylation [3] and subsequent degradation by the proteasome [4]. However, if
DNA damage is detected, the activity of Mule is inhibited by the
accumulation of ARF [5], which then generates more active Pol β that is
able to freely enter the nucleus to participate in DNA repair.

allow cells time to accomplish DNA repair.[25] Therefore, ARF plays a key role
in DNA damage signal transduction and, as expected, has strong links to
cancer development.

3.3.3   ARF in Cancer

ARF-null mice develop spontaneous tumours by two months of age, revealing
a fundamental role of the ARF protein in preventing tumour formation
in vivo.[26] It was reported that 28 out of 39 ARF-null mice died within one year
of birth following tumour formation, 43% of the ARF-null mice developed
sarcomas, 29% developed lymphoma and, interestingly, three mice developed
gliomas which rarely develop in normal mice, or even in p53-null mice. It was
also shown that ARF-heterozygous mice, which may arise due to the loss of
function of the second allele through deletion or methylation silencing of the
wild-type gene copy, also develop tumours. Immunostaining of papillomas
derived from ARF-null mice revealed an increase in the phosphorylation of

histone H2AX and a dramatic decrease in p53 protein levels, demonstrating that ARF protein is important in the regulation of p53 basal levels *in vivo.*[27]

There are multiple examples of ARF inactivation or deletion in several different human cancers, such as in colon carcinoma,[28] oligodendroglioma,[29] breast cancer,[30] oral squamous cell carcinoma, glioma,[31,32] primary colorectal cancer[33] and in primary gastric carcinoma.[34] Furthermore, alterations in ARF expression levels have been observed in lung cancer,[35,36] in breast carcinoma[37] and in colorectal tumours.[38] For example, from an analysis of breast cancer samples, it has been revealed that 26% expressed no detectable or very low levels of p14 *arf* mRNA and 17% expressed high levels of p14 *arf* mRNA.[39] Interestingly, the majority of genomic alterations have been found in haematological malignancies. Therefore, deletions of ARF and genomic instability are frequent events in non-Hodgkin's lymphomas,[40] and in several lymphomas and adult leukaemia.[41] ARF has also been reported to be strongly involved in melanoma predisposition, with a cluster of five different germ line mutations at the splice donor site of exon 1β being revealed, of which three of the variants were demonstrated to result in aberrant splicing of the p14*arf* mRNA.[42] So it is quite clear that ARF inactivation leads to cell transformation, however since ARF regulates both DNA repair and cell cycle progression, the relative contribution of these processes to cancer development is poorly understood.

3.3.4   ARF Protein Structure, Stability and Expression

ARF was originally identified as an alternative transcript of the INK4 locus which also encodes p16 protein, an inhibitor of cyclin dependent kinases. The two different transcripts are regulated by two distinct promoters and they also have two different first exons.[23] The ARF protein (132 amino acids, 14 kDa) has an unusual amino acid composition and is composed of more than 20% arginine residues making it highly basic (pI > 12). It has been proposed that ARF is probably unstructured and needs to form complexes with other molecules, both to be folded and for its charge to be neutralised at physiological pH.[25] Indeed, ARF has the ability to bind a range of different proteins, such as nucleophosmin (NPM/B23), the E3 ubiquitin ligases Mdm2 and Mule,[43,44] and some other proteins.[45] However, it is still not established if ARF is able to bind nucleic acids directly and there is no bioinformatics evidence of the presence of DNA binding domains. Although ARF does contain numerous potential sites of phosphorylation, these again are still uncharacterised.

In normal unstressed cells, the expression of ARF protein is very low, although the relative contribution of transcriptional control and proteasomal degradation of newly synthesised proteins in maintaining the normal low levels of ARF is unclear. However, it has been demonstrated in a number of cancer cell lines that transcription of ARF is repressed by promoter hypermethylation, although this may be a specific mechanism of ARF inactivation employed by cancer but not normal cells.[46,47]

The ubiquitin proteasome system is widely known as a mechanism that regulates cellular protein levels, although the ubiquitylation and subsequent proteasomal degradation of ARF was not considered for a long time since the human protein does not contain any lysine residues in its structure for ubiquitin conjugation, although the mouse ARF protein has a single lysine residue.[23] However, polyubiquitylation of ARF has been demonstrated to occur in mammalian cells and the N-terminus of the ARF protein was confirmed as the site of ubiquitylation.[48] These data indicate that the N-terminus is important for the stability of the ARF protein and subsequently its regulation in cells. More recently, the E3 ubiquitin ligase ULF was identified as a major enzyme involved in the ubiquitylation and subsequent degradation of ARF.[49] Most probably, down-regulation of ARF protein levels is accomplished through proteasomal degradation of newly synthesised ARF protein. However, the mechanism regulating ULF activity on ARF in response to DNA damage remains unknown.

3.3.5 ARF Roles in the Cellular Response to DNA Damage

It has been shown that ARF-null cells are unable to correctly respond to DNA damage induced by γ-irradiation and continue cell cycle progression, which supports the role of ARF in the cellular DNA damage response.[50–52] However, the precise role of the ARF protein in this process is still under discussion. One of the major reasons for this is that the majority of early studies on the role of ARF in the DNA damage response were performed with transformed cancer cells, where ARF is already expressed at high levels and sequestered in the nucleolus by nucleophosmin. The reason for this nucleolar localisation is still unclear, although it has been shown that interaction of ARF with nucleophosmin is critical for ARF nucleolar localisation and that this interaction is disrupted after DNA damage.[24,53,54] However, in unstressed normal cells, ARF is usually expressed at very low levels,[49] but they have been shown to increase following DNA damage.[55] The working model on the role of ARF in DNA damage responses is summarised in Figure 3.3.3. As long as DNA damage is identified (by an as yet unknown mechanism) ARF protein accumulates and the increased concentration of ARF has multiple cellular effects. Firstly, ARF inhibits the E3 ubiquitin ligase Mdm2 leading to an accumulation of p53, which either causes a subsequent delay in cell cycle progression or induces apoptosis if the DNA damage is too extensive. Secondly, ARF inhibits the E3 ubiquitin ligase Mule, which leads to an up-regulation in the cellular levels of Pol β and subsequently causes an increase in the rate of DNA repair. Since the increased DNA repair rate reduces the cellular DNA damage, this will down-regulate the cellular levels of ARF and thus causes a negative feedback loop. This cycle will be repeated continuously which results in transient fluctuations in the amounts of all proteins involved and provides a balanced cellular response to DNA damage.

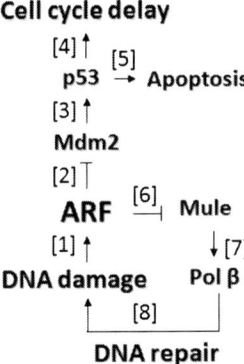

Figure 3.3.3   ARF coordinates the cellular response to DNA damage. Following the detection of cellular DNA damage, ARF protein accumulates [1], which has two major cellular effects. Firstly, ARF inhibits Mdm2 [2] which leads to an accumulation of p53 [3] that causes either a delay in cell cycle progression [4] or an induction of apoptosis, if the DNA damage is too extensive [5]. Secondly, ARF inhibits Mule [6] leading to an up-regulation in the cellular levels of Pol β [7] and consequently an increase in the rate of DNA repair [8].

3.3.6   Conclusions and Future Prospective

It is clear that ARF is an essential protein involved in tumour suppression in humans, with its importance akin to p53 which has received substantially more interest, and that ARF plays crucial roles in the cellular response to DNA damage. What still remains unclear is the signalling mechanism(s) that links the levels of cellular DNA damage to the increased levels of ARF protein. It is also important to understand if the regulation of the protein is coordinated by an increase of its expression, *via* a dose-dependent mechanism, or through a post-translational mechanism, such as the inhibition of ubiquitylation-dependent degradation. Since post-translational mechanisms of regulation of the ARF protein remain poorly understood, these need to be investigated in further detail. Currently, only ubiquitylation of ARF on the N-terminus of the protein has been reported and being a very basic protein without any predicted functional domains, it is probable that the only way in which ARF is able to exert its functions is through a direct interaction with its target proteins. In this scenario, how ARF chooses and consequently modulates the function of the many different cellular targets will be the challenge of future studies. Therefore, a better understanding of the mechanism(s) of regulation of ARF protein, and of the signalling pathway(s) that trigger its accumulation in response to DNA damage, will be a future challenge and this could have important implications in the treatment of human diseases, such as cancer.

References

1. I. I. Dianova, K. M. Sleeth, S. L. Allinson, J. L. Parsons, C. Breslin, K. W. Caldecott and G. L. Dianov, *Nucleic Acids Res.*, 2004, **32**, 2550–2555.
2. J. L. Parsons, I. I. Dianova, S. L. Allinson and G. L. Dianov, *Biochemistry*, 2005, **44**, 10613–10619.
3. G. L. Dianov and J. L. Parsons, *DNA Repair (Amst.)*, 2007, **6**, 454–460.
4. G. Frosina, P. Fortini, O. Rossi, F. Carrozzino, G. Raspaglio, L. S. Cox, D. P. Lane, A. Abbondandolo and E. Dogliotti, *J. Biol. Chem.*, 1996, **271**, 9573–9578.
5. A. J. Podlutsky, I. I. Dianova, V. N. Podust, V. A. Bohr and G. L. Dianov, *EMBO J.*, 2001, **20**, 1477–1482.
6. D. C. Cabelof, Z. Guo, J. J. Raffoul, R. W. Sobol, S. H. Wilson, A. Richardson and A. R. Heydari, *Cancer Res.*, 2003, **63**, 5799–5807.
7. D. C. Cabelof, Y. Ikeno, A. Nyska, R. A. Busuttil, N. Anyangwe, J. Vijg, L. H. Matherly, J. D. Tucker, S. H. Wilson, A. Richardson and A. R. Heydari, *Cancer Res.*, 2006, **66**, 7460–7465.
8. T. Lang, M. Maitra, D. Starcevic, S. X. Li and J. B. Sweasy, *Proc. Natl. Acad. Sci. U. S. A.*, 2004, **101**, 6074–6079.
9. K. Ochs, J. Lips, S. Profittlich and B. Kaina, *Cancer Res.*, 2002, **62**, 1524–1530.
10. K. Och, R. W. Sobol, S. H. Wilson and B. Kaina, *Cancer Res.*, 1999, **59**, 1544–1551.
11. R. W. Sobol, D. E. Watson, J. Nakamura, F. M. Yakes, E. Hou, J. K. Horton, J. Ladapo, B. Van Houten, J. A. Swenberg, K. R. Tindall, L. D. Samson, and S. H. Wilson, *Proc. Natl. Acad. Sci. U. S. A.*, 2002, **99**, 6860–68605.
12. J. B. Sweasy, T. Lang, D. Starcevic, K. W. Sun, C. C. Lai, D. Dimaio and S. Dalal, *Proc. Natl. Acad. Sci. U. S. A.*, 2005, **102**, 14350–14355.
13. K. W. Caldecott, S. Aoufouchi, P. Johnson and S. Shall, *Nucleic Acids Res.*, 1996, **24**, 4387–4394.
14. A. Rasouli-Nia, F. Karimi-Busheri and M. Weinfeld, *Proc. Natl. Acad. Sci. U. S. A.*, 2004, **101**, 6905–6910.
15. A. Unnikrishnan, J. J. Raffoul, H. V. Patel, T. M. Prychitko, N. Anyangwe, L. B. Meira, E. C. Friedberg, D. C. Cabelof and A. R. Heydari, *Free Radic. Biol. Med.*, 2009, **46**, 1488–1499.
16. M. R. Albertella, A. Lau and M. J. O'Connor, *DNA Repair (Amst.)*, 2005, **4**, 583–593.
17. G. Frosina, *Eur. J. Biochem.*, 2000, **267**, 2135–2149.
18. T. Lindahl, *Nature*, 1993, **362**, 709–715.
19. J. L. Parsons, P. S. Tait, D. Finch, I. I. Dianova, S. L. Allinson and G. L. Dianov, *Mol. Cell*, 2008, **29**, 477–487.
20. K. I. Nakayama and K. Nakayama, *Nat. Rev.*, 2006, **6**, 369–381.
21. J. L. Parsons, P. S. Tait, D. Finch, I. I. Dianova, M. J. Edelmann, S. V. Khoronenkova, B. M. Kessler, R. A. Sharma, W. G. McKenna and G. L. Dianov, *EMBO J.*, 2009, **28**, 3207–3215.

22. D. Chen, N. Kon, M. Li, W. Zhang, J. Qin and W. Gu, *Cell*, 2005, **121**, 1071–1083.

23. S. J. Gallagher, R. F. Kefford and H. Rizos, *Int. J. Biochem. Cell Biol.*, 2006, **38**, 1637–1641.

24. C. Lee, B. A. Smith, K. Bandyopadhyay and R. A. Gjerset, *Cancer Res.*, 2005, **65**, 9834–9842.

25. C. J. Sherr, *Nat. Rev. Cancer*, 2006, **6**, 663–673.

26. T. Kamijo, F. Zindy, M. F. Roussel, D. E. Quelle, J. R. Downing, R. A. Ashmun, G. Grosveld and C. J. Sherr, *Cell*, 1997, **91**, 649–659.

27. T. Kamijo, S. Bodner, E. van de Kamp, D. H. Randle and C. J. Sherr, *Cancer Res.*, 1999, **59**, 2217–2222.

28. N. Burri, P. Shaw, H. Bouzourene, I. Sordat, B. Sordat, M. Gillet, D. Schorderet, F. T. Bosman and P. Chaubert, *Lab. Invest.*, 2001, **81**, 217–229.

29. T. Watanabe, M. Nakamura, Y. Yonekawa, P. Kleihues and H. Ohgaki, *Acta Neuropathol.*, 2001, **101**, 185–189.

30. G. H. Ho, J. E. Calvano, M. Bisogna, Z. Abouezzi, P. I. Borgen, C. Cordon-Cardo and K. J. van Zee, *Breast Cancer Res. Treat.*, 2001, **65**, 225–232.

31. K. Ichimura, M. B. Bolin, H. M. Goike, E. E. Schmidt, A. Moshref and V. P. Collins, *Cancer Res.*, 2000, **60**, 417–424.

32. S. Shintani, Y. Nakahara, M. Mihara, Y. Ueyama and T. Matsumura, *Oral Oncol.*, 2001, **37**, 498–504.

33. M. Esteller, S. Tortola, M. Toyota, G. Capella, M. A. Peinado, S. B. Baylin and J. G. Herman, *Cancer Res.*, 2000, **60**, 129–33.

34. S. Iida, Y. Akiyama, T. Nakajima, W. Ichikawa, Z. Nihei, K. Sugihara and Y. Yuasa, *Int. J. Cancer*, 2000, **87**, 654–8.

35. S. Gazzeri, V. Della Valle, L. Chaussade, C. Brambilla, C. J. Larsen and E. Brambilla, *Cancer Res.*, 1998, **58**, 3926–3931.

36. S. Vonlanthen, J. Heighway, M. P. Tschan, M. M. Borner, H. J. Altermatt, A. Kappeler, A. Tobler, M. F. Fey, N. Thatcher, W. G. Yarbrough and D. C. Betticher, *Oncogene*, 1998, **17**, 2779–2785.

37. A. J. Brenner, A. Paladugu, H. Wang, O. I. Olopade, M. H. Dreyling and C. M. Aldaz, *Clin. Cancer Res.*, 1996, **2**, 1993–1998.

38. S. Zheng, P. Chen, A. McMillan, A. Lafuente, M. J. Lafuente, A. Ballesta, M. Trias and J. K. Wiencke, *Carcinogenesis*, 2000, **21**, 2057–2064.

39. J. Silva, G. Dominguez, J. M. Silva, J. M. Garcia, I. Gallego, C. Corbacho, M. Provencio, P. Espana and F. Bonilla, *Oncogene*, 2001, **20**, 4586–4590.

40. M. Herranz, M. Urioste, J. Santos, C. Rivas, B. Martinez, J. Benitez and J. Fernandez-Piqueras, *Leukemia*, 1999, **13**, 808–810.

41. S. Takemoto, R. Trovato, A. Cereseto, C. Nicot, T. Kislyakova, L. Casareto, T. Waldmann, G. Torelli and G. Franchini, *Blood*, 2000, **95**, 3939–3944.

42. M. Harland, C. F. Taylor, P. A. Chambers, K. Kukalizch, J. A. Randerson-Moor, N. A. Gruis, F. A. de Snoo, J. A. ter Huurne, A. M. Goldstein, M. A. Tucker, D. T. Bishop and J. A. Bishop, *Oncogene*, 2005, **24**, 4604–4608.

43. C. J. Sherr and J. D. Weber, *Curr. Opin. Genet. Dev.*, 2000, **10**, 94–99.

44. S. N. Brady, Y. Yu, L. B. Maggi Jr. and J. D. Weber, *Mol. Cell Biol.*, 2004, **24**, 9327–9338.
45. V. Tompkins, J. Hagen, V. P. Zediak and D. E. Quelle, *Cell Cycle*, 2006, **5**, 641–646.
46. K. D. Robertson and P. A. Jones, *Mol. Cell Biol.*, 1998, **18**, 6457–6473.
47. V. Badal, S. Menendez, D. Coomber and D. P. Lane, *Cell Cycle*, 2008, **7**, 112–119.
48. M. L. Kuo, W. den Besten and C. J. Sherr, *Cell Cycle*, 2004, **3**, 1367–1369.
49. D. Chen, J. Shan, W. G. Zhu, J. Qin and W. Gu, *Nature*, 2010, **464**, 624–627.
50. S. H. Khan, J. Moritsugu and G. M. Wahl, *Proc. Natl. Acad. Sci. U. S. A.*, 2000, **97**, 3266–3271.
51. K. K. Khanna and M. F. Lavin, *Oncogene*, 1993, **8**, 3411–3416.
52. Y. Xu and D. Baltimore, *Genes Dev.*, 1996, **10**, 2401–2410.
53. C. Korgaonkar, J. Hagen, V. Tompkins, A. A. Frazier, C. Allamargot, F. W. Quelle and D. E. Quelle, *Mol. Cell Biol.*, 2005, **25**, 1258–1271.
54. S. Moulin, S. Llanos, S. H. Kim and G. Peters, *Oncogene*, 2008, **27**, 2382–2389.
55. S. Khan, C. Guevara, G. Fujii and D. Parry, *Oncogene*, 2004, **23**, 6040–6046.

CHAPTER 3.4

Lesion Sensing and Decision Points in the DNA Damage Response

PHILIP C. HANAWALT

Department of Biology, Stanford University, 371 Serra Mall, Stanford, California 94305-5020, USA
Email: hanawalt@stanford.edu

3.4.1 Introduction

The essential genetic blueprint for all organisms is entrusted to the sequences of nucleotides in double-stranded cellular DNA. However, the genomic DNA is susceptible to deleterious alterations that can cause mutations, which in turn can lead to cancer or other disabilities in multicellular organisms such as humans. These alterations include spontaneous DNA modifications that are intrinsic to the chemical instability of DNA (*e.g.* cytosine deamination and depurination),[1] as well as damage inflicted by various chemical and physical agents. Some, like reactive oxygen species (ROS) are primarily endogenous, while other genotoxic chemicals and radiations are threats from the external environment. Some naturally occurring nucleotide sequences can adopt non-canonical DNA structures, which then become hotspots for deleterious alterations and genomic instability.[2] A complex set of surveillance and repair mechanisms has evolved to remove or nullify genomic damage. Several of the DNA repair pathways are so important that life cannot be sustained without

Issues in Toxicology No. 13
The Cellular Response to the Genotoxic Insult: The Question of Threshold for Genotoxic Carcinogens
Edited by Helmut Greim and Richard J. Albertini
© The Royal Society of Chemistry 2012
Published by the Royal Society of Chemistry, www.rsc.org

them. Furthermore, a number of human hereditary diseases with predisposition to cancer are linked to deficiencies in DNA repair. There have been many remarkable advances since the publication in 2006 of a textbook on *DNA Repair and Mutagenesis* that otherwise provides a comprehensive treatment of this broad field.[3]

This chapter summarizes the enzymatic pathways for repairing or tolerating various sorts of DNA damage. It considers the mechanisms by which DNA lesions are detected and the orchestration of multiple DNA damage processing modes to achieve desired biological endpoints. Examples of overlap between these modes are discussed, as stages at which common intermediates may be processed by several different pathways. Thus, there are crossover points from one excision repair pathway to another, as each step creates another lesion and the overall process is not complete until a repair patch has been synthesized and ligated to the contiguous DNA strand. Genetic diseases in which DNA damage processing is defective, causing sensitivity to particular genotoxins and usually, but not always, an enhanced risk of carcinogenesis, are also described. Our understanding of the regulation of DNA repair is related to the question of whether there are thresholds for cellular responses to genotoxic carcinogens.

3.4.2   Mechanisms for Processing Lesions in DNA

3.4.2.1   Direct Reversal of Damage

Some types of deleterious DNA alterations can be directly reversed by a single enzyme. Thus, an inappropriate methylation at the 0^6-position on guanine (causing that base to code as if it were adenine) can be removed by an alkyltransferase, which simply transfers the methyl to itself. Surprisingly, this is a 'suicide enzyme' that can only operate once. The biological importance of this specialized pathway is documented by the evident necessity of synthesizing an entire protein for the sole purpose of removing one offending methyl group from one guanine.

Many organisms possess a photolyase, which reverses the dimerization of adjacent pyrimidines in DNA caused by short-wavelength ultraviolet light (UV). The enzyme can bind to the predominant photoproduct, the cyclobutane–pyrimidine dimer (CPD), in the dark and then, upon activation by light in the blue spectral region, it can split the dimerized bases to restore the single pyrimidines without otherwise affecting the DNA structure. This mechanism does not operate in humans or other mammals, but the light-sensing component of photolyase (from an early stage of evolution) has been retained to regulate the circadian clock. A few organisms have a specialized photolyase to reverse the structure-distorting 6-4 pyrimidine–pyrimidone photoproduct (6-4PP), which is formed in genomic DNA with a somewhat lower quantum yield than that for the CPD.

Single-strand breaks in DNA, which arise frequently in the course of replication and as intermediates in excision repair, can be sealed directly by polynucleotide ligase when the respective abutting ends are 3′ hydroxyl and 5′ phosphate. However, ionizing radiation and ROS produce breaks with chemically heterogeneous ends that must be processed before ligation is possible.[4] Also, the gap created by deleted nucleotides at the site of a strand break requires several enzymes for its repair, as discussed below.

3.4.2.2 Excision Repair Pathways

The most ubiquitous and versatile modes of DNA repair are those in which a relatively short section of a strand containing a lesion, or an incorrect nucleotide, is excised and the resulting gap is patched by repair replication using the complementary strand as template. Thus, the redundancy of information in duplex DNA ensures that each of the complementary strands can serve as a template for accurate repair of the other strand as well as for semiconservative replication. The prototype mode, now called nucleotide excision repair (NER), was discovered in the early 1960s through basic research on the effects[5–8] of UV on DNA synthesis in the bacterium, *Escherichia coli*. Shortly after the subsequent documentation of NER in human cells, the first example of a DNA repair deficient disease, Xeroderma pigmentosum (XP), was reported. Victims of this rare autosomal recessive disease, present striking predisposition to cancer in sun-exposed skin; their cultured epidermal cells are UV-sensitive and defective in repair replication.[9]

3.4.2.2.1 Nucleotide Excision Repair (NER)

Lesions throughout the genome can be recognized for NER, although bound proteins on DNA and the chromatin structure generally impede this global genomic repair (GGR) mode. The access to lesions is improved in regions of active transcription, in which the chromatin is 'opened up', but there is an additional complication when the damage is encountered by a translocating RNA polymerase (RNAP), which can sensitively detect lesions in DNA. Many types of lesions arrest transcription and the blocked RNAP then prevents access of repair enzymes to the damage. Furthermore, the RNAP remains tightly bound at lesion sites such as CPDs and prolonged transcription arrest can trigger apoptosis (see Chapter 4.3). A dedicated sub-pathway of NER, transcription-coupled repair (TCR), can displace the arrested RNAP and recruit NER enzymes to remove the encumbrance so that transcription may resume (for a review see ref. 10).In all reported examples of TCR, the lesion that arrests the RNAP is on the transcribed DNA strand,[11] so TCR is thought to be specific for damage in the transcribed strands of expressed genes. However, alternative DNA secondary structures, such as the G4 quadruplex, arrest transcription when they are formed in the non-transcribed strand.[12,13] Some types of damage, including the DNA adducts of aristocholic acid and the monofunctional adducts formed by irofulven (6-hydroxymethylacylfulvene), are transparent to recognition by the GGR pathway, but they can be sensitively detected by translocating RNAP, resulting in their efficient repair by TCR.[14,15] Of course, only the transcribed strands of expressed genes are being repaired in these cases; mutation-generating lesions persist in most of the genome. The recently sequenced genomes in cells from human tumors have revealed the 'signatures' of the likely causal agents (*i.e.* UV in a melanoma and polycyclic hydrocarbons in a lung tumor), as well as the strand specificities at the mutation sites in an expressed gene, thereby implicating TCR in lesion processing during the development of these tumors.[16,17] It is anticipated that this approach will contribute importantly to genetic toxicology.

In addition to GGR and the targeted repair of transcribed DNA strands by TCR, there is another NER pathway, termed 'domain associated repair' (DAR), that is operative on both DNA strands in genomic domains containing expressed genes in terminally differentiated cells. DAR requires the same set of enzymes as GGR and it masks the observation of TCR in the domains in which it operates.[18,19] It is thought to be important for long-lived, non-dividing cells like neurons, in order to maintain a set of undamaged DNA strands for use as potential templates during TCR.

3.4.2.2.2 Mismatch Repair (MMR)

Genetic evidence for a second type of excision repair, mismatch repair (MMR), was reported in 1976 and soon confirmed through biochemical studies.[20] MMR recognizes and removes mismatched nucleotides in the otherwise complementary DNA strands, arising from DNA replication errors and recombination, as well as from some types of base modifications. The importance of MMR is highlighted by the observation that the frequency of spontaneous mutations is greatly enhanced in its absence. Defects in mismatch repair genes are linked to hereditary non-polyposis colon cancer (HNPCC) as well as to sporadic cancers, which exhibit genomic instability in regions containing short repetitive nucleotide sequences.

3.4.2.2.3 Base Excision Repair (BER)

Inappropriate bases in DNA (*e.g.* uracil) and a subset of base alterations, including those due to ROS (*e.g.* 8-oxo-guanine), are subject to a variant of the excision repair theme, termed base excision repair (BER), in which an initial step of base removal precedes incision of the DNA backbone.[21,22] Thus, BER is usually initiated by a specific *N*-glycosylase that recognizes and cleaves the damaged or inappropriate base from its sugar moiety. The resulting abasic site can then be processed by an apurinic/apyrimidinic (AP) endonuclease to cut the DNA backbone and facilitate synthesis of a repair patch.

There are several BER pathways, of which one employs DNA polymerase β to introduce a patch as short as one nucleotide, while another requires DNA polymerases δ or ε in association with proliferating cell nuclear antigen (PCNA) to complete a longer patch (but not nearly as long as the typical 23–30 nucleotide patch for NER) (see Chapter 3.1). Each glycosylase recognizes a particular form of base damage or a particular inappropriate base. For example, uracil-DNA glycosylase (UDG) removes uracil that has been mis-incorporated during replication or generated by deamination of cytosine.

The fact that mice deficient in enzymes required for BER typically succumb to early embryonic death attests to the necessity of repairing DNA lesions from endogenous causes during development. The rarity of human cancer prone diseases characterized by defects in BER genes is likely a consequence of the essential nature of this pathway. The term 'base excision repair' has also been applied to the repair of single-strand breaks, when this requires additional processing before ligation to complete the repair process. Evidence has been presented in support of a transcription-coupled BER pathway,[23] but a mechanism for coupling BER, instead of NER, to transcription arrest is not yet clear. A plausible scenario, discussed below, implicates TCR of an abasic site intermediate, following the removal of the affected base by a glycosylase.

Some organisms (but not human cells) contain a specialized *N*-glycosylase-AP-lyase that cleaves DNA at the sites of CPDs to initiate excision repair. An important example is the T4 bacteriophage CPD glycosylase, called T4 endonuclease V (TEV), which can serve as a quantitative probe for this particular photoproduct.[24] A complex strand break is formed, wherein the pyrimidine on the 5′ side is cleaved from its deoxyribose moiety, which remains at the 3′ end of a backbone incision introduced at the site between the dimerized pyrimidines, leaving the dimer hanging off the 5′ end at the break. A few organisms can deal with the excision of UV photoproducts more simply, by utilizing an endonuclease that introduces an ATP-independent incision just 5′ of a CPD (or a 6-4PP) to generate 3′-OH and 5′-P termini, so that no end-processing is required before DNA polymerase can begin to synthesize a repair patch while displacing the strand containing the photo-product. Ironically, it was this type of activity that was originally hypothe-sized as the first step in excision repair of CPDs in *E. coli*, before the detailed enzymology of NER was elucidated.[25,26]

3.4.2.3   Repair of Clustered Damage on both DNA Strands

Excision repair is always strand-specific, in the sense that it can only operate in duplex DNA, where there is a complementary strand to serve as a template for repair replication. The most serious type of DNA damage is arguably that in which both DNA strands have been damaged at sites within a few nucleotides of each other, so that there is no intact complementary strand for excision repair. This is always a problem for the repair of interstrand crosslinks in DNA, for which at least two repair modes (usually NER and homologous recombination) are required to operate in coordination. The extreme case is the double-strand break, by which the continuity of the duplex DNA is disrupted (a single-strand break on each strand within 5 or 6 nucleotides of each other will easily evolve to a double-strand break). Ionizing radiation is an important source of double-strand breaks because of the frequency of clustered strand breaks, but these are also generated naturally in the course of genetic recombination. A particularly complicated sort of double-strand break is formed when a replication fork encounters a single-strand break, releasing one arm of the fork without generating an abutting DNA duplex for rejoining.

Homologous recombination is the principal mechanism for accurate rejoining of double-strand breaks but this requires an available homologous DNA duplex to provide a genetic 'splint' across the ends to be rejoined. If such homology is not available, then the alternative pathway of non-homologous end joining must be employed, but this inevitably results in loss of information, since the abutting ends must be recessed by exonucleases in a search for 'microhomology' so that overlapping single strand extensions can be paired before rejoining can take place. A double-strand break is a highly lethal event and as few as one double-strand break in the entire genome is thought to be sufficient to signal cell cycle checkpoints that delay attempted DNA synthesis or cell division until repair has been completed. It is not clear how one double-strand break within the complex chromatin structure of the entire human genome might not evade detection by surveillance systems.

3.4.2.4 Tolerance of Persisting Damage by Translesion DNA Synthesis

Although this is not a repair pathway *per se,* the bypass of lesions in the template DNA during replication is an essential mode for responding to genomic alterations and maintaining cellular viability (*cf.* ref. 27). It is now known that human cells contain 15 or more DNA polymerases, and many of these appear to have dedicated roles for carrying out DNA synthesis over particular lesions that otherwise pose hurdles to the replicase machinery.[28] This mode usually operates at the cost of high mutagenicity since the active sites of the relevant DNA polymerases must be amorphous enough to accommodate adducted bases and/or dimerized bases as well as the normal ones.

Following translesion synthesis there is an opportunity for repair of the bypassed lesions before the next round of genomic DNA replication. This of course is a trade-off, as it permits cells to survive at the expense of low fidelity DNA synthesis, which could contribute to genomic instability and cancer. In a mouse model system it has been shown that spontaneous tumors arise upon loss of DNA polymerase ζ, one of the essential translesion DNA polymerases.[29] The clinical importance of translesion DNA synthesis is highlighted by the fact that one class of XP patients (the so-called XP variant) is deficient in DNA polymerase η, while retaining normal levels of NER. Pol η appears to have evolved expressly for the task of bypassing CPDs, since it replicates DNA containing these lesions with higher fidelity than it achieves on undamaged DNA. Similar clinical features in XP result from either the NER deficiency or loss of Pol η, leading one to conclude that ineffective bypass of CPDs is just as serious as the inability to remove them from genomic DNA.

3.4.3   Overlap between DNA Damage Processing Pathways

The various DNA repair pathways may sometimes compete with each other for the same lesion. Also, as noted above, each step in a multistep DNA repair pathway creates an intermediate that might be susceptible to intervention by enzymes from another pathway. Overall damage processing, and the allocation of cellular resources can then be viewed in terms of successive stages of repair, with a decision point at each stage until DNA integrity is finally re-established. Sometimes the first protein to bind might be a transcription factor or some other protein that is not part of any repair process. The outcome for the cell, and the organism of which it is a part, may depend upon which protein encounters the lesion first.[30]

There have been a number of excellent reviews on crosstalk between different DNA repair mechanisms, including the interplay of glycosylases and mismatch repair proteins among multiple pathways.[31] An enzyme from one pathway may compete with or stimulate the activity of an enzyme from another pathway. It is beyond the scope of this chapter to detail these many interactions. However, a few examples of crossover between DNA repair pathways are given below.

The repair of abasic sites, which arise by spontaneous depurination or as an intermediate in BER, is accomplished either through backbone cleavage by an AP endonuclease, an AP lyase activity inherent in some glycosylases or through recognition by the NER pathway, and this probably includes TCR, since an abasic site is a strong block to transcription.[32] In a genetic study in yeast it was established that the abasic site intermediate in repair of DNA containing uracil was removed by NER-dependent TCR, when the uracil was in the transcribed DNA strand of a highly expressed gene, while uracil in the non-transcribed strand was repaired by BER.[33] These results imply that the abasic site intermediate may persist for an indeterminate period and that it is not necessarily tightly coupled to the subsequent steps in BER, so that it becomes possible for it to be encountered by RNAP and diverted to the TCR sub-pathway of NER.

The mismatch repair gene product, MutS, binds to O^6-methyl guanine in competition with the alkyltransferase mentioned earlier. However, the action of MutS on O^6-methyl guanine paired with cytosine results in a repeated futile cycle of mismatch repair that is cytotoxic.[34] There is also a class of alkyltransferase-like proteins (ATLs) that can bind O^6-methyl guanine but which lack the cysteine to which the methyl must be transferred. However, the ATLs can then serve as antennae to attract the GGR sub-pathway of NER for removing the lesion. Thus, the ATLs serve as 'molecular switches' between several DNA repair pathways.[35] Similarly the *E. coli* photolyase, when bound to a CPD in the dark, can serve as an antenna to attract the NER enzymes to repair that lesion. Interestingly, this mode is species specific; the bound *E. coli* photolyase will not enhance CPD repair in yeast, while the yeast photolyase will only serve as an antenna for NER of a CPD in yeast, and not in *E. coli*.[36]

Direct physical interaction between components of MMR and NER has been documented using a yeast two-hybrid screen with MSH2 as bait. Mutations in

MSH2 also increased UV sensitivity of NER-deficient yeast strains and the msh2 mutations were epistatic with a mutator phenotype in these strains. The MMR proteins functionally recognize some damage usually repaired by NER and *vice versa*; some base–base mispairings are recognized by NER. It was suggested that MMR and NER proteins may co-exist in a complex to facilitate their operation in some common functions.[37]

Overlap between alternative pathways is also important in relation to the options for dealing with arrested replication forks at lesions, in which the choice of a particular solution to the problem, such as recombination, might preclude the selection of a more appropriate one, such as translesion synthesis.[38] Little is known about how the respective options for processing blocked replication forks are coordinated overall. However, it is intuitive that fork regression followed by excision repair of the revealed lesion would be the option most likely to maintain genomic stability. The value of effectively utilizing excision repair is further maximized if the overall rate of replication can be reduced or if replication can be curtailed until the genomic stress has been alleviated, so that replication forks are less likely to encounter the lesions in the first place. Otherwise, transient replacement of the replicase with another DNA polymerase, adapted for lesion bypass, would seem to be the best option. It remains an exercise for researchers to learn how the cellular priorities are actually set, including the choice, from a large corral of translesion polymerases, of the one most likely to achieve lesion bypass with minimal error.

3.4.4 Regulation of Damage Recognition for Nucleotide Excision Repair

The NER pathway has historically and traditionally been thought to deal primarily with bulky adduct DNA damage. However, the spectrum of lesion recognition for NER spans the immense range from the major structural distortion of the 6-4PP to the negligible distortion caused by a methylphosphonate or a phosphorothiolate in the DNA backbone. The abasic site, as mentioned earlier, is also an easily recognized lesion for NER, and this is not surprising since the recognition mechanism includes interrogation of the ease with which the DNA strands can be separated. The primary lesion recognition enzyme, XPC, actually binds the undamaged DNA strand rather than the one containing the lesion.[39–41] This helps to explain how NER can be so versatile in recognizing such a large variety of structurally unrelated DNA lesions.

Upon arbitrarily assigning 100% to the highly-distorting 6-4PP, the comparative sensitivities to NER-based DNA incisions in human cell extracts have been determined for various model substrates. The CPD was 18% on that scale, while methylphosphonate was 5% and the phosphorothiolate was 3%; DNA judged to be undamaged, by a variety of criteria, was incised at 1%.[42,43] Therefore, the 'gratuitous' incision activities in undamaged genomic DNA could be quite significant, especially when one considers the vast excess of undamaged DNA. It should not be surprising to learn that the cell normally maintains the levels of lesion-recognition enzymes for NER at low concentrations, which can then be increased as warranted by environmental conditions or genomic stress.

Cell cycle analyses have revealed no variations in the efficiency of TCR for the dihydrofolate reductase (*DHFR*) gene that is uniformly expressed throughout the cycle in Chinese hamster ovary (CHO) cells and there are also no major changes over the cycle for GGR.[44] Following the initial lesion recognition steps, which differ for TCR and GGR, there is convergence to a common set of steps required to complete the NER pathway. XPA is a critical protein at this stage of lesion recognition and verification, and it is known to be rate limiting for NER. It has been shown that NER activity in mouse skin, liver and brain undergoes daily oscillations as a consequence of the transcriptional regulation of the *xpa* gene by the circadian clock. Thus, there is temporal compartmentalization of the DNA repair response that is coupled to the 24-hour daily cycle, with high expression levels of *xpa* in the early evening and low levels early in the morning. The carcinogenic potential of UV-exposed mice varied in the corresponding predictable manner. It was also shown that loss of the core circadian clock proteins Cry1 and Cry2 sensitized tumor cells to apoptosis induced by DNA damage, an effect that was exacerbated if the cells were p53 deficient.[45]

GGR, but not TCR, is also modulated through up-regulation of lesion recognition enzymes under genomic stress. It is well known that the activation of p53 in response to DNA damage can lead to arrest of the cell cycle or apoptosis, and more complex details of the signaling pathways are still being elucidated.[46] The cell cycle checkpoint is thought to provide a window-of-opportunity for repair of the damage before the cell attempts to divide or initiate a new round of DNA replication. p53 activation is also required for efficient GGR of UV-induced CPDs in human fibroblasts.[47] This is analogous to the classic SOS damage-inducible stress response in *E. coli* that is critical for efficient GGR (but not TCR) for certain types of lesions such as CPDs.[48,49] Skin fibroblasts derived from tumors in Li–Fraumeni syndrome patients, homozygous for mutations in p53, are remarkably defective in GGR of CPDs compared with that in heterozygous mutants and normal cells.[50] These observations were confirmed in human fibroblasts in which the expression level of p53 could be controlled to establish that the effect is specifically due to the p53 gene product.[51] The further finding that repair of CPDs is much more dependent upon p53 than is repair of 6-4PPs suggested that there could be heterogeneity in the requirement of p53 for GGR of different types of DNA lesions.

We investigated repair of adducts formed by benzo(a)pyrene7,8-diol-9,10-epoxide (BPDE), a reactive metabolite of the potent carcinogen benzo(a)pyrene that binds predominantly to the exocyclic amino position on guanine. Using human fibroblasts in which p53 expression was controlled experimentally, the ultrasensitive technique of ^{32}P-post-labeling was employed to measure BPDE–DNA adducts after low level exposure to BPDE. The adducts (roughly 50 adducts per 10^8 nucleotides, as found in smokers' lung cells) persisted for at least three days in cells deficient in p53 expression, while in cells expressing p53 they were largely repaired.[52] These findings were confirmed by strand-specific repair analyses, which additionally showed that the efficiency of TCR is not reduced in p53-deficient cells exposed to BPDE.[53]

What is the mechanism of the p53 effect on GGR efficiency? *In vitro* studies had revealed no direct effect of p53 upon NER. The p53 effect on GGR of CPDs is mediated in large part through p48, a protein involved in DNA damage recognition that is missing in XPE cell lines, as well as in CHO cells. The p48 protein (also known as DDB2) is a component of the UV-damaged DNA binding complex, UV-DDB, and expression of the p48 gene is up-regulated in UV-irradiated human cells in a p53-dependent manner.[54] Transfection of the human p48 gene into CHO cells enhances the removal of CPDs from the genomic DNA by GGR. Thus, p48 is a link between p53 and efficient GGR in mammalian cells.[55] However, p53 is also involved in trans-activation of other genes required for the early steps of NER. Thus, the XPC gene is UV inducible in human WI38 fibroblasts and in HCT116 colorectal cancer cells with normal p53 expression. In contrast, no significant up-regulation of XPC was detected in p53 deficient derivatives of those cells.[56]

Some tumor viruses can result in abrogation of p53 and reduced GGR efficiency. When p53 deficiency was conferred in human primary fibroblasts by enhancing p53 degradation through expression of the papillomavirus E6 gene, a major reduction in GGR of CPDs was observed.[57] In SV40 transformed human fibroblasts, in which the large T-antigen interferes with p53 function, there was also a very significant reduction in GGR of CPDs. That deficiency was documented by three different assays and in several different SV40 transformed cell lines, and was in striking contrast to the proficient GGR in non-transformed parental cells.[58] A third example is the hepatitis B virus X gene (HBx). The HBx protein physically interacts with and sequesters p53 in the cytoplasm. Unlike cells from most rodent tissues, hepatocytes in mice exhibit proficient GGR of CPDs and so it is likely that p53 transactivation of p48 and inducible UV-DDB activity is normal in these cells. Yet, repair of CPDs was strikingly diminished in hepatocytes in which HBx was expressed.[59] The enhanced genomic instability and carcinogenesis in each of these examples is a likely consequence of compromised GGR in the virus-infected cells.

3.4.5 Hereditary Diseases in which DNA Damage Processing is Defective

The susceptibility to cancer involves a complex interplay between intrinsic hereditary factors, unique sequences of nucleotides and persisting DNA damage. The discovery of microsatellite instability in some hereditary colorectal cancers provided the first indication that their etiology might involve a deficiency in correcting errors introduced during DNA replication.[60] Since repetitive sequences have a tendency to form strand-slipped structures, with mispairings during replication, small deletions can be generated that give rise to frameshift mutations. These errors are normally corrected by MMR, so a defect in this pathway might be expected to result in microsatellite instability. The discovery of MMR gene defects in patients with HNPCC established that such defects could be the cause of the enhanced incidence of cancer. The correspondence between a defect in mismatch repair and susceptibility to cancer provides validation of the original hypothesis of Lawrence Loeb, that tumorigenesis is promoted by a 'mutator' phenotype.[61]

3.4.5.1   Xeroderma pigmentosum (XP)

As discussed earlier, XP was the first example of a DNA repair deficient disease in humans. Most of the seven classic XP complementation groups exhibit defects in the steps that are common to GGR and TCR.[62] However, several of the gene products, XPC and XPE (DDB2), operate only as recognition factors for GGR while those of three genes, CSA, CSB and UVSSA, are unique to TCR and not required for initiating GGR, as described below.

3.4.5.2   Cockayne syndrome (CS) and UV-sensitive syndrome (UV^{S}S)

Cockayne syndrome (CS) is another rare autosomal recessive genetic disease in which the victims are severely sensitive to sunlight.[63] However, in striking contrast to XP, CS patients do not develop skin cancers. Instead, these patients present serious developmental problems and neurological deficiencies. Most of the known CS patients are defective in one of two genes, CSA or CSB, but there are several examples of overlap between CS and XP. Thus, all three known XPB patients have CS/XP, two out of the 52 known XPD patients have CS/XP, and the six most severely afflicted XPG patients also have combined CS and XP.

   The DNA repair defect that is common to all CS patients is a deficiency in TCR. This can account for sunlight sensitivity (*i.e.* severe sunburn) because of epidermal cell death by apoptosis, due to transcription arrest at unrepaired DNA photoproducts. The lack of cancer susceptibility may derive from the fact that 'dead cells don't form tumors' and because the surviving cells exhibit normal global NER, so that most of the potential cancer-initiating lesions can be removed. One model for the severe developmental problems in CS is that endogenous oxidative damage (or the abasic site intermediate in BER) in metabolically active cells (*e.g.* neurons) is blocking transcription. The lack of TCR could then cause high levels of apoptosis in these essential cells. An alternative model derives from the realization that both the XPB and XPD gene products are components of the essential transcription initiation factor, TFIIH, which serves a dual role in NER and in transcription. It has been suggested that CS could be a 'transcription disease' in which certain essential genes cannot be transcribed at adequate frequencies.

   The only clinical presentation in UV^{S}S is a severe sensitivity to sunlight, leading to freckles and sunburn.[64] As with CS, no cancers have been reported in

UV^{S}S patients, although it must be noted that there have been very few victims identified. There are undoubtedly many more UV^{S}S individuals in the general population, but it is likely that dermatologists encountering such mild characteristics simply advise their patients to stay out of direct sunlight, but they do not register the problem as a hereditary disease. There are three complementation groups in UV^{S}S: UVSSA, certain mutations in CSB, and at least one known mutation in CSA. With respect to repair of UV-induced damage, UV^{S}S and CS are indistinguishable.[65] Therefore, TCR of bulky lesions, such as CPDs, cannot be responsible for the developmental and neurological problems in CS. Studies have shown, however, that CS cells are much more sensitive to oxidative damage than are UV^{S}S cells, which behave essentially like normal wild-type cells in this regard.[66]

3.4.5.3 Li–Fraumeni syndrome (LFS)

The rare autosomal dominant Li–Fraumeni syndrome (LFS) results in increased risk of developing a number of common tumors at an early age due to an inherited germ line defect in one allele of the p53 gene. Although the heterozygote 'carriers' of a defective p53 allele have no significant NER deficiencies, when the second allele has been mutated or lost, the absence of functional p53 results in severe problems, as noted earlier. The lack of the p53 controlled pathway of apoptosis permits severely damaged cells to survive. The loss of p53 dependent up-regulation of NER and the absence of p53 participation in cell cycle checkpoints also contributes to the genomic instability leading to carcinogenesis.

3.4.5.4 Ataxia Telangiectasia (AT)

Ataxia telangiectasia was originally identified in patients who exhibited a severe sensitivity to ionizing radiation while undergoing radiation therapy for tumors. AT also affects the immune system and the Purkinje cells in the cerebellum, causing progressive loss of motor control and leading to lack of coordination and balance. Respiratory infections develop in AT inflicted children as a consequence of the immunological deficiency. The single gene, ATM, is responsible for the surprisingly diverse symptoms of this disease that also include predisposition to lymphoma and leukemia. Over 10% of AT patients develop cancer at an early age and the heterozygotes, comprising 1% of the general population, also appear to have some predisposition to cancer.

A curious hallmark of AT is X-ray resistant DNA synthesis, but AT cells are not significantly sensitive to UV. ATM is a key element in controlling the cell cycle and specifically in delaying the initiation of DNA replication following DNA damage that involves strand breaks. The ATM protein exerts its regulatory role as a kinase that phosphorylates a number of other important proteins, such as the tumor suppressor p53, and proteins implicated in double-strand break repair. It also serves a regulatory role in controlling the number of developmentally programmed double-strand breaks permitted to initiate meiotic recombination, which promotes pairing and segregation of homologous chromosomes.[67]

3.4.5.5 Fanconi Anemia (FA)

Fanconi Anemia (FA) is the most frequent of rare diseases involving bone marrow failure, with several thousand cases known worldwide. It is characterized by severe life-threatening aplastic anemia and developmental abnormalities. The cause was originally characterized as a deficiency in the repair of interstrand DNA crosslinks, but it is now thought to reflect a more general defect in the processing of arrested replication forks and coordination of the repair of double-strand breaks, which can also be generated as intermediates in the repair of interstrand crosslinks.[68] At least 13 genes are implicated and these appear to operate in a common signaling pathway. The breast cancer susceptibility genes BRCA1 and BRCA2, shown to have roles in the control of RAD51 in homologous recombination and double-strand break repair, are among the genes that participate in the FA repair pathway. The 'core complex' of eight proteins constituting an E3 ligase monoubiquinates FANCD2 to activate the pathway. The monoubiquinated FANCD2 binds BRCA1 and the RAD51 recombinase, and associates with FANCD1, which has turned out to be BRCA2. Remarkable is the fact that mutations compromising function of any one of the proteins involved in the large multisubunit complex causes the same phenotype of FA. The FA model provides a preview of the complexities that lie ahead for a thorough understanding of the roles of ubiquitination and other post-translational protein modifications in the control of genomic stability.

3.4.5.6 Other Diseases Involving Deficient Processing of DNA

At least three cancer-prone hereditary diseases result from defects in homologues of the *recQ* gene, originally discovered in *Escherichia coli*.[69] The product of *recQ* is a helicase, which in *E. coli* is involved in processing nascent DNA at arrested replication forks.[70,71] The RecQ helicases in eukaryotes have been implicated in the processing of unusual DNA structures, such as the G-4 quadruplexes in telomeres.[72] There are five homologues of recQ in human cells and several have been localized to telomeres as well as to sites of replication.[73]

In Bloom's syndrome, the BLM helicase deficiency is characterized by an extremely high frequency of sister chromatid exchanges that leads to genomic instability and cancer. BLM also interacts with FA proteins during chromosomal segregation in meiosis.[74] In Werner's syndrome, the deficiency in WRN helicase results in remarkable features of premature aging as well as cancer predisposition.[75–77] In yet another recQ homologue defect, Rothmund–Thompson syndrome, the RTS helicase defect causes growth deficiencies and cancer predisposition.[78] Cells from RTS patients have elevated levels of fragile chromosomal ends and human cells depleted for RTS accumulate sister chromosome exchanges and double-strand breaks in telomeres. RTS resolves telomeric D-loop structures in association with the shelterin proteins TRF1, TRF2 and POT1. There also appears to be a functional interaction of RTS with WRN during D-loop unwinding, thereby implicating RTS helicase in telomere maintenance.[79] It is a continuing challenge to sort out the unique roles of these three recQ homologues that have functions in telomeres, but in which their deficiencies result in rather distinct clinical phenotypes.

Now that we have access to the completely sequenced human genome it is likely, indeed certain, that additional genes will be implicated in the diseases listed above as well as in new genetic diseases. In addition, there are increasing examples of overlap between the relevant damage response pathways and gene functions. Thus, as noted earlier, the genes BRCA1 and BRCA2 that predispose to breast and ovarian cancer when mutated, are involved in the repair of double-strand breaks. The Nijmegen Breakage Syndrome is also related to double-strand break repair and the implicated gene, NBS1, is one of the phosphorylation targets of ATM. There is yet a third disease called 'AT-like disorder' in which the responsible gene, MRE11, is also required for double-strand break repair.[80] There is a recently reported unique genetic disorder, called Warsaw breakage syndrome, in which defects in sister chromatid cohesion are a consequence of a deficiency in the ChlR1 helicase.[81]

The intricate web of overlapping DNA damage surveillance and repair pathways poses an exciting challenge for researchers and clinicians. It is now important to learn all the gene products involved in the different pathways, in addition to their relevant interactions with each other, so that susceptibility to the respective genetic diseases may be more fully assessed. The knowledge to be gained will be essential for the design of targeted therapeutic strategies to improve the prognosis for victims of these hereditary diseases.

3.4.6 Conclusions Relevant to Threshold Effects for Genotoxic Carcinogens

Issues to consider in relation to a possible threshold for carcinogenicity due to a particular genotoxic agent include the following:

(1) The large and diverse set of chemical carcinogens and several forms of radiation to which humans are exposed lead to a heterogeneous mix of lesions in genomic DNA, for which repair is effected by a number of pathways with overlapping specificities. Thus, the determination of the responsible lesions for carcinogenesis is complex, although general characteristics can be surmised from genomic sequencing of cells from the tumor.

(2) Some environmental agents produce the very same lesions that originate from endogenous ROS and simply add to the steady state level of those lesions. There could not be a threshold for deleterious effects of those lesions that could be set any lower than the existing background.

The estimated steady state level of endogenous DNA lesions per cell is ~50 000, of which most (~30 000) are abasic sites.[82]

(3) Cellular responses that are induced by genomic stress can reset the pattern and efficiency of DNA repair for particular lesions, as well as activating cell cycle check points and triggering apoptosis. This is elaborated below.

(4) Humans contain a variety of specialized cells in different growth states and in different organs. Some cells are rapidly proliferating while others are terminally differentiated and/or senescent. Thus, the sensitivities to particular genotoxins will vary widely among different tissues.

(5) Some damage will occur in sequences of nucleotides that can form non-canonical DNA structures, within which the lesions may be refractory to repair; other damage may be hidden in tightly-packed transcriptionally-silent chromatin and totally inaccessible to repair.

Perhaps most important, however, are the relationships between the genomic damage inflicted from environmental sources to the steady state level of endogenous damage. Figure 3.4.1 illustrates a greatly oversimplified view of the 'DNA damage response'. The steady state levels of endogenous lesions in all cells will undoubtedly account for some 'spontaneous mutagenesis' as replication forks pass through unrepaired lesions. Those mutations can contribute to the progression toward cancer and we can confidently state that the steady-state level is below any threshold for the genomic stress response because it otherwise would be triggering cell cycle checkpoints, apoptosis and up-regulated expression of repair enzymes. However, there is evidence that, under non-stressed conditions, there are periodic pulses of p53 activation thought to be triggered by random double-strand breaks produced through the cell cycle.[83] Upon environmental exposures or through therapeutic regimens, cells receive additional

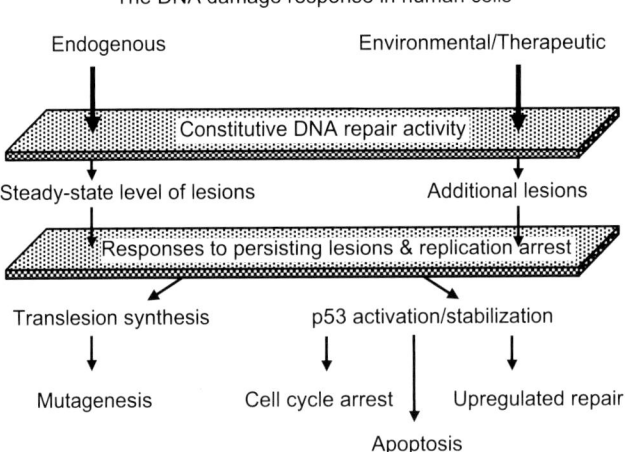

Figure 3.4.1 DNA damage response in human cells.

DNA damage, as well as lesions of several different types. If constitutive repair responses are unable to rapidly accommodate the new set of lesions, p53 will become further acetylated and the complex regulatory system will be activated. It is not clear how the options of cell cycle arrest and apoptosis are allocated within the cell population, but the transcriptional activation of DNA damage recognition factors should result in more efficient repair of some of the offending lesions. Once triggered by any type of lesion, the process may then be much more efficient in dealing with a variety of other lesions as well. In that case one might anticipate a dip in the carcinogenicity of a genotoxic exposure immediately following activation of the stress response. A meaningful threshold could be the level at which a particular genotoxin causes significant levels of apoptosis, but even that is complicated by the nature of the lesions and whether the problem is one of overwhelming DNA repair capacities, or of blocking replication forks and transcription complexes, as detailed in Chapter 4.3.

It will be nearly impossible to provide an unambiguous answer to the question of whether there is a threshold for carcinogenicity of genotoxins.

Finally, a serious issue in genetic toxicology is the important distinction between occupational exposures to genotoxins (*e.g.* the worker who is spraying a pesticide) and the levels of those same genotoxins to which individuals among the public may be exposed (*e.g.* by eating fruit from the treated orchard). It is still the 'dose that makes the poison', whether the risks are subject to thresholds or simply linear with increasing exposures.

Acknowledgements

Research in the author's laboratory was supported from 1987 to 2001 by an Outstanding Investigator Grant from the National Cancer Institute, 5R35 CA44349. It is currently supported by several NIH grants; 2RO1-CA77712 and 1RO1-ES018834. In addition to the seminal contributions from many outstanding researchers internationally, this field has benefited over the past 50 years from the ideas and dedicated efforts of talented students and postdoctoral associates from the author's laboratory. Most of the historical documentation for this chapter is covered in ref. 3; sources for statements not otherwise referenced will be provided upon request.

References

1. T. Lindahl, Instability and decay of the primary structure of DNA, *Nature*, 1993, **362**, 709–715.
2. K. M. Vasquez and P. C. Hanawalt, Intrinsic genomic instability from naturally occurring DNA structures: An introduction to the special issue, *Mol. Carcinog.*, 2009, **48**, 271–272.
3. E. C. Friedberg, G. C. Walker, W. Siede, R. D. Wood, R. A. Schultz and T. Ellenberger, *DNA Repair and Mutagenesis*, ASM Press, Washington DC, 2nd edn, 2006.

4. K. W. Caldecott, Single-strand break repair and genetic disease, *Nat. Rev. Genet.*, 2008, **9**, 619–631.
5. R. B. Setlow and W. L. Carrier, The Disappearance of thymine dimers from DNA: An error-correcting mechanism, *Proc. Natl. Acad. Sci., U. S. A.*, 1964, **51**, 226–231.
6. R. P. Boyce and P. Howard-Flanders, Release of ultraviolet light-induced thymine dimers from DNA in *E.coli* K-12, *Proc. Natl. Acad. Sci., U. S. A.*, 1964, **51**, 293–231.
7. D. Pettijohn and P. C. Hanawalt, Deoxyribonucleic acid replication in bacteria following ultraviolet irradiation, *Biochim. Biophys. Acta.*, 1963, **72**, 127–129.
8. D. Pettijohn and P. C. Hanawalt, Evidence for repair-replication of ultraviolet damaged DNA in bacteria, *J. Mol. Biol.*, 1964, **9**, 395–410.
9. J. E. Cleaver, Defective repair replication of DNA in Xeroderma pigmentosum, *Nature*, 1968, **218**, 652–656.
10. P. C. Hanawalt and G. Spivak, Transcription-coupled DNA repair: two decades of progress and surprises, *Nat. Rev. Mol. Cell Biol.*, 2008, **9**, 958–970.
11. S. Tornaletti, Transcription arrest at DNA damage sites, *Mutat. Res.*, 2005, **577**, 131–145.
12. S. Tornaletti, S. Park-Snyder and P. C. Hanawalt, G4-forming sequences in the non-transcribed DNA strand pose blocks to T7 RNA polymerase and mammalian RNA polymerase II, *J. Biol. Chem.*, 2008, **283**, 2756–12762.
13. C. Broxson, J. Beckett and S. Tornaletti, Transcription arrest by a G quadruplex forming-trinucleotide repeat sequence from the human c-myb gene, *Biochemistry*, 2011, **50**, 4162–4172.
14. V. S. Sidorenko, J. E. Yeo, R. R. Bonala, F. Johnson, O. D. Schärer and A. P. Grollman, Lack of recognition by global-genome nucleotide excision repair accounts for the high mutagenicity and persistence of aristolactam-DNA adducts, *Nucleic Acids Res.*, 2011, Nov 25 [Epub ahead of print].
15. A. E. Escargueil, V. Poindessous, D. G. Soares, A. Sarasin, P. R. Cook and A. K. Larsen, Influence of irofulven, a transcription-coupled repair-specific antitumor agent, on RNA polymerase activity, stability and dynamics in living mammalian cells, *J. Cell Sci.*, 2008, **121**, 1275–1283.
16. E. D. Pleasance, R. K. Cheetham, P. J. Stephens, D. J. McBride, S. J. Humphray, C. D. Greenman, I. Varela, M. Lin, G. R. Ordoñez, G. R. Bignell, K. Ye, J. Alipaz, M. J. Bauer, D. Beare, A. Butler, R. J. Carter, L. Chen, A. J. Cox, S. Edkins, P. I. Kokko-Gonzales, N. A. Gormley, R. J. Grocock, C. D. Haudenschild, M. M Hims, T. James, M. Jia, Z. Kingsbury, C. Leroy, J. Marshall, A. Menzies, L. J. Mudie, Z. Ning, T. Royce, O. B. Schultz-Trieglaff, A. Spiridou, L. A. Stebbings, L. Szajkowski, J. Teague, D. Williamson, L. Chin, M. T. Ross, P. J. Campbell, D. R. Bently, P. A. Futreal and M. R. Stratton, A comprehensive catalogue of somatic mutations from a human cancer genome, *Nature*, 2010, **463**, 191–196.

17. E. D. Pleasance, P. J. Stephens, S. O'Meara, D. J. McBride, A. Meynert, D. Jones, M. Lin, D. Beare, K. W. Lau, C. Greenman, I. Varela, S. Nik-Zainal, H. R. Davies, G. R. Ordoñez, L. J. Mudie, C. Latimer, S. Edkins, L. Stebbings, L. Chen, M. Jia, C. Leroy, J. Marshall, A. Menzies, A. Butler, J. W. Teague, J. Mangion, Y. A. Sun, S. F. McLaughlin, H. E. Peckham, E. F. Tsung, G. L. Costa, C. C. Lee, J. D. Minna, A. Gazdar, E. Birney, M. D. Rhodes, K. J. McKernan, M. R. Stratton, P. A. Futreal and P. J. Campbell, A small-cell lung cancer genome with complex signatures of tobacco exposure, *Nature*, 2010, **463**, 184–190.

18. T. Nouspikel and P. C. Hanawalt, DNA repair in terminally differentiated cells, *DNA Repair*, 2002, **1**, 59–75.

19. T. Nouspikel, N. Hyka-Nouspikel and P. C. Hanwalt, Transcription domain-associated repair in human cells, *Cell Mol. Biol.*, 2006, **26**(23), 8722–8730.

20. R. Wagner and M. Meselson, Repair tracts in mismatched DNA heteroduplexes, *Proc. Natl. Acad. Sci., U. S. A.*, 1976, **73**, 4135–4139.

21. T. Lindahl, An N-glycosidase from *Escherichia coli* that releases free uracil from DNA containing deaminated cytosine residues, *Proc. Natl. Acad. Sci., U. S. A.*, 1974, **71**, 3649–3653.

22. J. Duncan, L. Hamilton and E. C. Friedberg, Enzymatic degradation of uracil-containing DNA, *J. Virol.*, 1976, **19**, 338–345.

23. D. Banerjee, S. M. Mandal, A. Das, M. L. Hedge, S. Das, K. K. Bhakat, I. Boldogh, P. S. Sarkar, S. Mitra and T. K. Hazra, Preferential repair of oxidized base damage in the transcribed genes of mammalian cells, *J. Biol. Chem.*, 2011, **286**, 6006–6016.

24. A. K. Ganesan, A method for detecting pyrimidine dimers in the DNA of bacteria irradiated with low doses of ultraviolet light, *Proc. Natl. Acad. Sci., U. S. A.*, 1973, **70**, 2753–2756.

25. P. W. Doetsch, What's old is new: An alternative DNA excision repair pathway, *Trends Biochem. Sci.*, 1995, **20**, 384–388.

26. P. C. Hanawalt, Repair of genetic material in living cells, *Endeavor*, 1972, **31**, 83–87.

27. G. Spivak and P. C. Hanawalt, Translesion DNA synthesis in the DHFR domain of UV-irradiated CHO cells, *Biochemistry*, 1992, **31**, 6794–6800.

28. S. S. Lange, K. Takata and R. D. Wood, DNA polymerases and cancer, *Nat. Rev. Cancer*, 2011, **11**, 96–110.

29. J. P. Wittschieben, V. Patil, V. Glushets, L. J. Robinson, D. F. Kusewitt and R. D. Wood, Loss of DNA polymerase zeta enhances spontaneous tumorigenesis, *Cancer Res.*, 2010, **70**, 2770–2778.

30. S. D. Cline and P. C. Hanawalt, Who's on first in the cellular response to DNA damage? *Nat. Rev. Mol. Cell Biol.*, 2003, **4**, 361–372.

31. I. V. Kuvtun and C. T. McMurray, Crosstalk of DNA glycosylases with pathways other than base excision repair, *DNA Repair*, 2007, **6**, 517–529.

32. S. Tornaletti, L. S. Maeda and P. C. Hanwalt, Transcription arrest at an abasic site in the transcribed strand of template DNA, *Chem. Res. Toxicol.*, 2006, **19**, 1215–1220.

33. N. Kim and S. Jinks-Robertson, Abasic sites in the transcribed strand of yeast DNA are removed by transcription-coupled nucleotide excision repair, *Mol. Cell Biol.*, 2010, **30**, 3206–3215.

34. L. J. Rasmussen and L. Samson, The *Escherichia coli* mutS DNA mismatch binding protein specifically binds O^6-methylguanine DNA lesions, *Carcinogenesis*, 1996, **17**, 2085–2088.

35. J. L. Tubbs and J. A. Tainer, Alkyltransferase-like proteins: molecular switches between DNA repair pathways, *Cell. Mol. Life Sci.*, 2010, **67**, 3749–3762.

36. G. B Sancar and F. W. Smith, Interactions between yeast photolyase and nucleotide excision repair proteins in *Saccharomyces cerevisiae* and *Escherichia coli*, *Mol. Cell Biol.*, 1989, **9**, 4767–4776.

37. P. Bertrand, D. X. Tishkoff, N. Filosi, R. Dasgupta and R. D. Kolodner, Physical interaction between components of DNA mismatch repair and nucleotide excision repair, *Proc. Natl. Acad. Sci., U. S. A.*, 1998, **95**, 14278–1428.

38. P. C. Hanawalt, Paradigms for the three Rs: DNA replication, recombination, and repair, *Mol. Cell*, 2007, **28**, 702–707.

39. O. D. Scharer, Achieving broad substrate specificity in damage recognition by binding accessible nondamaged DNA, *Mol. Cell*, 2007, **28**, 184–186.

40. H. Naegeli and K. Sugasawa, The xeroderma pigmentosum pathway: Decision tree analysis of DNA quality, *DNA Repair*, 2011, **10**, 673–683.

41. Y. Liu, D. Reeves, K. Kropachev, Y. Cai, S. Ding, M. Kolbanovskiy, A. Kolbanovskiy, J. L. Bolton, S. Broyde, B. Van Houten and N. E. Geacintov, Probing for DNA damage with β-hairpins: Similarities in incision efficiencies of bulky DNA adducts by prokaryotic and human nucleotide excision repair systems *in vitro*, *DNA Repair*, 2011, **10**, 684–696.

42. M. E. Branum, J. T. Reardon and A. Sancar, DNA repair excision nuclease attacks undamaged DNA, *J. Biol. Chem.*, 2001, **276**, 25421–25426.

43. J. Huang, D. S. Hsu, A. Kazantsev and A. Sancar, Substrate spectrum of human excinuclease: Repair of abasic sites, methylated bases, mismatches, and bulky adducts, *Proc. Natl. Acad. Sci., U. S. A.*, **91**, 12213–12217.

44. L. Lommel, C. Carswell-Crumpton and P. C. Hanawalt, Preferential repair of the transcribed DNA strand in the dihydrofolate reductase gene throughout the cell cycle in UV-irradiated human cells, *Mutat. Res. DNA Repair*, 1995, **336**, 181–192.

45. S. Gaddameedhi, C. P. Selby, W. K. Kaufmann, R. C. Smart and A. Sancar, Control of skin cancer by the circadian rhythm, *Proc. Natl. Acad. Sci., U. S. A.*, 2011, **108**, 18790–18795.

46. J. W. Harper and S. J. Elledge, The DNA damage response: Ten years after, *Mol. Cell*, 2007, **28**, 739–745.

47. J. M. Ford, Regulation of DNA damage recognition and nucleotide excision repair: another role for p53, *Mutat. Res.*, 2005, **577**, 195–202.

48. J. Courcelle, A. Khodursky, B. Peter, P. O. Brown and P. C. Hanawalt, Comparative gene expression profiles following UV exposure in wild type and SOS deficient *Escherichia coli, Genetics*, 2001, **158**, 41–64.

49. D. J. Crowley and P. C. Hanawalt, Induction of the SOS response increases the efficiency of global nucleotide excision repair of cyclobutane pyrimidine dimers but not 6-4 photoproducts in UV-irradiated *Escherichia coli, J. Bacteriol.*, 1998, **180**, 3345–3352.

50. J. M. Ford, P. C. Hanawalt and Li-Fraumeni, Syndrome fibroblasts homozygous for p53 mutations are deficient in global DNA repair but exhibit normal transcription-coupled repair and enhanced UV-resistance, *Proc. Natl. Acad. Sci., U. S. A.*, 1995, **92**, 8876–8880.

51. J. M. Ford and P. C. Hanawalt, Expression of wild type p53 is required for efficient global nucleotide excision repair in UV-irradiated human fibroblasts, *J. Biol. Chem.*, 1997, **272**, 28073–28080.

52. D. Lloyd and P. C. Hanawalt, p53-dependent global genomic repair of benzo[alpha]pyrene-7,8-diol-9,10-epoxide adducts in human cells, *Cancer Res.*, 2000, **60**, 517–521.

53. M. A. Wani, Q. Zhu, M. El-Mahdy, S. Venkatachalam and A. A. Wani, Enhanced sensitivity to anti-benzo(a)pyrene-diol-epoxide DNA damage correlates with decreased global genomic repair attributable to abrogated p53 function in human cells, *Cancer Res.*, 2000, **60**, 2273–2280.

54. B. J. Hwang, J. M. Ford, P. C. Hanawalt and G. Chu, Expression of the p48 Xeroderma pigmentosum gene is p53-dependent and involved in global genomic repair, *Proc. Natl. Acad. Sci., U. S. A.*, 1999, **96**, 424–428.

55. J. Tang, B. J. Huang, J. Ford, P. C. Hanawalt and G. Chu, Xeroderma pigmentosum p48 gene enhances global genomic repair and suppresses UV-induced mutagenesis, *Mol. Cell*, 2000, **5**, 737–744.

56. S. Adimoolam and J. M. Ford, p53 and DNA damage-inducible expression of the xeroderma pigmentosum group C gene, *Proc. Natl. Acad. Sci., U. S. A.*, 2002, **99**, 12985–12990.

57. J. M. Ford, E. L. Baron and P. C. Hanawalt, Human Fibroblasts expressing the human papillomavirus E6 gene are deficient in global genomic nucleotide excision repair and sensitive to UV-irradiation, *Cancer Res.*, 1998, **58**, 599–603.

58. K. Bowman, D. Sicard, J. Ford and P. C. Hanawalt, Reduced global genomic repair of UV light-induced cyclobutane pyrimidine dimers in Simian virus 40 transformed human cells, *Carcinogenesis*, 2000, **29**, 17–24.

59. S. Prost, J. M. Ford, C. Taylor, J. Doig and D. J. Harrison, Hepatitis Bx protein inhibits p53-dependent DNA repair in primary mouse hepatocytes, *J. Biol. Chem.*, 1998, **273**, 33327–33332.

60. S. N. Thibodeau, G. Bren and D. Schaid, Microsatellite instability in cancer of the proximal colon, *Science*, 1993, **260**, 816–819.

61. L. A. Loeb, Human cancers express mutator phenotypes: origin, consequences and targeting, *Nature*, 2011, **11**, 450–457.

62. J. E. Cleaver, E. T. Lam and I. Revet, Disorders of nucleotide excision repair: the genetic and molecular basis of heterogeneity, *Nature*, 2009, **10**, 756–768.

63. M. Nance and S. A. Berry, Cockayne syndrome: review of 140 cases, *Am. J. Med. Genet.*, 1992, **42**, 68–84.

64. G. Spivak, UV-sensitive syndrome, *Mutation Res.*, 2005, **577**, 162–169.

65. G. Spivak, T. Itoh, T. Matsunaga, O. Nikaido, P. Hanawalt and M. Yamaizumi, Ultraviolet-sensitive syndrome cells are defective in transcription-coupled repair of cyclobutane pyrimidine dimers, *DNA Repair*, 2002, **50**, 1–15.

66. G. Spivak and P. C. Hanawalt, Host cell reactivation of plasmids containing oxidative DNA lesions is defective in Cockayne syndrome but normal in UV-sensitive syndrome fibroblasts, *DNA Repair*, 2006, **5**, 13–22.

67. J. Lange, J. Pan, F. Cole, M. P. Thelen, M. Jasin and S. Keeney, ATM controls meiotic double-strand-break formation, *Nature*, 2011, **479**, 237–240.

68. Y. Kee and A. D. D'Andrea, Expanded roles of the Fanconi anemia pathway in preserving genomic stability, *Genes Dev.*, 2010, **24**, 1680–1694.

69. H. Nakayama, K. Nakayama, R. Nakayama, N. Irino, Y. Nakayama and P. C. Hanawalt, Isolation and genetic characterization of a thymineless death-resistant mutant of *Escherichia coli* K-12: Identification of a new mutation (*recQ1*) that blocks the RecF recombination pathway, *Mol. Gen. Genet.*, 1984, **195**, 474–480.

70. J. Courcelle and P. C. Hanawalt, RecQ and RecJ process blocked replication forks prior to the resumption of replication in UV-irradiated, *Escherichia coli, Mol. Gen. Genet.*, 1999, **262**, 543–551.

71. J. Courcelle and P.C. Hanawalt, RecA-dependent recovery of arrested DNA replication forks, *Ann. Rev. Genet.*, 2003, **37**, 611–646.

72. K. Paeschke, K. R. McDonald and V. A. Zakian, Telomeres: structures in need of unwinding, *FEBS Lett.*, 2010, **584**, 3760–3772.

73. R. J. Monnat Jr., Human RECQ helicases: roles in DNA metabolism, mutagenesis and cancer biology, *Semin. Cancer Biol.*, 2010, **20**, 329–339.

74. S. Ying and I. D. Hickson, Fanconi anaemia proteins are associated with sister chromatid bridging in mitosis, *Int. J. Hematol.*, 2011, **93**, 440–445.

75. S. A. Compton, G. Tolun, A. S. Kamath-Loeb, L. A. Loeb and J. D. Griffith, The Werner syndrome protein binds replication fork and holliday junction DNA as an oligomer, *J. Biol. Chem.*, 2008, **283**, 24478–24483.

76. A. S. Kamath-Loeb, L. Lan, S. Nakajima, A. Yasui and L. A. Loeb, Werner syndrome protein interacts functionally with translesion DNA polymerases, *Proc. Natl. Acad. Sci., U. S. A.*, 2007, **104**, 10394–10399.

77. M. L. Rossi, A. K. Ghosh and V. A. Bohr, Roles of Werner syndrome protein in protection of genome integrity, *DNA Repair*, 2010, **9**, 331–344.

78. L. Larizza, G. Roversi and L. Volpi, Rothmund-Thomson syndrome, *Orphanet J. Rare Dis.*, 2010, **5**(2), http://www.ojrd.com/content/5/1/2.

79. A. K. Ghosh, M. L. Rossi, D. K. Singh, C. Dunn, M. Ramamoorthy, D. L. Croteau, Y. Liu and V. A. Bohr, RECQL4, the protein mutated in Rothmund-Thomson syndrome, functions in telomere maintenance, *J. Biol. Chem.*, 2012, **287**, 196–209.
80. R. J. Michelson and T. Weinert, Closing the gaps among a web of DNA repair disorders, *Bioessays*, 2000, **22**, 966–969.
81. Y. Wu, J. A. Sommers, I. Khan, J. P. de Winter and R. M. Brosh, Jr., Biochemical characterization of Warsaw breakage syndrome helicase, *J. Biol. Chem.*, 2011, Nov 18 [Epub ahead of print].
82. J. A. Swenberg, K. Lu, B. C. Moeller, L. Gao, P. B. Upton, J. Nakamura and T. B. Starr, Endogenous *versus* exogenous DNA adducts: Their role in carcinogenesis, epidemiology, and risk assessment, *Toxicol. Sci.*, 2011, **120**(S1), S130–S145.
83. T. Sun, W. Yang, J. Liu and P. Shen, Modeling the basal dynamics of P53 system, *PLoS One*, 2011, **6**, e27882 (1–9).

4. Apoptosis

CHAPTER 4.1

Survival and Death Strategies in Cells Exposed to Genotoxins

BERND KAINA,* MARKUS CHRISTMANN,
MARCUS EICH AND WYNAND P. ROOS

Institute of Toxicology, University Medical Center, Johannes Gutenberg
University, Obere Zahlbacher Strasse 67, D-55131 Mainz, Germany
*Email: kaina@uni-mainz.de

4.1.1 Introduction

Chemical carcinogens, ionizing radiation and the 'classic' genotoxic anticancer
drugs attack DNA, which is at the very heart of their genotoxic and toxic
attributes. It is therefore important to shed light upon the processes taking place
on the level of damaged DNA in order to understand the effects of carcinogens
(and thus cancer development), as well as the effects of anticancer drugs (and
thus cancer treatment). Potentially lethal events for the cell are DNA double-
strand breaks (DSB) and any type of damage that prevents the replication of
DNA. Cells have developed sensors that detect these lesions. Sensor systems
recognize the damage and relay the signal *via* kinases to 'executors', which start
a process that either inhibits cell cycle progression and strengthens DNA repair,
or induces apoptosis that destroys the cell. The main 'players' in DNA damage
recognition are ATM, ATR and DNA-dependent protein kinase (DNA-PK),
which phosphorylate a multitude of proteins and thus induce the DNA damage
response (DDR), in which p53 and BRCA1/2 play important roles. Downstream are targets that either help cells to survive or destine them to undergo
cell death. Here, we discuss the main pathways of survival and death.

Issues in Toxicology No. 13
The Cellular Response to the Genotoxic Insult: The Question of Threshold for
Genotoxic Carcinogens
Edited by Helmut Greim and Richard J. Albertini
Published by the Royal Society of Chemistry, www.rsc.org

4.1.2   The DNA Damage Response

DNA double-strand breaks and replication blocking lesions such as DNA polymerase blocking DNA modifications and DNA interstrand crosslinks are lethal DNA lesions and, therefore, cells have to be equipped with sophisticated mechanisms to deal with them. The first step is sensing critical lesions. DNA damage recognition by these sensors starts a cascade that results finally in cell cycle arrest and DNA repair; if repair fails, cells die.

The most important sensors are ataxia-telangiectasia mutated (ATM) and ATM- and Rad3-related (ATR) proteins. While the ATM protein is mainly activated by DSBs, ATR is activated by DNA single-strand breaks (SSB) and gapped DNA, which is generated at stalled DNA replication forks. Consequently, genotoxins that cause the formation of DSBs activate ATM while genotoxins that inhibit DNA synthesis and cause stalled DNA replication forks activate ATR (Figure 4.1.1). ATM and ATR are implicated in three crucial cellular processes: regulation and stimulation of DNA repair, activation of cell cycle checkpoints, and signaling to apoptosis (Figure 4.1.1). Therefore, they are key nodes in making the decision between survival and death following

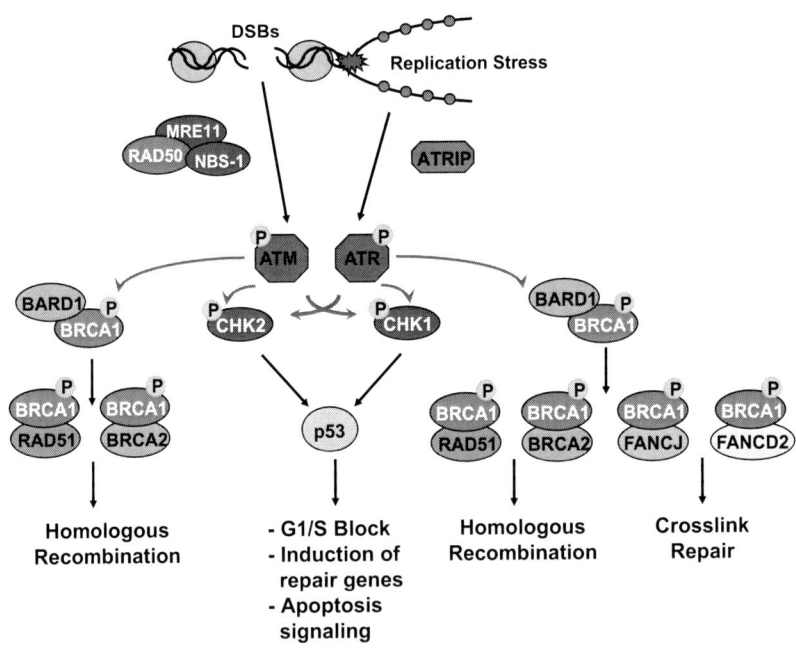

Figure 4.1.1   Signalling triggered by DNA double-strand breaks and replication blocking lesions. DSB are recognized by the MRN complex (MRE11, RAD50, NBN-1) and activate ATM as well as ATR. These PI3-kinases phosphorylate a large number of substrates that finally trigger downstream key players of homologous recombination and crosslink repair as well as p53 driven pathways resulting in cell cycle blockade, induction of repair genes (*e.g.* DDB2) or, if DNA damage exceeds a particular threshold, apoptosis.

genotoxin exposure. Although the signaling from ATM and ATR overlaps, the pathways are not identical and will therefore be discussed separately.

4.1.3 DSB Detection

DSBs are detected very rapidly, which in itself is remarkable considering the huge size of the genome. The detection and subsequent downstream signaling of DSBs requires the interplay of the MRN complex and ATM. It is possible that poly(ADP-ribose) polymerase (PARP-1) is also involved as indicated by some recent data.[1] PARP-1 catalyzes the formation of poly (ADP-ribose) chains, which in turn facilitate the docking of the MRN-complex to the DSB.[2,3] The MRN-complex consists of NBS1 (NBN or Nibrin), MRE11 and RAD50. MRE11, which exhibits exo- and endonuclease activity, and RAD50 have DNA binding capabilities while NBS1 is responsible for shuttling the MRN complex to the nucleus.[4] RAD50 holds the DSB ends together while MRE11 makes use of its nuclease activity to process the DSB ends. Once the MRN complex is localized to the DSBs, which can be visualized by immunofluorescence microscopy as small granular foci, ATM is recruited to the DSB.

ATM is a serine/threonine protein kinase and is a member of the phosphatidyl inositol 3-kinases-related (PI-3Ks) family of kinases. It is found in its inactive state as a homodimer, which upon recruitment to the DSB-PARP-MRN site, activates itself by phosphorylation.[5] This autophosphorylation occurs at Ser1981,[5] which leads to the disassociation of inactive ATM dimers into catalytic active monomers. Additional autophosphorylation sites on Ser367 and Ser1893 were also detected *in vitro* and *in vivo*. In humans, mutations in these phosphorylation sites abrogate ATM signaling upon ionizing radiation.[6] However, in mice, mutation in these phosphorylation sites does not influence ATM activity.[7] Interestingly, the *in vitro* activation of ATM kinase activity is not strictly dependent on autophosphorylation on Ser1981.[8] In the absence of the MRN complex, the catalytic activity of ATM and Ser1981 phosphorylation is reduced[9] and ATM is not recruited to the DSB.[10] The presence and absence of MRN also differentially affects the activation of several ATM targets.[11] ATM monomerization was also shown to occur by interaction of ATM with the MRN complex and single-stranded DNA (ssDNA).[10] *In vitro*, unwinding of DNA ends by the MRN complex, but not ATM autophosphorylation, is required for monomerization of ATM and ATM activation.[10] *In vivo*, active ATM monomers are responsible for DSB dependent signaling, as it phosphorylates a large number of proteins. Among them is NBN within the MRN complex, whose activity is in turn strengthened through phosphorylated NBN.

4.1.4 Detection of Blocked DNA Replication Forks

One of the essential processes of life is DNA synthesis. Normal DNA polymerases cannot synthesize new DNA if the template DNA contains modifications (*e.g.* depurinations) or replication blocking DNA adducts (*e.g.* N^3-methyladenine). These DNA lesions are normally bypassed by translesion synthesis (TLS) polymerases. However, if too many of these lesions are present in replicating DNA, the replicating fork is stalled and may lead to DSBs within the fragile DNA single-stranded regions. To combat the formation of these DSBs, a sensor system has evolved whose task it is to stabilize DNA in these areas and to initiate repair of the lesion-causing blockage in the replication process. This sensor system makes use of several components, namely replication protein A (RPA), the ATR-ATRIP complex, TopBP1 and the 9-1-1 complex. During normal DNA replication, RPA binds to single-stranded DNA and prevents it from winding back on itself or from forming secondary structures. In replication blocking situations, these long stretches of RPA-bound ssDNA persist and ATR is recruited to these sites by ATRIP (ATR-interacting protein).[12] Like ATM, ATR is a serine/threonine protein kinase and a member of the phosphatidylinositol 3-kinases-related (PI-3 kinase) family of kinases. ATR-ATRIP complex is bound to RPA-bound single stranded DNA by direct interaction of ATRIP with RPA[13–16] and oligomerization.[17] The recruitment of the 9-1-1 complex is a multi step process. The 9-1-1 complex consists of Rad9, Hus1 and Rad1.[18] Rad9 interacts with TopBP1[19] and recruits TopBP1 to the stalled replication fork.[20,21] TopBP1 directly activates the ATR-ATRIP complex, most likely *via* a conformational change of the ATR-ATRIP complex.[22] Similar to ATM, activated ATR phosphorylates many proteins (BRCA1 and BRCA2, Rad51 and FANCJ, among others), whose activity is modulated as a result. One of the main pathways activated by ATR is homologous recombination (Figure 4.1.1), which is required for the abolition of the replication blockage.

4.1.5 Phosphorylation of Histone 2AX

DSBs and stalled replication forks cause the phosphorylation of neighboring histone 2AX at Ser139 (H2AX for the unphosphorylated form and γH2AX for the phosphorylated form) *via* ATM and ATR respectively. For DSBs, the distribution of γH2AX within the nucleus is localized to the DSB. These γH2AX foci can be visualized by immunofluorescence microscopy, which has led to the ability to quantify the exact number of DSBs induced in the DNA by a given dose of genotoxin. Thus it has been shown that 1 Gy low linear energy transfer (LET) ionizing radiation induces 36 γH2AX foci that are supposed to correspond to 36 DSB.[23] These γH2AX foci already appear 10 min after DSB induction; their detection by immunocytochemistry suggests that a huge amount of H2AX is phosphorylated. Over time the amount and intensity of these γH2AX foci decreases due to the activity of the phosphatase PP4,[24,25] showing that DSB repair is occurring. The appearance and disappearance of γH2AX foci is used in many laboratories worldwide for the very accurate determination of DSB induction and repair on a cellular level following low dose genotoxin treatment. As stated above, stalled replication forks activate

ATR and ATR phosphorylates H2AX. The functional significance of γH2AX is assumed to be a signal that facilitates the repair of free DSB or DSB formed at stalled replication forks, presumably by causing the chromatin to be more accessible for DNA repair.

4.1.6 Signaling to Cell Cycle Checkpoints

Important phosphorylation targets of ATM and ATR are Chk1, Chk2 and p53. ATM phosphorylates Chk2 (checkpoint kinase-2) after the formation of DSBs at Thr68,[26,27] while ATR phosphorylates Chk1 (checkpoint kinase-1) at Ser345 upon stalled DNA replication forks.[28,29] In turn, Chk2 and Chk1 phosphorylate the transcription factor p53 at Ser20.[30] Phosphorylation of p53 at Ser20 does not activate the transcription factor; rather it prevents the proteosomal degradation of p53. Under unphosphorylated conditions, p53 is ubiquitinated by the ubiquitin E3 ligase MDM2 which targets it for proteosomal degradation. Phosphorylation of p53 by Chk1 and Chk2 prevents this process and results in stabilization of the p53 protein.[31–34] ATM and ATR can also directly phosphorylate p53 at Ser15, thereby increasing its transactivation activity.[35,36] Consequently, the nuclear localization of p53 and its DNA binding activity increases and its target genes are transcriptionally induced.

Signaling from ATM and ATR to the G1 checkpoint occurs *via* two mechanisms. First, cell cycle progression from G1 to S is governed by the cell cycle checkpoint. Cyclin D1 binds to Cdk4/6 (cyclin dependent kinase 4/6) and the Cyclin D1-Cdk4/4 complex phosphorylates Rb (retinoblastoma). Phosphorylated Rb releases E2F (a transcription factor) and E2F transcribes Cyclin E. Cyclin E binds to Cdk2 and the Cyclin E-Cdk2 complex drives the cells into S-phase. p53 prevents the entry of cells into S-phase by transcritionally causing the expression of p21, and p21 is a inhibitor of the Cyclin E-Cdk2 complex. Secondly, cell cycle progression is also regulated by the phosphorylation status of Cdk4 and Cdk2. When they are phosphorylated, they are inhibited. The dual specificity tyrosine and serine/threonine phosphatase Cdc25a (cell division cycle 25 homolog A) removes these inhibitory phosphate groups from Cdk2. Chk1 phosphorylates Cdc25a at Ser123, which leads to Cdc25a's ubiquitination and degradation,[37] thereby preventing the removal of the inhibitory phosphate groups from Thr14 and Tyr15 on Cdk2 and preventing the entry of cells into S-phase. ATM and ATR signaling therefore activate the G1 cell cycle arrest by the transcriptional activity of p53 and the direct phosphorylation of Cdc25a by Chk1.

The mechanism whereby ATM and ATR regulate the G2/M cell cycle block is as follows. Chk1 phosphorylates Cdc25c (cell division cycle 25 homolog C), also a dual specificity tyrosine and serine/threonine phosphatase, at Ser216 causes the binding of Cdc25c to the 14-3-3σ protein.[38,39] Within this complex, Cdc25c is transported out of the nucleus and is thereby unable to dephosphorylate/activate Cdk1 (cyclin dependent kinase 1), which finally results in G2/M arrest.[40]

Besides ATM and ATR causing the activation of G1/S and the G2/M checkpoint control, they can also cause an inter S-phase block. DNA replication in the S-phase of the cell cycle can be described using the following steps:[41–43]

(1) Binding of the origin recognition complex (ORC) to the origins of DNA replication.
(2) Formation of the pre-replication complex (preRC), which is activated by Cdk and Dbf4-dependent kinase (DDK). The preRC consists of ORC, Cdc6 (cell division cycle 6), Cdt1 (Cdc10-dependent target 1) and MCM6-7 (mini-chromosome maintenance 6-7).
(3) Formation of the initiation complex (IC) which activates the replicative helicases. The IC consists of RPA, Cdc45 and GINS (go-ichi-ni-san (5-1-2-3)).
(4) Loading of Pol-prim (DNA polymerase α-primase) onto the DNA by MCM10.
(5) Pol-prim synthesizes the RNA primer, which is recognized by replication factor C (RFC).
(6) RFC loads the replication sliding clamp proliferating cell nuclear antigen (PCNA) onto the DNA.
(7) RFC, PCNA and RPA mediate the polymerase switch from Pol-prim to the replicative polymerases δ and ε.

As stated above, modifications to the DNA prevents the DNA polymerases from replicating DNA and thereby causing the inhibition of S-phase progression. But how do the cells prevent new replicating forks from forming at other origins of DNA replication once ATR or ATM is activated? ATR does this by preventing IC function. ATR activation leads *via* Chk1 to the loss of interaction between Cdc45 and MCM7,[44] thereby preventing the IC from activating replicative helicases and the replication fork cannot form. An additional mechanism for preventing the initiation of DNA replication forks is *via* the ATM and ATR activation of Cdc25a, as already discussed. Cdc25a degradation prevents the activation of the preRC by preventing the phosphorylation by Cdks.[45]

4.1.7 Survival Strategy: DSB Repair

The DNA damage dependent signaling system has the task of recognizing critical DNA damage, preparing the cell for this 'emergency situation' and to stimulate repair. This is accomplished through three strategies:

(1) Inhibition of the cell cycle
(2) Increased synthesis of repair proteins, which are present only in small amounts
(3) Recruitment of DNA repair proteins (not all of which are enzymes) to the lesion.

The repair of DSB has been studied intensively over the past few years. It has to be seen as part of the cell's DNA damage response and a main survival strategy of the cell and is therefore dealt with here in this context.

There are two pathways of DSB repair: non-homologous joining of broken DNA ends (NHEJ) (Figure 4.1.2) and homologous recombination (HR). NHEJ occurs whenever there is only one copy of the DNA, *i.e.* in the G1-phase of the cell cycle when there is no homologous DNA strand. HR requires an unbroken homologous DNA strand and can thus only occur in the S- and G2-phases. HR plays a particularly significant role as resistance factor for geno-toxic anticancer drugs (such as alkylating agents), because it facilitates repair of DSBs, which are formed at blocked replication forks.

NHEJ is based on the interplay of the proteins Ku70, Ku80 and DNA-PK$_{cs}$. Ku70/Ku80 recognizes the DNA ends and is bound there, while DNA-PK$_{cs}$ (the catalytic subunit of DNA-PK) is eventually bound to these proteins, forming a complex that is described as DNA-PK.[46] This complex has kinase activity. Not only does the complex phosphorylate H2AX, but also XRCC4, ligase IV, artemis and XLF, which then remain at the lesion and bring about the (often faulty) joining of the broken DNA ends (Figure 4.1.2).

HR is different and more complex. It requires additional systems beside damage recognition, which find the intact homologous DNA strand, transport it to the lesion, partial denature the DNA, initiate and execute DNA synthesis at the intact matrix, separate the DNA strands, and restore the undamaged and intact DNA. Describing this complex process comprehensively requires some imagination, especially since the three-dimensional DNA and chromatin structure have to be considered. At this point we will refrain from discussing this process in detail. What is important is that ATM/ATR signaling activates

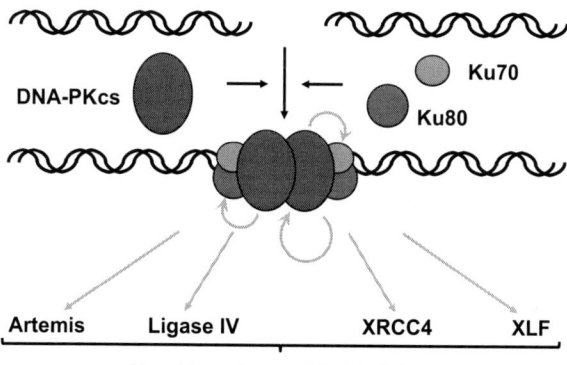

Figure 4.1.2 DNA double-strand breaks trigger DNA-PK activation. DSBs are recognized by Ku70 and Ku80 proteins, to which DNA-PK$_{CS}$ binds. The complex possesses kinase activity which phosphorylates several repair proteins that, in concert, mediate the ligation of broken DNA ends, a process called non-homologous end joining.

proteins through phosphorylation that are essential for HR. These include 'breast cancer gene 1' (BRCA1), 'breast cancer gene 2' (BRCA2), FANCD2, WRN and BLM. BRCA2 is responsible for the hereditary form of mamma carcinoma, as well as BRCA1. FANCD2 is one of the mutated complementation groups of the hereditary cancer predisposition syndrome 'Fanconi's anemia', BLM is the hereditary syndrome 'Bloom's Syndrome' and WRN is the hereditary premature ageing syndrome 'Werner's'. Both BLM and WRN are DNA damage repair helicases. All these syndromes have in common a high genomic instability, and a predisposition to develop tumors; the reason for which can be found in defective DNA repair through faulty or missing activation of the above-mentioned proteins or the absence of one of the components.

4.1.8 Survival Strategy: Autophagy

Autophagy is a catabolic process whereby cells break down intracellular components in lysosomes. It is normally activated under conditions of stress and starvation, where it cleans up damaged cellular components (including organelles) and breaks down misfolded proteins. Thus, not only does autophagy maintain cellular homeostasis by eliminating surplus or damaged proteins and organelles, but it also provide substrates for energy production and biosynthesis.[47] DNA damage triggers autophagy.[48–50] The influence of autophagy on cell death triggered by methylating agents is still unclear as whether it is pro-survival of pro-death is still controversial; the available literature shows opposing effects on overall survival—some papers demonstrate a protective role[49] while others a sensitization role.[48] For ionizing radiation it is clear that it has a protective role.[50] Ionizing radiation induces DSBs by causing localized reactive oxygen species (ROS) to form in close proximity to the DNA. These ROS are very reactive and break the phosphate backbone of the DNA, giving rise to DSBs. These and other lesions activate ATM and ATM that in turn activate TSC2 (tuberous sclerosis protein 2) tumor suppressor *via* the STK11 (serine/threonine kinase 11) and AMPK (AMP-activated protein kinase) metabolic pathway in the cytoplasm to repress the autophagy suppressor mTOR (mammalian target of rapamycin), which finally induces autophagy.[51,52]

4.1.9 Survival Signaling Mediated by NF-κB

The transcription factor NF-κB is found in an inactive state in the cytoplasm of cells[53] and becomes rapidly activated and translocated into the nucleus upon signaling from the receptor activator of nuclear factor κ B (RANK),[54] tumor necrosis factor receptor (TNFR)[55] or toll-like receptors (TLR)[56] on the cell membrane. NF-κB is kept inactive in the cytoplasm due to its binding to its inhibitor IκB (inhibitor of κB) that shields the nuclear localization signal (NLS) of NF-κB. For the activation of NF-κB, IκB is phosphorylated by IκB kinase (IKK) at Ser32 and Ser36 (human IκB), which leads to the ubiquitination and proteasomal degradation of IκB.[57] NF-κB then becomes nuclear localized and can mediate the transcription of its target genes. Apart from its activation by cell receptors, it can also be activated by DNA damage. DSB causes the nuclear localization of NEMO (NF-κB essential modulator)[58] followed by NEMO's phosphorylation by ATM.[58] NEMO and ATM form a complex and are shuttled to the cytoplasm.[59] In the cytoplasm, IκB kinase (IKK) binds to NEMO/ATM. IKK phosphorylates the NF-κB inhibitor IκBα that traps NF-κB in the cytoplasm. Phosphorylated IκBα is subsequently ubiquitinated and degraded *via* the 26S proteasome pathway, liberating NF-κB and NF-κB becomes nuclear localized where it can perform its transcriptional function. Transcriptional activation by NF-κB leads to changes in pro- and anti-apoptotic factors, in addition to others. NF-κB causes the expression of Bcl-xL, A1/Bfl-1,[60] c-IAP2[61] and TRAF1 and 2.[62] Bcl-xL and A1/Bfl-1 are both members of the Bcl-2 family of proteins that inhibit apoptosis by heterodimerization with the pro-apoptotic Bcl-2 family members.[63–65] Since c-IAP2, TRAF1 and TRAF2 are involved in the inhibition of the death receptor pathway, the induction of these proteins by NF-κB suppresses the activation of caspase-8 dependent apoptosis.[62] Interestingly, NF-κB also up-regulates MKP-1,[66] which counteracts the c-Jun *N*-terminal kinase (JNK) phosphorylation and JNK-driven apoptosis observed after sustained DNA damage induction. NF-κB has also been implicated in the regulation of DNA damage induced by topoisomerase II inhibitors, where NF-κB activation leads to increased topoII-inhibitor induced DNA damage.[67]

4.1.10 Survival Signaling Mediated by Akt

Akt is a serine/threonine kinase that plays a central role in multiple cellular processes. This section focuses on the influence that Akt has on apoptosis. Akt activation by phosphatidylinositol 3-kinase (PI3-kinase) is required for the suppression of apoptosis.[68] This inhibition of apoptosis by Akt leads to increased survival and has been characterized in many cancer cell systems.[69–72] Akt suppresses apoptosis by the following mechanisms:

(1) Akt phosphorylates BAD on Ser136.[73,74] BAD is a member of the Bcl-2 family proteins and acts by binding to Bcl-2 or Bcl-X$_L$, and thereby inhibiting their anti-apoptotic function.[65] When Akt phosphorylates BAD, BAD becomes sequestered by 14-3-3 proteins and is inactivated, as the phosphorylated form of BAD has a lower binding affinity for Bcl-X$_L$.[75] The release of Bcl-2 and Bcl-X$_L$ frees these anti-apoptotic proteins allowing them to fulfill their apoptosis suppression function.

(2) Akt phosphorylates human caspase-9 on Ser196[76] thereby inhibiting it. Caspase-9, along with Apaf-1 and cytochrome c form the apoptosome, which is required for mitochondrial mediated apoptosis.[77]

(3) Akt phosphorylates apoptosis signal-regulating kinase 1 (ASK1) on Ser83 and this results in the inhibition of apoptosis induced by ASK1.[78] ASK1 is upstream to JNK and p38 kinase, which in turn is upstream to Bid cleavage, Bax translocation and cytochrome c release.[79]

(4) Akt stimulates the activity of NF-κB by promoting the degradation of the NF-κB inhibitor IκB.[80]
(5) Akt prevents the nuclear localization of p53. Akt binds to and phosphorylates Mdm2, the ubiquitin E3 ligase, on Ser166 and Ser186 thereby inducing its nuclear import or increasing its ubiquitin ligase activity.[81,82] Mdm2 is greatly involved in the inactivation of p53, and this increase in nuclear Mdm2 level and activity can thereby inactivate the pro-apoptotic transcriptional function of p53.

One link between Akt and DNA damage is p53. Not only does p53 regulate apoptosis, it also plays a role in DNA damage signaling and repair.[83] The suppression of p53 by Akt following DNA damage may lead to a high non-tolerable level of unrepaired DNA lesions. The cell will then have to overcome the pro-survival signaling of Akt in order to undergo apoptosis. The consequences for the cell in failing to do this might lead to necrotic cell death, which in the case of chemotherapy or radiation therapy will lead to inflammation, or fixation of DNA damage, which might lead to mutation and/or tumorigenesis. This shows the complex interplay between multiple pathways in the DDR.

4.1.11   Death Strategy: Apoptosis

The current paradigm states that if DNA repair fails, cells undergo death by activating one of the programmed death pathways, *i.e.* apoptosis. This implies that, in cases of heavy DNA damage, the DNA damage response leads to elimination of the genetically damaged cells from the population in a given tissue. There are various experimental systems which show that defects in the execution of apoptosis increase cancer incidence.[84] Thus, apoptosis following DNA damage is a protective mechanism which counters carcinogenesis.

The pathways of apoptosis, following from DSBs, are shown in Figure 4.1.3. p53 is essential in the 'extrinsic' path, which works through activation of the death receptor Fas (alias Apo-1, CD95), as p53 regulates the transcription of the Fas-receptor—just like Bax and other genes. Phosphorylation and stability of anti-apoptotic proteins (such as Bcl-2) also play a key role. Bcl-2 degradation, in fact, is a key event in DNA damage-induced apoptosis.[85] The evidence that DNA damage is able to activate the apoptotic machinery is as follows:

(1) Cells that are not able to repair DNA damage are hypersensitive towards genotoxins, and this increased sensitivity is due to the induction of apoptosis. This has been shown for cells defective in O^6-methylguanine-DNA methyltransferase (MGMT), base excision repair (BER), nucleotide excision repair (NER), DSB repair and DNA crosslink repair.
(2) When cells incorporate modified nucleotides into their DNA (*e.g.* 6-thioguanine or gancyclovir), they undergo apoptosis.
(3) DSBs induced in the cellular genome by restriction enzymes trigger apoptosis.[86]

The apoptotic pathways activated by specific DNA lesions such as O^6-methylguanine (O^6MeG) and cyclobutane pyrimidine dimers (CPDs), as well as their interplay with DNA repair pathways have been studied in great detail (for reviews, see refs. 87 and 88). Thus it has been shown that O^6MeG does not block DNA replication; it mispairs with thymine, giving rise to mismatch repair substrates that are recognized by MutSα (MSH2 and MSH6) and MutLα (MLH1 and PMS2), which provoke a futile mismatch repair cycle leading to the formation of DSBs and apoptosis signaling.[87] CPDs are an example of bulky DNA lesions that block DNA replication and transcription. They have been shown to trigger apoptosis either by forming DSBs or blocking the transcription of pro- and anti-apoptotic genes.[85] A critical down-regulated gene is MAP kinase phosphatase 1 (MKP1), which has been shown to be blocked in expression upon exposure to ultraviolet (UV) C light. This is turn leads *via* dephosphorylation of JNK to down-regulation of the MAP kinase pathway, resulting in low AP-1 levels, which has an impact on the regulation of anti-apoptotic (*e.g.* repair genes) or pro-apoptotic (*e.g.* Fas-L) genes.[89]

The main apoptosis players activated by DNA damage are outlined in Figure 4.1.3. DNA damage triggered apoptosis relies on the interplay between the extrinsic (receptor based) and the intrinsic (mitochondrial based) apoptosis pathways. DNA damage causes the activation of ATM and ATR, which leads the stabilization of p53 as already discussed. A transcriptional target of p53 is the *fas* gene. DNA damage, therefore, leads to the activation of the extrinsic caspase-8 dependent apoptosis pathway. DNA damage can also activate

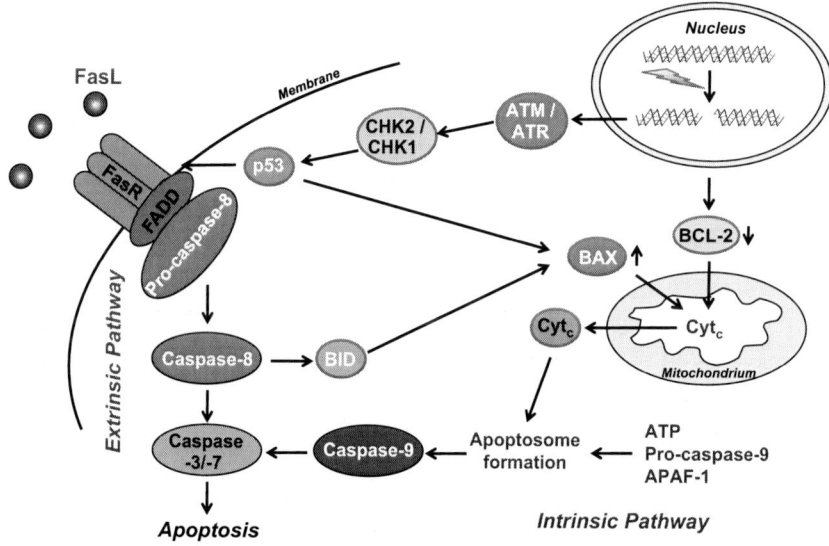

Figure 4.1.3 Induction of apoptosis triggered by DNA double-strand breaks *via* the intrinsic (mitochondria damage dependent) and extrinsic (FAS dependent) pathway. Only the main players of apoptosis are shown.

the intrinsic mitochondrial caspase-9 dependent apoptosis pathway in a p53 dependent and independent manner. The hallmark of this is the release of cytochrome c and the formation of the apoptosome, a complex consisting of Apaf-1, ATP, pro-caspase-9 and cytochrome c. The p53 independent mechanism relies on the hereto mechanistic undefined decrease in Bcl-2 protein levels. The p53 dependent mechanism relies on the p53 mediated increase in Noxa, Bax and Puma. Both mechanisms increase the leakiness of mitochondria, causing cytochrome c release and activation of the apoptosome. Caspase-9 in turn cleaves caspase-3 and -7, which degrades the inhibitor of caspase activated DNase (ICAD). The DNase then cleaves the DNA into the typical nucleosomal fragments. Activated caspase-3 and -7, being proteases, inactivate other proteins that play a role in cell cycle, DNA repair, signaling and cellular structure. There is crosstalk between the two pathways; caspase-8 cleaves Bid, which causes Bax and Bak mitochondrial localization. Important factors that modulate the effect of DNA damage are p53, Jun kinase/p38 kinase, caspase-2, NF-kB and Akt, which are discussed briefly below.

4.1.11.1   p53

As outlined above, p53 becomes activated in response of DNA damage. p53 transcriptionally regulates both pro- and anti-apoptotic genes. Pro-apoptotic genes are Fas-R, Bax, Puma, Noxa, Apaf-1 and Pidd all are apoptosis genes. Anti-apoptotic are DDB2, XPC, Fen1, MGMT and MSH2[90] all are DNA repair proteins. The transcription factor p53 therefore plays a protective role by increasing DNA repair and a sensitization role by increasing apoptosis execution. This dual role is illustrated by way of an example: In glioblastoma cells, p53 stimulates apoptosis following methylating drug treatment due to its up-regulation of Fas,[91] while causing resistance to chloroethylating and crosslinking agents due to is up-regulation of the DNA repair genes *ddb2* and *xpc*.[92] In this case, the decision of p53 to stimulate apoptosis or DNA repair is obviously dependent on the genotoxic agent. It may also be dependent on the dose level of a given genotoxin.

4.1.11.2   Sustained JNK and p38 Kinase Activation

DNA damaging agents have been shown to activate stress-activated protein kinase (SAPK)/JNK and p38 kinase, which results in an increase in c-Jun level and the activity of the transcription factor AP-1. This sustained activation of AP-1 is accompanied by increased Fas-L expression. NER repair defective mutants display a higher level of sustained JNK/p38 kinase activation, indicating DNA damage is responsible for the response.[93] Together with DNA damage-induced p53 up-regulation and subsequent Fas expression, the AP-1 mediated Fas-L expression drives the extrinsic apoptosis pathway upon DNA damage.

How is sustained activation of JNK/p38 kinase achieved? JNK/p38 kinases become activated by epidermal growth factor (EGF) receptor phosphorylation.

This is counteracted by the phosphatase MKP1 that dephosphorylates JNK and therefore down-regulates its activity.[94,95] Genotoxic agents provoke a reduction in the level of MKP1, which is likely to be due to transcription blockage due to bulky lesions in the DNA, and therefore ameliorates apoptosis by sustained up-regulation of JNK/p38 kinase.[89,93]

4.1.11.3 Caspase-2 Activation

The only pro-caspase constitutively present in the nucleus is caspase-2,[96,97] which is therefore a possible player in DNA damage triggered apoptosis. Caspase-2 has been shown to play a role in etoposide, doxorubicin, cisplatin and UV-light induced apoptosis.[98,99] Caspase-2 induced apoptosis requires caspase-9 activation[100] indicating that caspase-2 acts through the mitochondria damage pathway. The action of caspase-2 may activate the mitochondrial apoptotic pathway in the following manner:

(1) Caspase-2 acts upstream of the mitochondria by inducing Bid cleavage, Bax translocation and subsequent cytochrome c release.[98,100,101]
(2) Caspase-2 directly activates the mitochondrial pathway,[102] which is independent of its enzymatic activity.

Caspase-2 can also disrupt the interaction of cytochrome c with anionic phospholipids, notably cardiolipin, and thereby enhances the release of cytochrome c.[103] Recently, it has been shown that caspase-2 can directly initiate apoptosis following IR when Chk1 signaling is prevented.[104] Therefore, nuclear caspase-2 is able to directly activate apoptosis from DNA damage independent of caspase-3 under specific conditions.

4.1.12 Death Strategy: Necrosis

In contrast to apoptosis, necrosis has long been considered to be an unregulated, pathological mode of cell death. Recent studies revealed that, besides apoptosis, necrosis is also regulated and executed *via* different pathways. Necrosis is induced by DNA damage, but also by physico-chemical stress or different pathologies (*e.g.* ischemia). Furthermore, necrosis is described to occur during the early embryonic development[105] and is required for maintaining tissue homeostasis.[106,107]

Characteristic morphologies of necrosis are swelling of the cell and the subsequent loss of membrane integrity, which leads to the release of cytoplasmic compartments into the surrounding environment.[108] Since inflammatory compounds like HMGB1 (high mobility group box 1 protein),[109,110] heat shock protein 70 (HSP70)[111] and urea[112] are released, an immune response is induced and inflammation occurs in the surrounding tissue.

The uptake of necrotic cells by macrophages is less efficient than in the case of apoptosis.[113] The signaling pathways involved in necrosis are summarized

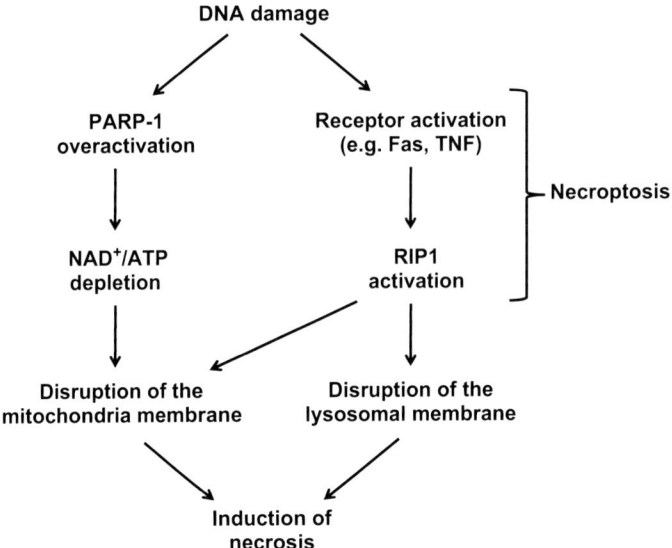

Figure 4.1.4   Induction of necrosis upon DNA damage. Necrosis requires energy depletion and an impairment of the integrity of mitochondria or lysosomal membranes.

in Figure 4.1.4. In these pathways the mitochondria are a central player.[114] Under stress conditions they undergo a process called 'mitochondrial permeability transition' (MPT). Under these conditions, specific protein channels, the mitochondrial permeability transition pores (MTPT), which connect the mitochondrial matrix with the cytosol, change their conformation and become permeable for a variety of molecules. As a consequence, the inner mitochondrial membrane depolarizes and the mitochondrial matrix swells, which leads to rupture of the outer mitochondria membrane.[115] Finally the concerned mitochondrion dies. If a critical threshold of dying mitochondria is reached, the cell suffers seriously from energy depletion and is unable to maintain basal survival functions, including maintenance of the integrity of the outer cell membrane: they undergoes necrotic cell death.

The induction of the MPT can be induced by different stimuli. A reasonable (simplified) model for the mechanism whereby DNA damage activates necrosis is as follows. DNA damage causes hyperactivation of PARP-1,[116] which provokes a drop in the NAD and the ATP level. This induces the MPT *via* the activation of RIP1 and JNK1.[108] Furthermore, when the cellular ATP level is depleted sufficiently, cells will initiate necrosis instead of apoptosis even though apoptosis was triggered, as apoptosis is an ATP dependent process. An analogue to apoptosis is necroptosis (programmed necrosis). Necroptosis (the term was first introduced in 2005[117]) is dependent on the activation of different receptors (*e.g.* FAS or TNF) and RIP1.[118,119] As DNA damage triggers necrosis *via* RIP1 activation, it can be concluded that DNA damage is responsible for inducing the process of necroptosis. Given the central role of ATP in both apoptosis (where it is required) and necrosis (where it is not required), it is reasonable to conclude that the cellular energy status is decisively involved in the decision of which cell death pathway will be activated following DNA damage.[120]

4.1.13 Conclusions

Genotoxic agents target not only DNA but also other components in the cell such as RNA, enzymes and proteins as well as cellular free nucleotides that may cause receptor activation/inactivation and signaling. An example is provided by the EGF receptor which can be activated by UV-C light. Nevertheless, damage to nuclear and mitochondrial DNA is at the very heart of the cellular responses that make the decision between survival and death. We have summarized the cell's strategies for survival and death following DNA damage and showed that DNA damage recognition and repair as well as damage tolerance and auto-phagy are major survival mechanisms. The current paradigm states that a high level of DNA damage (or residual DNA damage) activates one of the death programmes including apoptosis and necrosis. It is not well understood how the cell switches between these pathways, but it seems that 'clean' DSBs activate the apoptotic pathway almost exclusively, whereas it is DNA base damage that activates the BER machinery causing ATP depletion due to PARP activation triggers necrosis. The necrotic pathway following DNA damage and ATP depletion is less well understood. Since apoptosis requires ATP, it is reasonable to posit that a drop in the ATP level to a particular threshold determines the switch between apoptosis and necrosis. It is clear that further studies are required to clarify this important question.

Acknowledgements

We gratefully acknowledge Dr Christina Strauch for secretarial and editing work. The authors' work is supported by Deutsche Forschungsgemeinschaft (DFG KA724) and Deutsche Krebshilfe.

References

1. J. F. Hance, S. Kozlov, V. L. Dawson, T. M. Dawson, M. J. Hendzel, M. F. Lavin and G. G. Poirier, Ataxia telangiectasia mutated (ATM) signaling network is modulated by a novel poly(ADP-ribose)-dependent pathway in the early response to DNA-damaging agents, *J. Biol. Chem.*, 2007, **282**, 16441–16453.
2. O. K. Mirzoeva and J. H. Petrini, DNA damage-dependent nuclear dynamics of the Mre11 complex, *Mol. Cell. Biol.*, 2001, **21**, 281–288.

3. J. F. Haince, D. McDonald, A. Rodrigue, U. Dery, J. Y. Masson, M. J. Hendzel and G. G. Poirier, PARP1-dependent kinetics of recruitment of MRE11 and NBS1 proteins to multiple DNA damage sites, *J. Biol. Chem.*, 2008, **283**, 1197–1208.

4. J. P. Carney, R. S. Maser, H. Olivares, E. M. Davis, M. Le Beau, J. R. Yates 3rd, L. Hays, W. F. Morgan and J. H. Petrini, The hMre11/hRad50 protein complex and Nijmegen breakage syndrome: linkage of double-strand break repair to the cellular DNA damage response, *Cell*, 1998, **93**, 477–486.

5. C. J. Bakkenist and M. B. Kastan, DNA damage activates ATM through intermolecular autophosphorylation and dimer dissociation, *Nature*, 2003, **421**, 499–506.

6. S. V. Kozlov, M. E. Graham, C. Peng, P. Chen, P. J. Robinson and M. F. Lavin, Involvement of novel autophosphorylation sites in ATM activation, *EMBO J.*, 2006, **25**, 3504–3514.

7. M. Pellegrini, A. Celeste, S. Difilippantonio, R. Guo, W. Wang, L. Feigenbaum and A. Nussenzweig, Autophosphorylation at serine 1987 is dispensable for murine Atm activation *in vivo*, *Nature*, 2006, **443**, 222–225.

8. J. T. Powers, S. Hong, C. N. Mayhew, P. M. Rogers, E. S. Knudsen and D. G. Johnson, E2F1 uses the ATM signaling pathway to induce p53 and Chk2 phosphorylation and apoptosis, *Mol. Cancer Res.*, 2004, **2**, 203–214.

9. T. Uziel, Y. Lerenthal, L. Moyal, Y. Andegeko, L. Mittelman and Y. Shiloh, Requirement of the MRN complex for ATM activation by DNA damage, *EMBO J.*, 2003, **22**, 5612–5621.

10. J. H. Lee and T. T. Paull, ATM activation by DNA double-strand breaks through the Mre11-Rad50-Nbs1 complex, *Science*, 2005, **308**, 551–554.

11. R. Kitagawa, C. J. Bakkenist, P. J. McKinnon and M. B. Kastan, Phosphorylation of SMC1 is a critical downstream event in the ATM-NBS1-BRCA1 pathway, *Genes Dev.*, 2004, **18**, 1423–1438.

12. T. S. Byun, M. Pacek, M. C. Yee, J. C. Walter and K. A. Cimprich, Functional uncoupling of MCM helicase and DNA polymerase activities activates the ATR-dependent checkpoint, *Genes Dev.*, 2005, **19**, 1040–1052.

13. L. Zou and S. J. Elledge, Sensing DNA damage through ATRIP recognition of RPA-ssDNA complexes, *Science*, 2003, **300**, 1542–1548.

14. L. Zou, D. Liu and S. J. Elledge, Replication protein A-mediated recruitment and activation of Rad17 complexes, *Proc.. Natl. Acad. Sci. U. S. A.*, 2003, **100**, 13827–13832.

15. H. L. Ball, J. S. Myers and D. Cortez, ATRIP binding to replication protein A-single-stranded DNA promotes ATR-ATRIP localization but is dispensable for Chk1 phosphorylation, *Mol. Biol. Cell*, 2005, **16**, 2372–2381.

16. S. M. Kim, A. Kumagai, J. Lee and W. G. Dunphy, Phosphorylation of Chk1 by ATM- and Rad3-related (ATR) in Xenopus egg extracts requires binding of ATRIP to ATR but not the stable DNA-binding or coiled-coil domains of ATRIP, *J. Biol. Chem.*, 2005, **280**, 38355–38364.

17. H. L. Ball and D. Cortez, ATRIP oligomerization is required for ATR-dependent checkpoint signaling, *J. Biol. Chem.*, 2005, **280**, 31390–31396.
18. E. R. Parrilla-Castellar, S. J. Arlander and L. Karnitz, Dial 9-1-1 for DNA damage: the Rad9-Hus1-Rad1 (9-1-1) clamp complex, *DNA Repair (Amst.)*, 2004, **3**, 1009–1014.
19. M. Makiniemi, T. Hillukkala, J. Tuusa, K. Reini, M. Vaara, D. Huang, H. Pospiech, I. Majuri, T. Westerling, T. P. Makela and J. E. Syvaoja, BRCT domain-containing protein TopBP1 functions in DNA replication and damage response, *J. Biol. Chem.*, 2001, **276**, 30399–30406.
20. S. Delacroix, J. M. Wagner, M. Kobayashi, K. Yamamoto and L. M. Karnitz, The Rad9-Hus1-Rad1 (9-1-1) clamp activates checkpoint signaling *via* TopBP1, *Genes Dev.*, 2007, **21**, 1472–1477.
21. J. Lee, A. Kumagai and W. G. Dunphy, The Rad9-Hus1-Rad1 checkpoint clamp regulates interaction of TopBP1 with ATR, *J. Biol. Chem.*, 2007, **282**, 28036–28044.
22. A. Kumagai, J. Lee, H. Y. Yoo and W. G. Dunphy, TopBP1 activates the ATR-ATRIP complex, *Cell*, 2006, **124**, 943–955.
23. K. Rothkamm and M. Lobrich, Evidence for a lack of DNA double-strand break repair in human cells exposed to very low x-ray doses, *Proc. Natl. Acad. Sci. U. S. A.*, 2003, **100**, 5057–5062.
24. S. Nakada, G. I. Chen, A. C. Gingras and D. Durocher, PP4 is a gamma H2AX phosphatase required for recovery from the DNA damage checkpoint, *EMBO Rep.*, 2008, **9**, 1019–1026.
25. D. Chowdhury, X. Xu, X. Zhong, F. Ahmed, J. Zhong, J. Liao, D. M. Dykxhoorn, D. M. Weinstock, G. P. Pfeifer and J. A. Lieberman, PP4-phosphatase complex dephosphorylates gamma-H2AX generated during DNA replication, *Mol. Cell*, 2008, **31**, 33–46.
26. B. B. Zhou, P. Chaturvedi, K. Spring, S. P. Scott, R. A. Johanson, R. Mishra, M. R. Mattern, J. D. Winkler and K. K. Khanna, Caffeine abolishes the mammalian G(2)/M DNA damage checkpoint by inhibiting ataxia-telangiectasia-mutated kinase activity, *J. Biol. Chem.*, 2000, **275**, 10342–10348.
27. S. Matsuoka, G. Rotman, A. Ogawa, Y. Shiloh, K. Tamai and S. J. Elledge, Ataxia telangiectasia-mutated phosphorylates Chk2 *in vivo* and *in vitro*, *Proc. Natl. Acad. Sci. U. S. A.*, 2000, **97**, 10389–10394.
28. Q. Liu, S. Guntuku, X. S. Cui, S. Matsuoka, D. Cortez, K. Tamai, G. Luo, S. Carattini-Rivera, F. DeMayo, A. Bradley, L. A. Donehower and S. J. Elledge, Chk1 is an essential kinase that is regulated by Atr and required for the G(2)/M DNA damage checkpoint, *Genes Dev.*, 2000, **14**, 1448–1459.
29. Z. Guo, A. Kumagai, S. X. Wang and W. G. Dunphy, Requirement for Atr in phosphorylation of Chk1 and cell cycle regulation in response to DNA replication blocks and UV-damaged DNA in Xenopus egg extracts, *Genes Dev.*, 2000, **14**, 2745–2756.
30. S. Y. Shieh, J. Ahn, K. Tamai, Y. Taya and C. Prives, The human homologs of checkpoint kinases Chk1 and Cds1 (Chk2) phosphorylate

p53 at multiple DNA damage-inducible sites, *Genes Dev.*, 2000, **14**, 289–300.

31. N. H. Chehab, A. Malikzay, E. S. Stavridi and T. D. Halazonetis, Phosphorylation of Ser-20 mediates stabilization of human p53 in response to DNA damage, *Proc. Natl. Acad. Sci. U. S. A.*, 1999, **96**, 13777–13782.

32. N. H. Chehab, A. Malikzay, M. Appel and T. D. Halazonetis, Chk2/hCds1 functions as a DNA damage checkpoint in G(1) by stabilizing p53, *Genes Dev.*, 2000, **14**, 278–288.

33. A. Hirao, Y. Y. Kong, S. Matsuoka, A. Wakeham, J. Ruland, H. Yoshida, D. Liu, S. J. Elledge and T. W. Mak, DNA damage-induced activation of p53 by the checkpoint kinase Chk2, *Science*, 2000, **287**, 1824–1827.

34. T. Unger, T. Juven-Gershon, E. Moallem, M. Berger, R. Vogt Sionov, G. Lozano, M. Oren and Y. Haupt, Critical role for Ser20 of human p53 in the negative regulation of p53 by Mdm2, *EMBO J.*, 1999, **18**, 1805–1814.

35. S. Banin, L. Moyal, S. Shieh, Y. Taya, C. W. Anderson, L. Chessa, N. I. Smorodinsky, C. Prives, Y. Reiss, Y. Shiloh and Y. Ziv, Enhanced phosphorylation of p53 by ATM in response to DNA damage, *Science*, 1998, **281**, 1674–1677.

36. C. E. Canman, D. S. Lim, K. A. Cimprich, Y. Taya, K. Tamai, K. Sakaguchi, E. Appella, M. B. Kastan and J. D. Siliciano, Activation of the ATM kinase by ionizing radiation and phosphorylation of p53, *Science*, 1998, **281**, 1677–1679.

37. N. Mailand, J. Falck, C. Lukas, R. G. Syljuasen, M. Welcker, J. Bartek and J. Lukas, Rapid destruction of human Cdc25A in response to DNA damage, *Science*, 2000, **288**, 1425–1429.

38. C. Y. Peng, P.R. Graves, R. S. Thoma, Z. Wu, A. S. Shaw and H. Piwnica-Worms, Mitotic and G2 checkpoint control: regulation of 14-3-3 protein binding by phosphorylation of Cdc25C on serine-216, *Science*, 1997, **277**, 1501–1505.

39. Y. Sanchez, C. Wong, R. S. Thoma, R. Richman, Z. Wu, H. Piwnica-Worms and S. J. Elledge, Conservation of the Chk1 checkpoint pathway in mammals: linkage of DNA damage to Cdk regulation through Cdc25, *Science*, 1997, **277**, 1497–1501.

40. S. N. Dalal, C. M. Schweitzer, J. Gan and J. A. DeCaprio, Cytoplasmic localization of human cdc25C during interphase requires an intact 14-3-3 binding site, *Mol. Cell. Biol.*, 1999, **19**, 4465–4479.

41. D. Y. Takeda and A. Dutta, DNA replication and progression through S phase, *Oncogene*, 2005, **24**, 2827–2843.

42. J. J. Blow and A. Dutta, Preventing re-replication of chromosomal DNA, *Nat. Rev. Mol. Cell. Biol.*, 2005, **6**, 476–486.

43. H. P. Nasheuer, R. Smith, C. Bauerschmidt, F. Grosse and K. Weisshart, Initiation of eukaryotic DNA replication: regulation and mechanisms, *Prog. Nucleic Acid Res. Mol. Biol.*, 2002, **72**, 41–94.

44. P. Liu, L. R. Barkley, T. Day, X. Bi, D. M. Slater, M. G. Alexandrow, H. P. Nasheuer and C. Vaziri, The Chk1-mediated S-phase checkpoint

targets initiation factor Cdc45 *via* a Cdc25A/Cdk2-independent mechanism, *J. Biol. Chem.*, 2006, **281**, 30631–30644.

45. R. Broderick and H. P. Nasheuer, Regulation of Cdc45 in the cell cycle and after DNA damage, *Biochem. Soc. Trans.*, 2009, **37**, 926–930.

46. G. C. Smith and S. P. Jackson, The DNA-dependent protein kinase, *Genes Dev.*, 1999, **13**, 916–934.

47. H. Y. Chen and E. White, Role of autophagy in cancer prevention, *Cancer Prev. Res. (Phila.)*, 2011, **4**, 973–983.

48. T. Kanzawa, I. M. Germano, T. Komata, H. Ito, Y. Kondo and S. Kondo, Role of autophagy in temozolomide-induced cytotoxicity for malignant glioma cells, *Cell Death Differ.*, 2004, **11**, 448–457.

49. M. Katayama, T. Kawaguchi, M. S. Berger and R. O. Pieper, DNA damaging agent-induced autophagy produces a cytoprotective adenosine triphosphate surge in malignant glioma cells, *Cell Death Differ.*, 2007, **14**, 548–558.

50. S. L. Lomonaco, S. Finniss, C. Xiang, A. Decarvalho, F. Umansky, S. N. Kalkanis, T. Mikkelsen and C. Brodie, The induction of autophagy by gamma-radiation contributes to the radioresistance of glioma stem cells, *Int. J. Cancer*, 2009, **125**, 717–722.

51. A. Alexander, J. Kim and C. L. Walker, ATM engages the TSC2/mTORC1 signaling node to regulate autophagy, *Autophagy*, 2010, **6**.

52. A. Alexander, S. L. Cai, J. Kim, A. Nanez, M. Sahin, K. H. MacLean, K. Inoki, K. L. Guan, J. Shen, M. D. Person, D. Kusewitt, G. B. Mills, M. B. Kastan and C. L. Walker, ATM signals to TSC2 in the cytoplasm to regulate mTORC1 in response to ROS, *Proc. Natl. Acad. Sci. U. S. A.*, 2010, **107**, 4153–4158.

53. J. A. Molitor, W. H. Walker, S. Doerre, D. W. Ballard and W. C. Greene, NF-kappa B: a family of inducible and differentially expressed enhancer-binding proteins in human T cells, *Proc. Natl. Acad. Sci. U. S. A.*, 1990, **87**, 10028–10032.

54. B. G. Darnay, V. Haridas, J. Ni, P. A. Moore and B. B. Aggarwal, Characterization of the intracellular domain of receptor activator of NF-kappaB (RANK). Interaction with tumor necrosis factor receptor-associated factors and activation of NF-kappab and c-Jun N-terminal kinase, *J. Biol. Chem.*, 1998, **273**, 20551–20555.

55. S. Schutze, T. Machleidt and M. Kronke, The role of diacylglycerol and ceramide in tumor necrosis factor and interleukin-1 signal transduction, *J. Leukoc. Biol.*, 1994, **56**, 533–541.

56. G. Zhang and S. Ghosh, Molecular mechanisms of NF-kappaB activation induced by bacterial lipopolysaccharide through Toll-like receptors, *J. Endotoxin Res.*, 2000, **6**, 453–457.

57. M. Karin and Y. Ben-Neriah, Phosphorylation meets ubiquitination: the control of NF-[kappa]B activity, *Annu. Rev. Immunol.*, 2000, **18**, 621–663.

58. T. T. Huang, S. M. Wuerzberger-Davis, Z. H. Wu and S. Miyamoto, Sequential modification of NEMO/IKKgamma by SUMO-1 and ubiquitin mediates NF-kappaB activation by genotoxic stress, *Cell*, 2003, **115**, 565–576.

59. K. Brzoska and I. Szumiel, Signalling loops and linear pathways: NF-kappaB activation in response to genotoxic stress, *Mutagenesis*, 2009, **24**, 1–8.

60. H. H. Lee, H. Dadgostar, Q. Cheng, J. Shu and G. Cheng, NF-kappaB-mediated up-regulation of Bcl-x and Bfl-1/A1 is required for CD40 survival signaling in B lymphocytes, *Proc. Natl. Acad. Sci. U. S. A.*, 1999, **96**, 9136–9141.

61. Z. L. Chu, T. A. McKinsey, L. Liu, J. J. Gentry, M. H. Malim and D. W. Ballard, Suppression of tumor necrosis factor-induced cell death by inhibitor of apoptosis c-IAP2 is under NF-kappaB control, *Proc. Natl. Acad. Sci. U. S. A.*, 1997, **94**, 10057–10062.

62. C. Y. Wang, M. W. Mayo, R. G. Korneluk, D. V. Goeddel and A. S. Baldwin Jr., NF-kappaB antiapoptosis: induction of TRAF1 and TRAF2 and c-IAP1 and c-IAP2 to suppress caspase-8 activation, *Science*, 1998, **281**, 1680–1683.

63. H. Zhang, S. W. Cowan-Jacob, M. Simonen, W. Greenhalf, J. Heim and B. Meyhack, Structural basis of BFL-1 for its interaction with BAX and its anti-apoptotic action in mammalian and yeast cells, *J. Biol. Chem.*, 2000, **275**, 11092–11099.

64. A. B. Werner, E. de Vries, S. W. Tait, I. Bontjer and J. Borst, Bcl-2 family member Bfl-1/A1 sequesters truncated bid to inhibit is collaboration with pro-apoptotic Bak or Bax, *J. Biol. Chem.*, 2002, **277**, 22781–22788.

65. E. Yang, J. Zha, J. Jockel, L. H. Boise, C. B. Thompson and S. J. Korsmeyer, Bad, a heterodimeric partner for Bcl-XL and Bcl-2, displaces Bax and promotes cell death, *Cell*, 1995, **80**, 285–291.

66. A. Zhou, S. Scoggin, R. B. Gaynor and N. S. Williams, Identification of NF-kappa B-regulated genes induced by TNFalpha utilizing expression profiling and RNA interference, *Oncogene*, 2003, **22**, 2054–2064.

67. S. Karl, Y. Pritschow, M. Volcic, S. Hacker, B. Baumann, L. Wiesmuller, K. M. Debatin and S. Fulda, Identification of a novel pro-apopotic function of NF-kappaB in the DNA damage response, *J. Cell. Mol. Med.*, 2009, **13**, 4239–4256.

68. R. Yao and G. M. Cooper, Requirement for phosphatidylinositol-3 kinase in the prevention of apoptosis by nerve growth factor, *Science*, 1995, **267**, 2003–2006.

69. H. G. Wendel, E. De Stanchina, J. S. Fridman, A. Malina, S. Ray, S. Kogan, C. Cordon-Cardo, J. Pelletier and S. W. Lowe, Survival signalling by Akt and eIF4E in oncogenesis and cancer therapy, *Nature*, 2004, **428**, 332–337.

70. S. Bao, G. Ouyang, X. Bai, Z. Huang, C. Ma, M. Liu, R. Shao, R. M. Anderson, J. N. Rich and X. F. Wang, Periostin potently promotes metastatic growth of colon cancer by augmenting cell survival *via* the Akt/PKB pathway, *Cancer Cell*, 2004, **5**, 329–339.

71. Q. Shi, S. Bao, J. A. Maxwell, E. D. Reese, H. S. Friedman, D. D. Bigner, X. F. Wang and J. N. Rich, Secreted protein acidic, rich in cysteine (SPARC), mediates cellular survival of gliomas through AKT activation, *J. Biol. Chem.*, 2004, **279**, 52200–52209.

72. D. Gupta, N. A. Syed, W. J. Roesler and R. L. Khandelwal, Effect of overexpression and nuclear translocation of constitutively active PKB-alpha on cellular survival and proliferation in HepG2 cells, *J. Cell. Biochem.*, 2004, **93**, 513–525.

73. L. del Peso, M. Gonzalez-Garcia, C. Page, R. Herrera and G. Nunez, Interleukin-3-induced phosphorylation of BAD through the protein kinase Akt, *Science*, 1997, **278**, 687–689.

74. S. R. Datta, H. Dudek, X. Tao, S. Masters, H. Fu, Y. Gotoh and M. E. Greenberg, Akt phosphorylation of BAD couples survival signals to the cell-intrinsic death machinery, *Cell*, 1997, **91**, 231–241.

75. J. Zha, H. Harada, E. Yang, J. Jockel and S. J. Korsmeyer, Serine phosphorylation of death agonist BAD in response to survival factor results in binding to 14-3-3 not BCL-X(L), *Cell*, 1996, **87**, 619–628.

76. M. H. Cardone, N. Roy, H. R. Stennicke, G. S. Salvesen, T. S. Franke, E. Stanbridge, S. Frisch and J. C. Reed, Regulation of cell death protease caspase-9 by phosphorylation, *Science*, 1998, **282**, 1318–1321.

77. I. Budihardjo, H. Oliver, M. Lutter, X. Luo and X. Wang, Biochemical pathways of caspase activation during apoptosis, *Annu. Rev. Cell. Dev. Biol.*, 1999, **15**, 269–290.

78. A. H. Kim, G. Khursigara, X. Sun, T. F. Franke and M. V. Chao, Akt phosphorylates and negatively regulates apoptosis signal-regulating kinase 1, *Mol. Cell. Biol.*, 2001, **21**, 893–901.

79. V. V. Sumbayev and I. M. Yasinska, Regulation of MAP kinase-dependent apoptotic pathway: implication of reactive oxygen and nitrogen species, *Arch. Biochem. Biophys.*, 2005, **436**, 406–412.

80. L. P. Kane, V. S. Shapiro, D. Stokoe and A. Weiss, Induction of NF-kappaB by the Akt/PKB kinase, *Curr. Biol.*, 1999, **9**, 601–604.

81. T. M. Gottlieb, J. F. Leal, R. Seger, Y. Taya and M. Oren, Cross-talk between Akt, p53 and Mdm2: possible implications for the regulation of apoptosis, *Oncogene*, 2002, **21**, 1299–1303.

82. L. D. Mayo and D. B. Donner, A phosphatidylinositol 3-kinase/Akt pathway promotes translocation of Mdm2 from the cytoplasm to the nucleus, *Proc. Natl. Acad. Sci. U. S. A.*, 2001, **98**, 11598–11603.

83. M. Christmann, M. T. Tomicic, W. P. Roos and B. Kaina, Mechanisms of human DNA repair: an update, *Toxicology*, 2003, **193**, 3–34.

84. S. Wirtz, G. Nagel, L. Eshkind, M. F. Neurath, L. D. Samson and B. Kaina, Both base excision repair and O6-methylguanine-DNA methyltransferase protect against methylation-induced colon carcinogenesis, *Carcinogenesis*, 2010, **31**, 2111–2117.

85. W. P. Roos and B. Kaina, DNA damage-induced cell death by apoptosis, *Trends Mol. Med.*, 2006, **12**, 440–450.

86. J. Lips and B. Kaina, DNA double-strand breaks trigger apoptosis in p53-deficient fibroblasts, *Carcinogenesis*, 2001, **22**, 579–585.

87. B. Kaina, M. Christmann, S. Naumann and W. P. Roos, MGMT: key node in the battle against genotoxicity, carcinogenicity and apoptosis induced by alkylating agents, *DNA Repair (Amst.)*, 2007, **6**, 1079–1099.

88. L. F. Batista, B. Kaina, R. Meneghini and C. F. Menck, How DNA lesions are turned into powerful killing structures: Insights from UV-induced apoptosis, *Mutat Res.*, 2009, **881**, 197–208.

89. M. Christmann, M. T. Tomicic, D. Aasland and B. Kaina, A role for UV-light-induced c-Fos: Stimulation of nucleotide excision repair and protection against sustained JNK activation and apoptosis, *Carcinogenesis*, 2007, **28**, 183–190.

90. M. Christmann, G. Fritz and B. Kaina, Induction of DNA repair genes in mammalian cells in response to genotoxic stress, in *Genome Dynamics and Stability*, ed. D. Lankenau, Springer Verlag, Berlin-Heidelberg, 2007, vol. 1, pp. 383–398.

91. W. P. Roos, L. F. Batista, S. C. Naumann, W. Wick, M. Weller, C. F. Menck and B. Kaina, Apoptosis in malignant glioma cells triggered by the temozolomide-induced DNA lesion O(6)-methylguanine, *Oncogene*, 2007, **26**, 186–197.

92. L. F. Batista, W. P. Roos, M. Christmann, C. F. Menck and B. Kaina, Differential sensitivity of malignant glioma cells to methylating and chloroethylating anticancer drugs: p53 determines the switch by regulating xpc, ddb2, and DNA double-strand breaks, *Cancer Res.*, 2007, **67**, 11886–11895.

93. M. Hamdi, J. Kool, P. Cornelissen-Steijger, F. Carlotti, H. E. Popeijus, C. van der Burgt, J. M. Janssen, A. Yasui, R. C. Hoeben, C. Terleth, L. H. Mullenders and H. van Dam, DNA damage in transcribed genes induces apoptosis *via* the JNK pathway and the JNK-phosphatase MKP-1, *Oncogene*, 2005, **24**, 7135–7144.

94. C. C. Franklin and A. S. Kraft, Conditional expression of the mitogen-activated protein kinase (MAPK) phosphatase MKP-1 preferentially inhibits p38 MAPK and stress-activated protein kinase in U937 cells, *J. Biol. Chem.*, 1997, **272**, 16917–16923.

95. D. D. Hirsch and P. J. Stork, Mitogen-activated protein kinase phosphatases inactivate stress-activated protein kinase pathways *in vivo*, *J. Biol. Chem.*, 1997, **272**, 4568–4575.

96. M. Mancini, C. E. Machamer, S. Roy, D. W. Nicholson, N. A. Thornberry, L. A. Casciola-Rosen and A. Rosen, Caspase-2 is localized at the Golgi complex and cleaves golgin-160 during apoptosis, *J. Cell Biol.*, 2000, **149**, 603–612.

97. B. Zhivotovsky, A. Samali, A. Gahm and Orrenius, Caspases: their intracellular localization and translocation during apoptosis, *Cell Death Differ.*, 1999, **6**, 644–651.

98. P. Lassus, X. Opitz-Araya and Y. Lazebnik, Requirement for caspase-2 in stress-induced apoptosis before mitochondrial permeabilization, *Science*, 2002, **297**, 1352–1354.

99. L. Bergeron, G. I. Perez, G. Macdonald, L. Shi, Y. Sun, A. Jurisicova, S. Varmuza, K. E. Latham, J. A. Flaws, J. C. Salter, H. Hara, M. A. Moskowitz, E. Li, A. Greenberg, J. L. Tilly and J. Yuan, Defects in regulation of apoptosis in caspase-2-deficient mice, *Genes Dev.*, 1998, **12**, 1304–1314.

100. Y. Guo, S. M. Srinivasula, A. Druilhe, T. Fernandes-Alnemri and E. S. Alnemri, Caspase-2 induces apoptosis by releasing proapoptotic proteins from mitochondria, *J. Biol. Chem.*, 2002, **277**, 13430–13437.
101. J. D. Robertson, M. Enoksson, M. Suomela, B. Zhivotovsky and S. Orrenius, Caspase-2 acts upstream of mitochondria to promote cytochrome c release during etoposide-induced apoptosis, *J. Biol. Chem.*, 2002, **277**, 29803–29809.
102. J. D. Robertson, V. Gogvadze, A. Kropotov, H. Vakifahmetoglu, B. Zhivotovsky and S. Orrenius, Processed caspase-2 can induce mitochondria-mediated apoptosis independently of its enzymatic activity, *EMBO Rep.*, 2004, **5**, 643–648.
103. M. Enoksson, J. D. Robertson, V. Gogvadze, P. Bu, A. Kropotov, B. Zhivotovsky and S. Orrenius, Caspase-2 permeabilizes the outer mitochondrial membrane and disrupts the binding of cytochrome c to anionic phospholipids, *J. Biol. Chem.*, 2004, **279**, 49575–49578.
104. S. Sidi, T. Sanda, R. D. Kennedy, A. T. Hagen, C. A. Jette, R. Hoffmans, J. Pascual, S. Imamura, S. Kishi, J. F. Amatruda, J. P. Kanki, D. R. Green, A. A. D'Andrea and A. T. Look, Chk1 suppresses a caspase-2 apoptotic response to DNA damage that bypasses p53, Bcl-2, and caspase-3. *Cell*, 2008, **133**, 864–877.
105. M. Chautan, G. Chazal, F. Cecconi, P. Gruss and P. Golstein, Interdigital cell death can occur through a necrotic and caspase-independent pathway, *Curr. Biol.*, 1999, **9**, 967–970.
106. H. I. Roach and N. M. Clarke, Physiological cell death of chondrocytes *in vivo* is not confined to apoptosis. New observations on the mammalian growth plate, *J. Bone Joint Surg. Br.*, 2000, **82**, 601–613.
107. D. H. Barkla and P. R. Gibson, The fate of epithelial cells in the human large intestine, *Pathology*, 1999, **31**, 230–238.
108. N. Festjens, T. Vanden Berghe and P. Vandenabeele, Necrosis, a well-orchestrated form of cell demise: signalling cascades, important mediators and concomitant immune response, *Biochim. Biophys. Acta*, 2006, **1757**, 1371–1387.
109. P. Scaffidi, T. Misteli and M. E. Bianchi, Release of chromatin protein HMGB1 by necrotic cells triggers inflammation, *Nature*, 2002, **418**, 191–195.
110. M. E. Bianchi and A. Manfredi, Chromatin and cell death, *Biochim. Biophys. Acta*, 2004, **1677**, 181–186.
111. S. Basu, R. J. Binder, R. Suto, K. M. Anderson and P. K. Srivastava, Necrotic but not apoptotic cell death releases heat shock proteins, which deliver a partial maturation signal to dendritic cells and activate the NF-kappa B pathway, *Int. Immunol.*, 2000, **12**, 1539–1546.
112. Y. Shi, J. E. Evans and K. L. Rock, Molecular identification of a danger signal that alerts the immune system to dying cells, *Nature*, 2003, **425**, 516–521.
113. G. Brouckaert, M. Kalai, D. V. Krysko, X. Saelens, D. Vercammen, M. Ndlovu, G. Haegeman, K. D'Herde and P. Vandenabeele, Phagocytosis of necrotic cells by macrophages is phosphatidylserine dependent and

does not induce inflammatory cytokine production, *Mol. Biol. Cell*, 2004, **15**, 1089–1100.

114. N. Vanlangenakker, T. Vanden Berghe, D. V. Krysko, N. Festjens and P. Vandenabeele, Molecular mechanisms and pathophysiology of necrotic cell death, *Curr. Mol. Med.*, 2008, **8**, 207–220.
115. S. Javadov and M. Karmazyn, Mitochondrial permeability transition pore opening as an endpoint to initiate cell death and as a putative target for cardioprotection, *Cell. Physiol. Biochem.*, 2007, **20**, 1–22.
116. H. C. Ha and S. H. Snyder, Poly(ADP-ribose) polymerase is a mediator of necrotic cell death by ATP depletion, *Proc. Natl. Acad. Sci. U. S. A.*, 1999, **96**, 13978–13982.
117. A. Degterev, Z. Huang, M. Boyce, Y. Li, P. Jagtap, N. Mizushima, G. D. Cuny, T. J. Mitchison, M. A. Moskowitz and J. Yuan, Chemical inhibitor of nonapoptotic cell death with therapeutic potential for ischemic brain injury, *Nat. Chem. Biol.*, 2005, **1**, 112–119.
118. L. Galluzzi and G. Kroemer, Necroptosis: a specialized pathway of programmed necrosis, *Cell*, 2008, **135**, 1161–1163.
119. G. Kroemer, L. Galluzzi, P. Vandenabeele, J. Abrams, E. S. Alnemri, E. H. Baehrecke, M. V. Blagosklonny, W. S. El-Deiry, P. Golstein, D. R. Green, M. Hengartner, R. A. Knight, S. Kumar, S. A. Lipton, W. Malorni, G. Nunez, M. E. Peter, J. Tschopp, J. Yuan, M. Piacentini, B. Zhivotovsky and G. Melino, Classification of cell death: recommendations of the Nomenclature Committee on Cell Death 2009, *Cell Death Differ.*, 2009, **16**, 3–11.
120. M. Leist, B. Single, A. F. Castoldi, S. Kuhnle and P. Nicotera, Intracellular adenosine triphosphate (ATP) concentration: a switch in the decision between apoptosis and necrosis, *J. Exp. Med.*, 1997, **185**, 1481–1486.

CHAPTER 4.2

Different Modes of Cell Death Induced by DNA Damage

OLGA SUROVA[1] AND BORIS ZHIVOTOVSKY[1,2,*]

[1] Institute of Environmental Medicine, Division of Toxicology, Karolinska Institute, Box 210, SE-171 77 Stockholm, Sweden; [2] MV Lomonosov Moscow State University, 119991, Moscow, Russia
*Email: Boris.Zhivotovsky@ki.se

4.2.1 Introduction

Genomic integrity and stability is constantly threatened by DNA damage, which may arise from numerous environmental and intrinsic sources, including reactive oxygen species (ROS), spontaneous replication errors, ionizing radiation (IR), ultraviolet (UV) light, therapeutic genotoxins and oncogene expression (*i.e.* c-Myc, viral oncogenes). These sources induce different types of DNA damage, such as DNA single- and double-strand breaks (SSB and DSB, respectively) (induced by IR), DNA crosslinks (*e.g.* induced by cisplatin), base modifications (*e.g.* induced by doxorubicin), and stalling of replication forks (*e.g.* caused by etoposide). Failure to repair DNA damage can cause mutations, genomic instability, premature ageing, mental retardation and other developmental disorders as well as cancer. Consequently eukaryotes, from yeast to mammals, have developed a comprehensive and complex set of mechanisms to sense different types of DNA damage and to safeguard the genome—the so-called DNA damage response (DDR). Depending on the cell type and the severity and extent of DNA damage, which determines whether or not damage

Issues in Toxicology No. 13
The Cellular Response to the Genotoxic Insult: The Question of Threshold for Genotoxic Carcinogens
Edited by Helmut Greim and Richard J. Albertini
© The Royal Society of Chemistry 2012
Published by the Royal Society of Chemistry, www.rsc.org

is reparable, the activated DDR can elicit different cellular responses. Mild DNA damage is normally managed through induction of cell cycle arrest mediated by the up-regulation of cyclin-dependent kinase inhibitors, such as p21, and subsequent repair of the lesions. In response to more severe and irreparable DNA damage, the cellular response shifts towards induction of the senescence or cell death programme such as apoptosis, mitotic catastrophe, autophagy and necrosis. The molecular basis underlying the decision-making process is currently the subject of intense investigation. Here we review current knowledge about the DDR and signalling with particular attention to cell death induction and the molecular switches between different cell death modalities following DNA damage.

4.2.2   DNA-damage Sensing: ATM and p53 System

The cellular response to DNA damage is a complex process that comprises a multiple network of signal transduction pathways. It is initiated by sensing or detecting damage, followed by signal transmission to effectors through activation of a cascade of phosphorylation events. This results in the initiation of a number of cellular responses such as cell cycle transition, DNA replication, DNA repair and cell death (Figure 4.2.1). Even though significant progress has been made in recent years in elucidating the molecular machinery of cellular responses to DNA damage, the exact mechanism by which DNA damage is sensed in cells still remains a matter of great interest and intense research.

Localization of DDR factors to DNA damage sites is initiated by special sensor proteins that directly recognize specific DNA lesions and activate the DDR. A group of proteins, including mediator of DNA damage checkpoint protein 1 (MDC1/NFBD1), 53BP1, Mre11-Rad50-NBS1 (MRN complex) and the Rad9-Rad1-Hus1 (9-1-1) complex have been implicated as DNA damage sensors, and recognize both DSBs and SSBs as well as regions of replication stress.[1–4] Binding of the MRN and 9-1-1 complexes to the sites of DNA damage in chromatin activates the signal transducing kinase ataxia-telangiectasia mutated (ATM), the ATM and Rad3-related (ATR) kinase, and the DNA-dependent protein kinase (DNA-PK), all of which are members of the phosphoinositide 3-kinase-related kinase family. The transducer kinases respond to different types of stimuli. DNA-PK and ATM are preferentially activated in response to DSBs and the disruption of chromatin structure, while ATR primarily responds to SSBs and replication forks.

4.2.2.1   ATM/ATR Activation and Kinase Cascade

ATM and ATR are conserved serine/threonine kinases characterized by a *C*-terminal catalytic motif containing a phosphatidylinositol 3-kinase domain. They share many targets that are phosphorylated by ATM immediately after the appearance of DSBs, and later by ATR, which probably maintains these phosphorylations for extended periods of time.[5] Once activated, ATR and

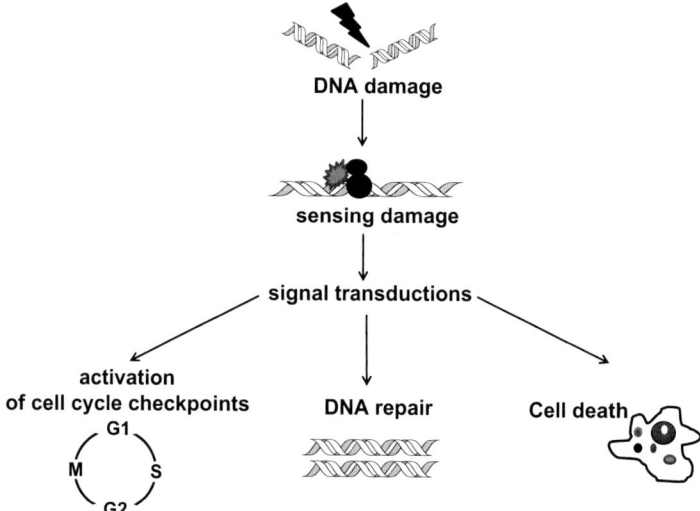

Figure 4.2.1 Schematic representation of DNA damage response. DNA damage results from numerous environmental and intrinsic sources, and activates a complex set of reactions called DNA damage response. It is initiated by sensing or damage detection, followed by signal transmission to effectors, and later results in initiation of number of cellular responses such as induction of cell cycle arrest, repair of the lesions or in case of more severe damage leads to activation of cell death programme. The activation of cell cycle checkpoints and DNA repair act as an immediate response to DNA damage to provide protection and recovery of damaged cells, whereas activation of cell death happens much later and aims to eliminate the irreversibly damaged cell.

ATM amplify the damage signal by phosphorylating various substrates, including key DNA damage response proteins H2AX, BRCA1, NBS1, SMC1, MDC1 and 53BP1, as well as repair proteins Ku70/80, artemis and Rad-like proteins.[6,7] Through interaction with ATR and the MRN complex, ATM mediates DNA damage repair through two main mechanisms—non-homologous end joining (NHEJ) and homologous recombination (HR). In addition, in order to allow time for DNA repair, ATM temporarily arrests the cell cycle by phosphorylation and activation of cell cycle checkpoints. Both G1/S and G2/M arrests can be mediated by ATM either through the ATM-p53/Chk2-p21-CDK or ATM-ATR-Chk1-mediated pathways, respectively.

Following DSBs ATM phosphorylates serine residue 15 (Ser15) of the tumour suppressor p53, which is involved in enhancing the transcriptional transactivation activity of p53.[8] ATM also regulates p53 activity through phosphorylation of its negative regulator mouse double minute 2 (MDM2) on Ser395, which inhibits p53 nuclear export and subsequent degradation.[9] Another control mechanism involves phosphorylation of the Chk2 kinase by ATM, which in turn phosphorylates p53 on Ser20, thereby blocking its interaction with MDM2. Upon activation p53 transcriptionally up-regulates the cyclin-dependent kinase inhibitor p21$^{waf1/cip1}$, which suppresses cyclin E/CDK2 kinase activity and prevents G1-to-S phase progression.[10] Phosphorylation of Chk2 or Chk1 by ATM and ATR, respectively, leads to the phosphorylation and subsequent ubiquitination of Cdc25A, and in this way prevents the activation of CDK2, resulting in prevention of DNA polymerase recruitment to replication origins and thereby blocking the transcription of genes required for progression of the S-phase. Progression of cells from the G2-phase into mitosis is also under the control of ATM in cells that were in earlier phases of the cell cycle at the time of DNA damage, and depends to a higher degree on ATR at the late G2 checkpoint.[11]

A recently conducted genome-wide RNA interference (RNAi) screen in *Drosophila* cells identified a list of 64 genes, of which more than 90% were not previously known to play a role in the G2–M checkpoint induced by DNA DSBs. These results provide insight into the diverse mechanisms that link DNA damage and the checkpoint signalling pathway.[12] Once the process of DNA repair has been successfully completed, cells can re-enter the cell cycle, whereas irreparable DNA damage can lead to irreversible cell cycle arrest known as senescence or induction of cell death.

4.2.2.2   Key Role of p53 in Genotoxic Stress Response

As one of the main targets of ATM/ATR, p53 is a universal sensor of genotoxic stress known as the 'guardian of the genome' and plays a pivotal role in maintaining its stability. *p53* is the most frequently mutated gene in all human cancers and its inactivation has been reported in more than 50% of all cancers. The p53 protein plays a key role in cellular responses to DNA damage, determining the fate of cells towards either survival, which involves cell cycle delay accompanied by repair of DNA damage or cell death.[13] p53 can be activated by UV or gamma irradiation, DNA-damaging drugs or cellular oxidative stress at the transcriptional level as well as through a number of post-transcriptional modifications, including phosphorylation, methylation and acetylation.

In addition to the DNA damage-dependent signalling pathway that causes p53 protein activation by ATM kinase selective phosphorylation, DNA-PK, casein kinase 1, p38 mitogen-activated kinase (MAPK) and Jun *N*-terminal kinase (JNK) are also involved in p53 phosphorylation at different serine or tyrosine residues, leading to its stabilization.[14–16] p53 modifications and chromatin remodelling through interaction of p53 with transcriptional co-activators further activate p53 and increase its DNA binding specificity, allowing selective transcription of p53-dependent genes.

The *p21$^{waf1/cip1}$* gene product is the critical downstream effector of *p53* in DNA damage-induced G1 arrest. The identified p21 was shown to bind various cyclin–CDK complexes that regulate both G1–S and G2–M transitions, and to inhibit their action.[17,18] The molecular mechanism of p53 regulation of the G2 checkpoint is still a matter of debate. Besides p21, another p53 effector gene, cyclin G, is believed to regulate G2 arrest, suggesting the ambiguous role of p53 as a tumour suppressor. The accumulated evidence suggests that the deregulated p53-dependent checkpoint control can cause the propagation of mutations and chromosomal aberrations, leading to carcinogenesis.[19]

The role of p53 in promoting cell death is critical to its role in tumour suppression. Upon activation p53 induces up-regulation of various pro-apoptotic proteins such as Bax, Bid, Apaf-1, CD95, PIDD, PUMA, Noxa, Scotin and PIG3.[20]

4.2.3 Cell Senescence in Response to DNA Damage

Cellular senescence (Figure 4.2.2) is one of the consequences of severe or irreparable DNA damage, especially DNA DSBs, and involves an irreversible growth arrest of cells in the G1 phase. Cellular senescence limits the proliferation of DNA-damaged cells that are at risk of neoplastic transformation. However, these cells remain metabolically active, showing a greatly altered

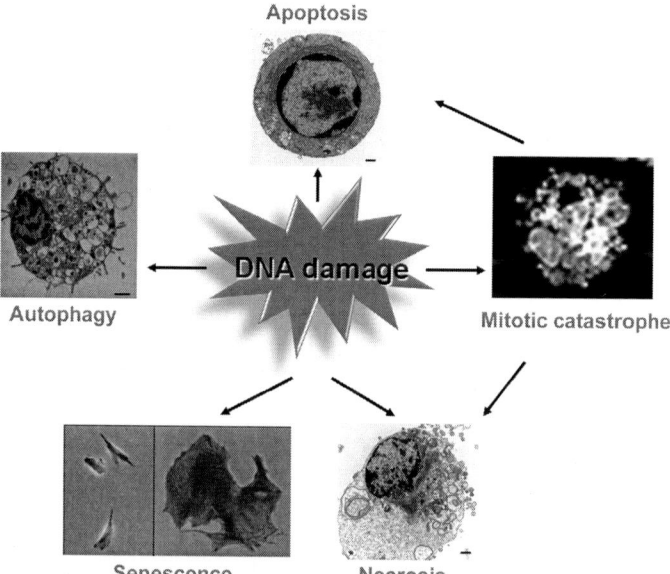

Figure 4.2.2 Different modes of cell death induced by DNA damage. Depending on cellular context as well as on type and the severity of stimulus, DNA damage response shifts towards induction of the senescence or cell death programme such as apoptosis, mitotic catastrophe, autophagy and necrosis. The apoptotic response is generally orchestrated by p53, whereas the necrosis pathway is mediated by PARP activation. The mitotic catastrophe is initiated when the cell cycle checkpoints are compromised and cells enter mitosis prematurely before the completion of DNA repair; commonly it serves as a pre-stage of apoptosis or necrosis. Depending on the cellular context, type and magnitude of stress, autophagy may enhance cell survival or result in autophagic cell death, having tight crosstalk with the apoptotic pathway.

pattern of gene expression. Most senescent cells acquire characteristic morphological changes, becoming flat, increasing in volume up to ten-fold and displaying a vacuole-rich cytoplasm, plus an increased nucleus to cytoplasm ratio and mitochondrial mass.[21] There are two main types of senescence—replicative (Hayflick limit) and premature senescence, both of which share a common signal, now recognized as DNA strand breaks. In replicative senescence, telomere destruction and shortening yield DNA DSBs at chromosome ends that initiate DNA damage responses. In premature senescence, oncogenic activation leads to elevated intracellular levels of reactive species, augmented numbers of active replicons, alterations in DNA replication fork progression, and the appearance of DNA SSBs and DSBs that initiate DNA damage responses.

4.2.3.1   Molecular Pathways of DNA Damage-Induced Senescence

How does cellular senescence result from DNA damage? First it should be noted that the outcomes of cellular senescence following DNA damage are highly dependent on the cell type and physiological context. For instance, cells of epithelial and stromal origin undergo senescence following severe DNA damage, while lymphoid cells undergo apoptosis. In most cell types, activation of p53 is crucial for initiating the senescence response following DNA damage, and in certain cells p53 is also important for maintaining the senescent growth arrest. As mentioned above, the p53 pathway is regulated at multiple points. p21[wafl/cip1] is a crucial target of p53 and is also a mediator of senescence. It inhibits a spectrum of cyclin-dependent kinases that regulate phosphorylation of the retinoblastoma protein pRb; the dephosphorylated form of pRb binds to the transcription factor E2F, leading to transactivation of the genes necessary for the G1–S transition. On the other hand, p21 can mediate transient DNA damage-induced growth arrest. It is still unclear what determines whether cells senesce or arrest transiently. It is likely that rapid DNA repair will quickly shut down p53–p21 signalling, whereas incomplete or incorrect repair results in prolonged signalling and senescence.[22–24]

Stimuli that produce DDR can also engage the p16–pRb pathway. It was shown that, in the terminal stages of growth arrest in the senescence of human diploid fibroblasts, the p16 mRNA and cellular protein levels were up-regulated almost 40-fold. p16 was also shown to be complexed to both CDK4 and CDK6. Immunodepletion analysis of p21 and p16 from the senescent fibroblast extracts revealed that p16 is the major CDK inhibitor for both CDK4 and CDK6 kinases.[25] Inducible expression of p16 in various normal and tumour cell types resulted in a senescence response, which has been found to be, at least in part, dependent on functional Rb. Following treatment with cyclophosphamide *in vivo*, mouse lymphomas underwent cellular senescence rather than apoptosis in a p53- and p16-dependent manner.[26] Although both p16 and p21 are CDK inhibitors, they are not equivalent. Cells that undergo senescence solely due to activation of p53–p21 can restart growth after inactivation of the p53 pathway. However, cells that fully engage the p16–pRB pathway cannot resume growth even after inactivation of p53, pRB or p16.[27] Thus, both of these pathways can cause transient or reversible cell cycle arrest, and it remains to be clarified exactly how their activities are modified by senescence-inducing signals and how cells make decisions in response to damage signals.

4.2.3.2 Cellular Senescence—Dead End Or?

Despite the common definition of cellular senescence as a permanent growth-arrested state, technical difficulties remain with regard to discriminating irreversible conditions from the reversible long-term arrested state. *In vitro* studies indicate that senescent cells can re-enter the cell cycle, and that this can be achieved by selective inactivation of critical mediators of DNA damage checkpoints, which are required to establish and maintain the senescent growth arrest. It was shown that depletion of p53 in some replicatively senescent human cells completely reversed senescent growth arrest. Such experimental manipulations that abolish p53 function cause post-mitotic senescent cells to resume growth for many cycles until widespread severe telomere dysfunction drives them into crisis, characterized by acute genomic instability.[27]

Overall, the existing data suggest that a DNA damage signalling pathway must be constitutively active in order to keep cells in a senescent state. The ability of the cell to repair damaged DNA plays an essential role. Senescence is an important anti-carcinogenic programme and neoplastic transformation comprises a number of events that allow cells to bypass senescence by the inactivation of pathways associated with it. Nonetheless, many tumour cells retain the capacity to senesce when exposed to external stress. Most conventional anti-cancer therapies activate DNA damage signalling pathways in order to induce primarily apoptotic cell death, but treated cells often undergo growth arrest or senescence rather than dying. Currently it is not clear which specific signals force cells to undergo either senescence or initiate apoptosis.[21] Senescent cells, despite the arrest of growth, continue to display metabolic activity and synthesize different soluble factors with diverse biological activities, thus retaining the ability to influence neighbouring cells. Studies on colon carcinoma cells exposed to doxorubicin demonstrated that cells senescing in response to chemotherapy secrete proteins with anti-apoptotic, mitogenic and angiogenic activities, suggesting that drug-induced senescence is associated with paracrine tumour-promoting effects.[28]

While apoptotic cells have been shown to attract macrophages due to specific signals from apoptotic bodies that stimulate phagocytosis, very little is known about such mechanisms for the clearance of senescent cells. In some cases senescent cells have been found to persist in the human body for a long time.[29] It was also demonstrated that senescent sarcoma cells were cleared away by an innate immune response due to the production of pro-inflammatory molecules by the senescent cells themselves. Thus, senescent cells can be cleared *in vivo* by the coordinated action of the senescence programme and the innate immune system in the context of tumour suppression. Nevertheless, more *in vivo* data are necessary to fully understand and evaluate the consequences of cell senescence.

4.2.3.3   Overlap between Senescence and Apoptosis

Given the complex overlap between senescence and apoptosis, an important question remains. What is the crucial trigger that drives cell to senesce or die through apoptosis? Even though the severity of DNA damage has been described as the main driver, the exact molecular mechanisms that determine the choice between apoptosis and senescence have not been completely established. Since both events are under the control of p53, it is difficult to understand the switch between these two modes of cellular response upon DNA damage-induced activation of p53. A recent study on mouse liver attempted to analyse the effects of the loss of telomere protection by TRF2 (TATA box binding protein-related factor 2) inhibition in correlation with the p53 gene status. Higher telomere dysfunction provoked p53-independent apoptosis whereas cellular senescence was associated with low levels of telomere dysfunction.[30] It seems that normal cells, in which all signalling pathways are intact, have a preference toward senescence compared with transformed cells and that senescence is usually induced by lower levels of damage than those leading to apoptosis.[21] Overall, p53 stands at the centre of a complex signal transduction pathway following DNA damage, and at least 20 different sites are modified in the human p53 protein in response to various stress signals. For example, phosphorylation at Ser15, Thr18 and Ser376 and dephosphorylation at Ser392 have been detected in fibroblasts undergoing senescence.[31] In contrast, phosphorylation at Ser46 was shown to be associated with the induction of p53AIP1 (p53-regulated apoptosis-inducing protein 1) and apoptosis. A study on MCF-7 cells demonstrated the role of different phosphorylation sites (Ser15 and Ser46), induction of p21, and high or low ROS levels in determining the dosage of adriamycin that induced either apoptosis or senescence. Cells in these two states exhibited quite distinct time course profiles of the proteins p53, p21 and E2F1. As mentioned above, DNA damage can also activate RB through the p16 pathway to induce senescence. An *in vivo* study showed that, following cyclophosphamide treatment, a p53-dependent treatment effect can also occur in an apoptosis-independent fashion by inducing cellular senescence in a p53- and p16-dependent manner.[26] Thus, the genetic context, cell type or the specific activation of certain target genes upon DNA damage can favour either apoptosis or senescence.

4.2.4   Apoptosis as a Fast-acting Response to DNA Damage

Like cellular senescence, apoptosis is an extreme response to cellular stress (Figure 4.2.2); it is activated if repair of DNA damage is slow or incomplete, and represents an important tumour-suppressive mechanism. It is still not completely clear how cells decide whether to undergo senescence or apoptosis, and despite the fact that most cells are capable of both, these processes seem to be mutually exclusive. Although apoptosis is a fast-acting response mode, very little is known about the possibility that apoptosis-competent cells might be guided into senescence following DNA-damaging therapy, or whether senescent cells could ever undergo apoptosis upon an additional pro-apoptotic signal. Cell type appears to be a determinant in the decision-making process; for instance, lymphocytes preferentially undergo DNA-damage-induced apoptosis, while damaged epithelial cells and fibroblasts are more likely to undergo senescence. Numerous studies have clearly shown crosstalk between the processes of apoptosis and senescence, primarily at the level of the tumour suppressor protein p53.

4.2.4.1 Multiple Functions of p53 in DNA Damage-Induced Apoptosis

The tumour suppressor protein p53 is considered to be a major player in the apoptotic response to genotoxins. The phosphorylation of p53 at Ser46 was reported to be correlated with the induction of apoptosis, and thus is the most prominent in priming its apoptotic activity.[32] Subsequently, homeodomain-interacting protein kinase 2 (HIPK2), a serine/threonine nuclear kinase, was identified as the p53 Ser46 kinase.[33,34] Following severe damage, HIPK2 binds p53 and is recruited to promyelocytic leukaemia protein (PML) nuclear bodies, where PML acts as the main co-factor for efficient HIPK2-mediated p53 phosphorylation. HIPK2-mediated p53 Ser46 phosphorylation leads to enhanced acetylation (Lys382) of p53 by the CREB binding protein (CBP), which also interacts with HIPK2 and co-localizes in PML nuclear bodies. The latter results in complete transcriptional activation of p53, followed by potentiation of pro-apoptotic target gene expression.[33,34] p53 can up-regulate PML expression, thus potentiating the apoptotic signal. Depending on the extent of DNA damage, p53 itself can either potentiate HIPK2 activity or negatively regulate HIPK2 stability.[35,36]

Following activation, p53 initiates expression of its apoptotic target genes such as p53-up-regulated modulator of apoptosis (PUMA), p53AIP1 and Bax, and apoptotic protease activating factor 1 (Apaf-1), in this way regulating the intrinsic apoptotic pathway. Following DNA damage, p53 also promotes the extrinsic apoptotic pathway through up-regulation of the TRAIL receptors, death receptor-4 (DR4) and death receptor-5 (DR5, KILLER), as well as the CD95 (Fas/Apo1) receptor and the CD95 (Fas/Apo1) ligand.

Acting primarily as a nuclear transcription factor, p53 was also shown to contribute to apoptosis induction in the cytoplasm by transcription-independent mechanisms. It was found to translocate to the cytoplasm in response to different stress signals where it stimulates mitochondrial outer membrane permeabilization and caspase activation, as well as antagonizing anti-apoptotic proteins Bcl-2 and Bcl-X_L by their direct binding.[37]

4.2.4.2   p53-independent Response to DNA Damage

Although p53 plays a central role in DNA damage-induced apoptosis, p53-mutated cells do not completely lose their ability to undergo apoptosis; they still die, but less efficiently. There are, obviously, some backup mechanisms that can be engaged upon DNA damage *via* a p53-independent pathway. The fact that p53 is very frequently mutated in tumour cells makes these mechanisms a matter of great interest and intense research. Clearly, one of the first mechanisms involves p53 homologous proteins such as p73 and p63. Following etoposide-induced DNA damage, Chk1 and Chk2 are activated by ATM and/or ATR, which leads to activation of the p73 transcription factor E2F1, followed by an increased level of p73 protein. Although p53 requires p63 and p73 to induce apoptosis, p73 is pro-apoptotic itself even in the absence of p53.[38] p73 elicits apoptosis *via* the mitochondrial pathway using PUMA and Bax as mediators, followed by the release of cytochrome *c*. Additionally, p73 shares many pro-apoptotic target genes with p53 such as Noxa, caspase-6 and CD95. Interestingly, p73 is also a substrate for caspases and, after processing, its truncated versions localize in the mitochondria and amplify apoptosis in response to treatment with TRAIL.[39] Nevertheless, it remains unclear whether this function of p73 is also activated following DNA damage. The cytoplasmic functions of p73 thus remain to be investigated further.

HIPK2 was also shown to be involved in DNA damage-induced cell death signalling independently of its fundamental role in p53-mediated apoptosis as discussed above. Expression of HIPK2 or exposure to UV irradiation *via* a proteosome-mediated pathway reduces levels of transcriptional co-repressor C-terminal binding protein (CtBP), which is a critical repressor of pro-apoptotic genes such as *bax* and *noxa*.[40] A recent study also showed that c-Jun N-terminal kinase 1 (JNK1) activation triggered CtBP phosphorylation on Ser422 and subsequent degradation, inducing p53-independent apoptosis in human lung cancer cells. JNK1 has previously been linked with UV-directed apoptosis. Expression of MKK7 (mitogen-activated protein kinase kinase 7)-JNK1 or exposure to UV irradiation reduces cellular levels of CtBP *via* a proteasome-mediated pathway.[41] Additionally, HIPK2 activates the JNK signalling pathway in hepatoma cells after treatment with TGF-β, and consequently JNK is capable of stimulating cell death *via* the mitochondrial pathway.[42] Interestingly, a report by Zhang and coworkers indicates that TGF-β mediates activation of ATM in 293 cells, which results in p53 Ser15 phosphorylation.[43] However, it remains unclear whether HIPK2 plays any role in this pathway.

The protein Nur77, also known as TR3 or nerve growth factor IB (NGFI-B), is a unique transcription factor belonging to the orphan nuclear receptor superfamily and can be activated by different stress stimuli, including DNA damage. To an even greater extent than p53, Nur77 can provoke the opposing biological activities of survival and death. Similarly to p53, the pro-apoptotic action of Nur77 is believed to be exerted through its nuclear transactivation of gene expression. Accumulating data suggest that translocation of Nur77 from the nucleus to the mitochondria represents an important mechanism by which

Nur77 initiates apoptosis. In response to differential apoptotic stimuli, Nur77 migrates from the nucleus to the cytoplasm, where it targets mitochondria, associated with the release of cytochrome *c* and induction of apoptosis. Remarkably, DNA-binding domain-deleted Nur77 constitutively resides in the mitochondria and is accompanied by massive cytochrome *c* release.[44] However, studies have shown the ability of Nur77 to inhibit apoptosis.[45] When over-expressed in the fibroblastic cell line HEK293, Nur77 promotes resistance to programmed cell death induced by death receptor engagement, DNA-damaging agents and endoplasmic reticulum stress. Nur77 overexpression leads to enhanced NF-κB activity, which contributes to Nur77 anti-apoptotic activity *via* induction of the anti-apoptotic gene cIAP1. These results show that crosstalk between Nur77 and other transcription factors contributes to cell fate in response to different apoptosis-inducing agents.[45]

Another pathway believed to trigger p53-independent apoptosis involves Bcl-2 degradation, which is a hallmark of genotoxin-induced apoptosis in many p53-mutated experimental systems, although the mechanism remains unclear. Some data suggest the involvement of stress kinases because phosphorylation of Bcl-2 at Ser87 by the MAP kinases protects against Bcl-2 degradation and apoptosis. Furthermore, dephosphorylation of Bad leads to sequestration of Bcl-2 and Bcl-X_L from the mitochondrial membrane, which is followed by apoptosis. However, the direct link between DNA damage and the regulation of Bcl-2 protein in the mitochondrial membrane remains to be elucidated.

4.2.4.3 Role of Caspase-2 in DNA Damage Responses

An additional mechanism that links DNA damage and the apoptotic pathway involves caspase-2, which is constitutively present in the nucleus and is required for etoposide-, cisplatin- and UV-light-induced apoptosis. Caspase-2-induced cell death is followed by cytochrome *c* release and requires caspase-9 activation, indicating that caspase-2 acts through the mitochondrial apoptotic pathway. Caspase-2 is cleaved and activated by a cytosolic protein complex PIDDosome, which contains PIDD (p53-induced protein with a death domain), an adaptor protein RAIDD (receptor-interacting protein-associated ICH-1 homologous protein with a death domain) and caspase-2. Increased PIDD expression results in spontaneous activation of caspase-2 and sensitization to apoptosis induced by genotoxic stimuli.[46] The prodomain of caspase-2 was shown to be required for its nuclear localization and activation, and further studies demonstrated that nuclear caspase-2 is cleaved rapidly in cells upon DNA damage.[48]

In addition to its contribution to apoptosis, caspase-2 is thought to be involved in other parts of the DNA damage response. One study identified a nuclear protein complex that activates caspase-2 in the cell nucleus.[47] This contains PIDD, caspase-2 and the catalytic subunit of DNA-PK (DNA-PKcs) but not RAIDD. In response to γ-radiation, caspase-2 is phosphorylated at S122 within its prodomain by DNA-PKcs, whose kinase activity is promoted by PIDD, leading to caspase-2 cleavage and activation. Characterization of this complex revealed unexpected involvement of caspase-2 in a G2/M DNA damage checkpoint and the repair of DNA DSBs mediated by the NHEJ pathway.[47] Thus, obtained data raise interesting issues with respect to the complexity of the G2/M checkpoint and NHEJ regulation mediated by a caspase that has long been implicated in the execution of apoptosis. However, how nuclear caspase-2 affects the G2/M checkpoint and NHEJ is yet to be elucidated.

Another study obtained data showing that Chk1 acts as a suppressor of the caspase-2 apoptotic response to DNA damage.[48] The ATM/ATR-caspase-2 pathway is triggered by DNA damage in cells in which Chk1 activity is simultaneously compromised. This pathway seems to be independent of p53 and requires the combination of IR and Chk1 inhibition in order to be triggered. Deletion of any subunit of the DNA-PK, similarly to Chk1, is able to rescue the p53-deficient cells from apoptosis. Since both p53 and Chk1 are downstream ATM/ATR substrates, inhibition of these proteins may affect DNA-PK signalling. Whereas DNA-PK mediates the activation of Chk2, its function as a kinase is regulated by Chk1, leading to the phosphorylation of p53. It is unclear whether p53 is regulated directly by DNA-PKcs or indirectly through the DNA-PKcs-mediated pathway. In fact, Chk1 and Ku70/80 are assembled in a complex that stimulates the DNA-PK-dependent end-joining process *in vitro*. Accordingly, Chk1 appears to influence caspase-2 activation *via* the regulation of DNA-PK.[48]

Overall, the existing data indicate the importance of caspase-2 activation for successful induction of apoptosis upon DNA damage, as well as for cell cycle checkpoints and DNA repair. However, the complex set of functions of this unique caspase is far from being completely identified and understood.

4.2.5 Cell Death by Mitotic Catastrophe

As mentioned above, to maintain genome integrity, cells respond to DNA damage by either a delay in cell cycle progression, allowing time for proper DNA repair, or the elimination of cells that are irreparably damaged. Checkpoints that monitor the cell cycle are capable of inducing arrest at multiple points, including the G1–S transition, S-phase progression and G2–M transition as well as progression through mitosis. When checkpoints are compromised, cells can enter mitosis prematurely before the completion of DNA repair and initiate mitotic catastrophe (MC) (Figure 4.2.2).[49]

Mitotic catastrophe is a process occurring either during or shortly after a dysregulated/failed mitosis and can be accompanied by morphological alterations including micronucleation (which often results from chromosomes and/or chromosome fragments that have not been distributed evenly between the daughter nuclei) and multinucleation (the presence of two or more nuclei of similar or heterogeneous sizes, derived from a deficient separation during cytokinesis). MC can be caused by chemical or physical stresses. In addition to mitotic failure caused by defective cell cycle checkpoints, it can be triggered by agents that influence the stability of microtubules, and various anticancer drugs. MC has also been described as the main form of cell death induced by ionizing radiation.[54] How does DNA damage induce MC and what is the link between MC and the apoptotic machinery?

It is well-known that tumours differ in their sensitivity to treatment with DNA damaging agents. Sensitive cells die during interphase, before entering mitosis. Even low doses of irradiation can induce interphase death in some tumours of haematopoietic origin. Cells that do not undergo death in interphase may become arrested in G1 and/or G2. If the cellular machinery is not able to repair the damage during arrest, various death-associated consequences may follow. For example, some tumour cells die by apoptosis after one or even repeated mitotic cycles. In other circumstances premature mitosis may lead to MC. However, depending on the severity of DNA damage, cells can also exit mitosis to form 4N G1 cells, and after the second arrest at G1 they die or survive. The extent of cell death and loss of clonogenicity is correlated with the duration of growth arrest.

Dying cells are characterized by mitotic abnormalities such as a multipolar meta- or anaphase, lagging telophase, the random distribution of condensed chromosomes throughout the cells, *etc.* The biochemical mechanisms responsible for these differences remain unclear. However, without prematurely entering mitosis, a cell cannot undergo MC. The damage during interphase is expanded during mitosis and activates the mitotic checkpoints. It is important to note that cells that are sensitive to apoptosis and die during interphase have very little or no likelihood of displaying features of MC.

Several experimental models have been utilized to clarify the biochemical changes associated with MC. Thus, the inability of HCT116 colon carcinoma cells to sequester cyclin B and Cdc2 in the cytoplasm following doxorubicin-induced DNA damage is a well-established MC model. However, other data indicate that Chk2 and not 14-3-3σ may be the main negative regulator of MC in this model.[50] Abrogation of the ATR-initiated checkpoint cascade mediated through Chk1 also directs cells into MC and has been analysed with respect to specific Chk1 phosphorylations. Other cell cycle regulating proteins whose normal function seems to inhibit entry into mitosis are the checkpoint mediators Brca1, p73 and p53. In the absence of p53 cells have the ability to activate alternative routes to halt the cell cycle. One of the proteins that may be involved in this process in addition to Chk1 is p38MAPK.[51] Depletion of p38MAPK/MK2 in p53-deficient cells leads to checkpoint deregulation and MC irrespective of the type of DNA damage induced. This two-hit model suggests that MC is the outcome of DNA damage only when multiple central checkpoint proteins are suppressed and other options for cell death and survival are minimized.

One of the most prominent morphological characteristics of MC is the formation of giant cells with abnormal nuclei.[52] Several possible scenarios for the death of giant cells have been suggested. According to one of these, death is a direct cause of MC and is distinct from apoptosis. This conclusion was based on the dissimilarities in morphology between MC and apoptotic cells. Despite their distinct morphology, some authors have suggested that the two processes share several biochemical hallmarks. Thus, following inhibition of Chk2 in syncytia generated by the fusion of asynchronous HeLa cells, metaphase-associated MC was accompanied by sequential caspase-2 activation, mitochondrial release of pro-apoptotic proteins, activation of caspase-3, DNA fragmentation and chromatin condensation.[53] However, it has also been reported that dying multinucleated giant cells are characterized by uncondensed chromatin and the absence of DNA ladder formation.[54]

It seems that the presence or absence of chromatin condensation during MC depends on different events, including the stage at which mitotic arrest takes place. There is no consensus concerning the requirement of caspases for MC, but it is likely that the progression of MC is caspase-independent since neither inhibition nor down-regulation of caspases influences the appearance of giant cells. However, caspases are essential for the termination of MC, suggesting that MC-related morphological changes are followed by activation of the apoptotic machinery. Thus, MC is not an ultimate manifestation of cell death but rather a process leading to apoptosis.[54]

Apoptosis, however, is not always required for MC lethality since some giant cells can undergo slow death in a necrosis-like manner. This conclusion is based on similarities between morphological changes during MC and necrosis (the loss of nuclear and plasma membrane integrities). Mitotic arrest in docetaxel-treated tumour cells was followed by massive cell destruction by means of cell lysis. Appearance of multinucleated giant cells that were terminated *via* a necrosis-like lysis was also observed in a cisplatin-treated ovarian carcinoma cell line, SKOV-3. Necrosis following MC could be an effect of genetic instability caused by aneuploidy and/or polyploidy. Taken together, it is important to note that cells that are facing a mitotic-linked cell death can in fact die by two separate mechanisms, either by apoptosis or necrosis, depending on the molecular profile of the cell.[54]

4.2.6   DNA Damage and Autophagy

Autophagy represents an evolutionarily conserved catabolic programme for the degradation of proteins and other subcellular elements through lysosomal proteolysis. This process plays a fundamental role in maintaining tissue homeostasis, promotes protein turnover and removes damaged proteins and organelles, intracellular pathogens and superfluous portions of the cytoplasm, as well as long-lived, aberrant or aggregated proteins.

Three types of autophagy have been described, which differ by the mode of cargo delivery to lysosomes. The most studied type is macroautophagy, which is characterized by autophagosome formation and is negatively regulated by the mammalian target of rapamycin (mTOR). The second type is micro-autophagy, in which engulfment is carried out directly by the lysosomal membrane; and the third type of self-eating is chaperone-mediated autophagy (CMA), during which cytosolic proteins that contain a specific lysosome-targeting motif are recognized by a complex of chaperone proteins and targeted to the lysosomal membrane.[55]

In most cases autophagy serves as a pro-survival mechanism, which adapts cells to stress conditions by providing metabolic precursors for cellular renewal and maintenance through the recycling of cellular components. On the other hand, substantial activation of autophagy in response to stress can lead to cell death.

4.2.6.1 Autophagic Cell Death Induced by DNA Damage

The cellular outcome of inducing autophagy in response to stress is complex and depends on the cellular context, type and magnitude of stress. In mild stress conditions, autophagy may enhance cell survival by recycling cellular components and allowing the cell to engage repair mechanisms and checkpoint activation; higher levels of damage over a prolonged period of time may trigger autophagic cell death, also known as type II programmed cell death (Figure 4.2.2). Studies conducted on MCF-7 cells treated with tamoxifen, which induces DNA adducts or chromosomal aberrations, revealed the induction of active self-destruction of cells, characterized by the autophagic degradation of cytoplasmic components including organelles prior to nuclear collapse, but the preservation of cytoskeletal elements until later stages.[56]

The typical DNA-damaging compounds 6-thioguanine (6-TG) and 5-fluorouracil (5-FU) were shown to induce autophagy. An important role in this process was suggested for BNIP3 (Bcl-2/adenovirus E1B 19 kDa protein-interacting protein 3), which mediates autophagy in a p53- and mTOR-dependent manner.[57] In fact, a number of studies have indicated the role of p53 as one of the important links between DNA damage and autophagy. Although the mechanisms of p53-dependent induction of autophagy remain incompletely understood, there are indications that both transcription-dependent (up-regulation of mTOR inhibitors, PTEN and tuberous sclerosis protein 1 (TSC1) or DRAM) and transcription-independent pathways (AMPK activation) are involved.[58,59] Interestingly, p53-induced autophagy may lead to cell death and this can be blocked by DRAM siRNA. However, under some circumstances, such as in c-Myc-driven lymphomas, p53-mediated autophagy can actually promote cell survival, as inhibition of autophagosomal maturation enhances p53-mediated tumour-cell death. Thus, p53-mediated autophagy can affect both life and death decisions depending on the cell type or type of stimulus, and reflects the activation of different sets of p53 signals.

It is important to note that the role of p53 as a positive regulator of autophagy has been questioned by some authors.[60] In particular, deletion, depletion or inhibition of p53 was shown to induce autophagy in human, mouse and nematode cells. Inhibition of p53 led to autophagy in anucleated cells and cytoplasmic, not nuclear, p53 is believed to repress the enhanced autophagy of p53$^{-/-}$ cells. Moreover, many inducers of autophagy (for example, starvation, rapamycin and toxins) stimulate proteasome-mediated degradation of p53 through a pathway relying on the HDM2. Inhibition of p53 degradation prevents the activation of autophagy in various cell lines in response to distinct stimuli.[60] One possible explanation for the dual action of p53 in the regulation of autophagy is that basal *versus* stress-induced levels of p53 might coordinate its regulatory outputs. Given the complexity of such regulation, the specific nature of the stress signal probably also makes a significant contribution to switching autophagy 'on' or 'off'. Overall, the opposing control of autophagy by p53 raises many important questions for future research.

Recent data described a new cytoplasmic pathway linking ATM and autophagy, suggesting that ATM can engage the TSC2/mTORC1 signalling pathway in order to regulate autophagy. Even though the molecular mechanisms regulating this phenomenon are not clear, it was indicated that a substantial cytoplasmic pool of ATM is phosphorylated in the presence of leptomycin B and could initiate signalling *via* LKB1 (serine/threonine kinase 11), AMPK (5′-adenosine monophosphate-activated protein kinase) and TSC2 in the cytoplasm to suppress mTORC1, resulting in increased autophagic flux.[61]

PARP has also been implicated in the cell's decision to commit to autophagy following DNA damage. Studies using PARP-1 wild-type and deficient cells revealed the formation of autophagic vesicles and the increased expression of genes involved in autophagy (bnip-3, cathepsin b and beclin 1) in wild-type cells treated with doxorubicin but not in PARP$^{-/-}$ cells or cells treated with a PARP inhibitor. Mechanistically the lack of autophagic features in PARP-1 deficient/ PARP inhibited cells is attributed to the prevention of ATP and NAD$^+$ depletion and to the activation of the key autophagy regulator mTOR.[62]

4.2.6.2 Controversial Role of Autophagy in Mechanisms of Cell Survival or Death

Despite the large amount of accumulated data concerning autophagic cell death, it remains controversial whether autophagy is protective or toxic for cells. In addition to its housekeeping function, autophagy has been considered a mechanism by which the cell protects itself from various stresses. There is extensive evidence for a cytoprotective function of autophagy both in normal and tumour cells. For example, induction of autophagy provokes an adaptive response, suppresses cisplatin-induced apoptosis, and prolongs cell survival during cisplatin injury of renal tubular epithelial cells.[63] In HCT-116 colon carcinoma cells, the promotion of autophagy was associated with resistance to TRAIL,[63] whereby a blockade of autophagy resulted in apoptosis. Similarly, in colon carcinoma cells exposed to the antitumour drug, 5-fluorouracil, blocking autophagy resulted in an increase in apoptosis. Autophagy was shown to delay the response to DNA damage in camptothecin-treated MCF-7 and inhibition of autophagy by eliminating the expression of beclin 1 and Atg7 proteins stimulates apoptotic pathways in DNA-damaged MCF-7 cells, suggesting that a post-mitochondrial caspase cascade is delayed as a result of early disposal of damaged mitochondria within autophagosomes.[64,66] This proposal is supported by evidence that allelic loss of beclin1 and defective autophagy activate the DNA damage response and promote genomic instability *in vitro* and in mammary tumours *in vivo*.

As described above, numerous studies have considered autophagy as the mode of cell death, especially in tumour cells. Many studies have been performed using MCF-7 breast tumour cells, which do not express caspase-3 and are characterized by reduced susceptibility to caspase-dependent apoptosis.[65] Resistance to apoptosis is believed to increase the possibility that cells may undergo autophagic cell death in response to chemotherapy and irradiation, promoting an alternative pathway for drugs and irradiation to facilitate tumour cell killing. An analogue of the microtubule poison, paclitaxel, was shown to promote autophagy in MCF-7 cells.[66] On the other hand, cervical carcinoma cells die in response to etoposide *via* both autophagy and apoptosis, and the blockade of either mode of cell death was associated with reduced drug toxicity.[67] Accumulated evidence supports the role of autophagy as a mode of radiosensitization rather than protection from radiation injury and cell death. Moreover, there are some reports indicating that autophagy is a primary response to radiation, although it is generally thought to be protective against radiation. Overall, despite data on both the cytoprotective and cytotoxic properties of autophagy, it should be emphasized that the conditions under which autophagy acts as a cytoprotective mechanism and the types of stress under which autophagy acts as a mode of cell death remain to be fully defined.

4.2.6.3 Switch between Autophagic and Apoptotic Pathways

The functional relationship between autophagy and apoptosis is complex, and was first recognized when autophagy-related gene Atg6 or beclin 1 was initially identified as a Bcl-2- interacting protein. Both autophagy ('self-eating') and apoptosis ('self-killing') are highly regulated processes, share many common inducers, and can display cross-inhibitory and activation interactions. Depending on the stimulus and cellular context, autophagy may be a prerequisite for apoptosis by preceding and later turning it on. On the other hand, there are some circumstances in which autophagy may antagonize or delay apoptosis, or in which these two processes may act exclusively and serve as reciprocal backup mechanisms.

Although it has become clear that cell fate in response to stress, including DNA damage, can be determined by the functional status and the interaction between apoptosis and autophagy, the precise mechanisms underlying this switch between these two stress-mitigating pathways remain obscure. Studies on double knock out mouse embryonic fibroblasts (MEFs) and DNA-damaging agents indicate that after treatment with etoposide Bax$^{-/-}$ Bak$^{-/-}$ MEFs fail to undergo apoptosis and instead manifest massive autophagy followed by delayed cell death. Knockdown of the essential autophagy gene products beclin 1 and Atg5 also reduces the etoposide-induced cell death of Bax$^{-/-}$ Bak$^{-/-}$ MEFs. In this cellular setting, no autophagy was induced when apoptosis was inhibited by other perturbations, such as Apaf-1$^{-/-}$ and caspase-9$^{-/-}$ MEFs or the addition of the caspase inhibitor zVAD-fmk to wild-type MEFs.[68] This could imply that Bcl-2-family proteins such as Bax and Bak may have direct or indirect roles in regulating autophagy that are independent of their role in modulating apoptosis. Furthermore, PUMA, a key mediator of p53-dependent apoptosis, was found to induce autophagy that leads to the selective removal of mitochondria. The study suggested that, under some circumstances, the selective targeting of mitochondria for autophagy can enhance apoptosis since inhibition of PUMA or Bax-induced autophagy leads to a reduced apoptotic response.[69] Data obtained with MCF-7 cells suggest that, depending on the functional status of caspase-3, MCF-7 cells may switch between autophagic and apoptotic features of cell death, supporting the notion that autophagy can substitute the apoptotic pathway when the apoptotic pathway is deregulated.

There is an interesting link between the p53 and JNK signalling pathways in terms of the regulation of autophagy. Ewing sarcoma cells treated with 2-methoxyestradiol (2-ME) are characterized by enhanced autophagy and apoptosis occurring through the activation of both the p53 and JNK pathways. In this context, p53 regulates, at least partially, JNK activation, which in turn modulates autophagy *via* two distinct mechanisms: the promotion of Bcl-2 phosphorylation resulting in the dissociation of the beclin 1-Bcl-2 complex, and the up-regulation of DRAM. The critical role of DRAM in 2-ME-mediated autophagy and apoptosis is underlined by the fact that its silencing efficiently prevents the induction of both processes. These findings not only demonstrate the interplay between JNK and p53 in the regulation of autophagy but also uncover the role of JNK activation in the regulation of DRAM, a pro-autophagic and pro-apoptotic protein.[70] Overall, many important questions concerning crosstalk between apoptosis and autophagy remain to be addressed, and the most significant involves identifying the exact biochemical switches that direct a cell towards apoptosis or autophagy.

4.2.7   Regulated Form of Necrotic Cell Death Induced by DNA Damage

Necrosis is an acute form of cell death usually initiated following energy loss. Until recently necrosis was considered to be an unregulated form of cell death; however, accumulating experimental data demonstrate that, except under extreme conditions, necrosis may be a well-controlled process. A protein kinase RIP1 was suggested to be a major player in the necrotic cascade; it can be activated by a number of stimuli such as TNF, TRAIL, oxidative stress or DNA damage. RIP1 kinase transduces a signal to the mitochondria directly or indirectly *via* another kinase JNK, which causes mitochondrial permeability transition. The collapse of mitochondria results in the activation of different proteases (*e.g.* calpains, cathepsin) and phospholipases, which leads to plasma membrane destruction, a hallmark of necrotic cell death. Various treatments, such as UV radiation, DNA-alkylating drugs, actinomycin D and cisplatin, which mainly induce apoptosis, are also capable of causing necrotic cell death (Figure 4.2.2).

4.2.7.1   PARP-mediated Necrosis

Poly(ADP-ribosyl)ation is essential for the maintenance of genomic integrity and has been shown to play a central role in modulating the cellular response to severe genotoxic stress and induction of cell death through necrotic mechanisms. PARP is a nuclear zinc-finger DNA-binding protein that detects DNA strand breaks. PARP substrates, such as histones H1, H2B and lamin B, can influence chromatin architecture; however, DNA replication factors and PARP itself may act in DNA metabolism.[71]

Among the targets of PARP are various proteins that are involved in DNA damage checkpoints including p53, p21$^{\text{waf1/cip1}}$, XPA (Xeroderma pigmentosum group A), MSH6, DNA ligase III, XRCC1, DNA polymerase-ε, DNA-PK$_{\text{CS}}$, Ku70, NF-kB, iNOS (inducible nitric oxide synthase), caspase-activated DNase and telomerase.[71] Because of the high negative charge on ADP-ribose polymers, poly(ADP-ribosylated) proteins lose their affinity for DNA and as a consequence their biological activities.

ATP depletion is the central mechanism by which PARP activation is suggested to participate in DNA damage-induced necrotic death and this was thought to be its sole mechanism of eliciting cell death following cell stress. PARP$^{-/-}$ fibroblasts were shown to resist necrotic death induced by *N*-methyl-*N'*-nitro-*N*-nitrosoguanidine (MNNG), and PARP$^{-/-}$ mice resist necrotic death during haemorrhagic shock. In thymocytes, kidney and pulmonary artery PARP inhibitors also prevent necrotic, but not apoptotic, cell death induced by oxidants.[72,73]

4.2.7.2 Cleavage of PARP and Connection with p53

Cleavage of PARP by caspases at a single site occurs at an early stage of apoptosis and leads to separation of the PARP DNA-binding domain from its catalytic subunit, thus inactivating the enzyme. PARP cleavage has been shown in almost all forms of apoptosis and can be blocked by caspase inhibitors or Bcl-2 overexpression, suggesting that the loss of PARP function is required for efficient accomplishment of apoptosis.[74] Caspase-mediated PARP cleavage promotes apoptosis due to the essential reduction of its repair function, thus preventing unnecessary DNA repair and also facilitating the access of endonucleases to chromatin. Caspase inhibition is known to cause the switch from apoptosis to necrosis, suggesting that the induction of apoptosis is a protective mechanism against necrotic damage.

Some experimental data suggest a role for poly(ADP-ribose) and PARP in p53-mediated pathways upon DNA damage, although opinions concerning the interaction of these two components are controversial. Studies on splenocytes, bone marrow cells and embryonic fibroblasts from PARP$^{-/-}$ mice showed an increase in p53 accumulation after treatment with an alkylating agent.[75] X-ray induced accumulation of p53 was found to increase upon PARP inhibition. It has been established that the PARP inhibitor 3-aminobenzamide delays the repair of both SSBs and DSBs caused by ionizing radiation, and that cells treated with inhibitor immediately after X-ray irradiation exhibit a prolonged p53 response.[76]

Later studies suggested that inactivation of PARP affects the duration but not the magnitude of p53 accumulation in irradiated Burkitt lymphoma cells.[81]

p53 accumulation is controlled by PARP-dependent and PARP-independent pathways but p53 activation is largely independent of PARP. It has also been reported that the activity of DNA-PK can be stimulated by PARP in the presence of NAD^+ in *in vitro* reactions and that DNA-PK ADP-ribosylation stimulates p53 phosphorylation.[82] Hence, the role of PARP in p53 activation is complex, involving not only direct activation, but also by acting through DNA-PK and possibly other proteins to transmit the DNA damage signal to p53. Making new phenotypes of mice, such as those with genetic defects of PARP and p53, would be a helpful tool in further understanding the complex interaction between these two pathways. Overall, the current data suggest a passive rather than active role of PARP in the apoptotic process.

4.2.7.3   AIF-mediated Programmed Necrosis

Energy collapse is not the sole mechanism by which PARP contributes to cell death: a 'death link' between PARP, mitochondria and apoptosis-inducing factor (AIF) has been established. AIF is a bifunctional NADH oxidase involved in mitochondrial respiration and caspase-independent apoptosis. AIF plays a pro-survival role through its redox activity in mitochondria, but a lethal role upon its translocation to the nucleus. Once translocated to the nucleus, AIF is involved in chromatinolysis and cell death. In order to be released from mitochondria, AIF has to be cleaved into tAIF by calpain or cathepsins.[77,78] Data indicate that PARP activation is required for AIF translocation during cell death initiated by MNNG and H_2O_2 in fibroblasts and *N*-methyl-D-aspartic acid (NMDA) in cortical neurons. Moreover, AIF appears to be essential for PARP-1–mediated cell death. It was also demonstrated that tAIF release depends on PARP-1-mediated calpain and Bax activation.[79]

The critical role of PARP in controlling AIF-mediated cell death is further supported by observations that NMDA-induced AIF translocation and cell death are blocked in neurons derived from PARP-null mice as well as in wild-type neurons in the presence of the pharmacological PARP inhibitor DPQ. Studies on isolated mouse liver mitochondria and HeLa cells showed that AIF activity is maintained in the presence of the caspase inhibitor zVAD-fmk.[83] Therefore, death mediated by PARP-activation and translocation of AIF is an example of caspase-independent cell death.

While a number of studies have shown the ability of PARP to regulate AIF translocation following different cytotoxic stimuli and in different cell types, very little is known about the precise mechanism by which AIF mediates DNA fragmentation upon its translocation to the nucleus and leads to the execution of cell death. Examination of the AIF crystal structure together with a structure function analysis has demonstrated that AIF can bind DNA in a sequence-independent manner. Structure-based mutagenesis revealed that DNA-binding defective mutants of AIF fail to induce cell death while retaining nuclear translocation.[80] AIF-interacting partners as well as components that promote nuclear chromatin breakdown in mammalian cells remain to be identified. It has recently been shown that AIF associates with histone H2AX in the nucleus *via* the *C*-terminal proline-rich binding domain (PBD), suggesting that this interaction regulates chromatinolysis and programmed necrosis by generating an active DNA-degrading complex with cyclophilin A (CypA).[84] This was confirmed by deletion or directed mutagenesis in the AIF C-terminal PBD, which abolished the AIF/H2AX interaction and AIF-mediated chromatinolysis. Furthermore, H2AX genetic ablation or CypA down-regulation was shown to confer resistance to programmed necrosis. The data obtained demonstrated that AIF fails to induce chromatinolysis in H2AX- or CypA-deficient nuclei. Additionally, AIF phosphorylation at Ser139 was established after treatment with MNNG and suggested to be critical for caspase-independent programmed necrosis. Nevertheless, this study seems to show a requirement for the synchronized presence of AIF, H2AX and CypA in the nucleus in order to provoke DNA degradation rather than a mechanism by which an association between H2AX and AIF might promote activation of latent nuclease CypA. Overall, although these data shed new light on the mechanisms regulating programmed necrosis, additional studies are needed to elucidate the key nuclear partner of AIF and clarify its mechanism of action in the nucleus.

4.2.8 Conclusions

Recent studies have emphasized that, depending on the severity of treatment and resulting DNA damage, cells either survive normally (when all damage is efficiently repaired), survive while carrying the mutation (inappropriate repair), or die. Technical and methodological progress during recent years (especially high throughput and life imaging technologies) has helped us understand how cells die, and how DNA might activate not only apoptotic or necrotic mechanisms but also induce senescence, or activate autophagy or mitotic catastrophe, which may then be followed by apoptosis or necrosis. Future research is necessary to delineate these pathways in further detail and to explore the role of DNA damage in the development of other, less well-defined cell death programmes. The potential importance of these signalling pathways in disease pathogenesis should also be further substantiated. Although we have already entered the era of developing new pharmaceuticals that target DNA in order to kill tumour cells, the clinical advantage as well as the potential adverse effects of such therapy must be critically evaluated.

Acknowledgements

The work in the author's laboratory was supported by grants from the Swedish Research Council, the Swedish and the Stockholm Cancer Societies, the Swedish Childhood Cancer Foundation, the Swedish Society for Medical Research, the Russian Ministry of High Education and Science (11.G34.31.0006), and the EC FP-6 (Chemores) and FP7 (Apo-Sys) programmes. O.S. was supported by a fellowship from the Swedish Institute and

Karolinska Institute. We apologize to those authors whose primary references could not be cited due to space limitations.

References

1. I. Ward and J. Chen, Early events in the DNA damage response, *Curr. Topics Dev. Biol.*, 2004, **63**, 1–35.
2. B. Wang, S. Matsuoka, P. B. Carpenter and S. J. Elledge, 53BP1, a mediator of the DNA damage checkpoint, *Science*, 2002, **298**, 1435–1438.
3. M. Stucki and S. P. Jackson, MDC1/NFBD1: a key regulator of the DNA damage response in higher eukaryotes, *DNA Repair*, 2004, **3**, 953–957.
4. T. T. Paull and J. H. Lee, The Mre11/Rad50/Nbs1 complex and its role as a DNA double-strand break sensor for ATM, *Cell Cycle*, 2005, **4**, 737–740.
5. Y. Shiloh, ATM (ataxia telangiectasia mutated): expanding roles in the DNA damage response and cellular homeostasis, *Biochem. Soc. Trans.*, 2001, **29**, 661–666.
6. A. G. Pallis and M. V. Karamouzis, DNA repair pathways and their implication in cancer treatment, *Cancer Metastasis Rev.*, 2010, **29**, 677–685.
7. A. K. Freeman and A. N. Monteiro, Phosphatases in the cellular response to DNA damage, *Cell Commun., Signaling*, 2010, **8**, 27–38.
8. S. Banin, L. Moyal, S. Shieh, Y. Taya, C. W. Anderson, L. Chessa, N. I. Smorodinsky, C. Prives, Y. Reiss, Y. Shiloh and Y. Ziv, Enhanced phosphorylation of p53 by ATM in response to DNA damage, *Science*, 1998, **281**, 1674–1677.
9. R. Maya, M. Balass, S. T. Kim, D. Shkedy, J. F. Leal, O. Shifman, M. Moas, T. Buschmann, Z. Ronai, Y. Shiloh, M. B. Kastan, E. Katzir and M. Oren, ATM-dependent phosphorylation of Mdm2 on serine 395: role in p53 activation by DNA damage, *Genes Dev.*, 2001, **15**, 1067–1077.
10. C. Deng, P. Zhang, J. W. Harper, S. J. Elledge, P. Leder, S. Banin, L. Moyal, S. Shieh, Y. Taya, C.W. Anderson, L. Chessa, N. I. Smorodinsky, C. Prives, Y. Reiss, Y. Shiloh and Y. Ziv, Mice lacking p21 CIP1/WAF1 undergo normal development, but are defective in G1 checkpoint control, *Cell*, 1995, **82**, 675–684.
11. B. Xu, S. T. Kim, D. S. Lim and M. B. Kastan, Two molecularly distinct G2/M checkpoints are induced by ionizing irradiation, *Mol. Cell. Biol.*, 2002, **22**, 1049–1059.
12. S. Kondo and N. Perrimon, A genome-wide RNAi screen identifies core components of the G2-M DNA damage checkpoint, *Sci. Signaling*, 2011, **4**, 1–10.
13. D. P. Lane, Cancer. p53, guardian of the genome, *Nature*, 1992, **358**, 15–16.
14. K. A. Boehme, R. Kulikov and C. Blattner, p53 stabilization in response to DNA damage requires Akt/PKB and DNA-PK, *Proc. Natl. Acad. Sci. U. S. A.*, 2008, **105**, 7785–7790.
15. A. S. Huart, N. J. MacLaine, D. W. Meek and T. R. Hupp, CK1alpha plays a central role in mediating MDM2 control of p53 and E2F-1 protein stability, *J. Biol. Chem.*, 2009, **284**, 32384–32394.

16. G. S. Wu, The functional interactions between the p53 and MAPK signaling pathways, *Cancer Biol. Ther.*, 2004, **3**, 156–161.

17. J. Wade-Harper, G. R. Adami, N. Wei, K. Keyomarsi and S. J. Elledge, The p21 CDK-interacting protein Cip1 is a potent inhibitor of G1 cyclin-dependent kinases, *Cell*, 1993, **75**, 805–816.

18. A. B. Niculescu III, X. Chen, M. Smeets, L. Hengst, C. Prives and S. I. Reed, Effects of p21$^{cip1/waf1}$ at both the G1/S and the G2/M cell cycle transitions: pRb is a critical determinant in blocking DNA replication and in preventing endoreduplication, *Mol. Cell. Biol.*, 1998, **18**, 629–643.

19. R. Paules, E. Levedakou, S. Wilson, C. Innes, M. Rhodes, T. Tisty, D. Galloway, L. Donehower, M. Tainsky and W. Kaufmann, Defective G2 checkpoint function in cells from individuals with familial cancer syndromes, *Cancer Res.*, 1995, **55**, 1763–1773.

20. K. Polyak, Y. Xia, J. L. Zweier, K. W. Kinzler and B. Vogelstein, A model for p53-induced apoptosis, *Nature*, 1997, **389**, 300–305.

21. G. Saretzki, Cellular senescence in the development and treatment of cancer, *Curr. Pharm. Des.*, 2010, **16**, 79–100.

22. J. Bartek and J. Lukas, Pathways governing G1/S transition and their response to DNA damage, *FEBS Lett.*, 2001, **490**, 117–122.

23. J. P. Brown, W. Wei and J. M. Sedivy, Bypass of senescence after disruption of p21CIP1/WAF1 gene in normal diploid human fibroblasts, *Science*, 1997, **277**, 831–834.

24. J. Campisi and F. d'Adda di Fagagna, Cellular senescence: when bad things happen to good cells, *Nat. Rev. Mol. Cell. Biol.*, 2007, **8**, 729–740.

25. D. A. Alcorta, Y. Xiong, D. Phelps, G. Hannon, D. Beach and J. C. Barrett, Involvement of the cyclin-dependent kinase inhibitor p16 (INK4a) in replicative senescence of normal human fibroblasts, *Proc. Natl. Acad. Sci. U. S. A.*, 1996, **93**, 13742–13747.

26. C. A. Schmitt, J. S. Fridman, M. Yang, S. Lee, E. Baranov and R. M. Hoffman, A senescence program controlled by p53 and p16INK4a contributes to the outcome of cancer therapy, *Cell*, 2002, **109**, 335–346.

27. C. M. Beausejour, A. Krtolica, F. Galimi, M. Narita, S. W. Lowe, P. Yaswen and J. Campisi, Reversal of human cellular senescence: roles of the p53 and p16 pathways, *EMBO J.*, 2003, **22**, 4212–4222.

28. B. D. Chang, M. E. Swift, M. Shen, J. Fang, E. V. Broude and I. B. Roninson, Molecular determinants of terminal growth arrest induced in tumor cells by a chemotherapeutic agent, *Proc. Natl. Acad. Sci. U. S. A.*, 2002, **99**, 389–394.

29. C. Michaloglou, L. C. Vredeveld, M. S. Soengas, C. Denoyelle, T. Kuilman and C. M. van der Horst, BRAFE600-associated senescence-like cell cycle arrest of human naevi, *Nature*, 2005, **436**, 720–724.

30. A. Lechel, A. Satyanarayana, Z. Ju, R. R. Plentz, S. Schaetzlein, C. Rudolph, L. Wilkens, S. U. Wiemann, G. Saretzki, N. P. Malek, M. P. Manns, J. Buer and K. L. Rudolph, The cellular level of telomere dysfunction determines induction of senescence or apoptosis in vivo, *EMBO Rep.*, 2005, **6**, 275–281.

31. K. Webley, J. A. Bond, C. J. Jones, J. P. Blaydes, A. Craig and T. Hupp, Posttranslational modifications of p53 in replicative senescence overlapping but distinct from those induced by DNA damage, *Mol. Cell. Biol.*, 2000, **20**, 2803–2808.

32. K. Oda, H. Arakawa, T. Tanaka, K. Matsuda, C. Tanikawa, T. Mori, H. Nishimori, K. Tamai, T. Tokino and Y. Nakamura, p53AIP1, a potential mediator of p53-dependent apoptosis, and its regulation by Ser-46-phosphorylated p53, *Cell*, 2000, **102**, 849–862.

33. T. G. Hofmann, A. Moller, H. Sirma, H. Zentgraf, Y. Taya, W. Droge, H. Will and M. L. Schmitz, Regulation of p53 activity by its interaction with homeodomain-interacting protein kinase-2, *Nat. Cell Biol.*, 2002, **4**, 1–10.

34. G. D'Orazi, B. Cecchinelli, T. Bruno, I. Manni, Y. Higashimoto, S. Saito, M. Gostissa, S. Coen, A. Marchetti and G. Del Sal, Homeodomain-interacting protein kinase-2 phosphorylates p53 at Ser 46 and mediates apoptosis, *Nat. Cell Biol.*, 2002, **4**, 11–19.

35. E. Gresko, A. Roscic, S. Ritterhoff, A. Vichalkovski, G. Del Sal and M.L. Schmitz, Autoregulatory control of the p53 response by caspase-mediated processing of HIPK2, *EMBO J.*, 2006, **25**, 1883–1894.

36. C. Rinaldo, A. Prodosmo, F. Mancini, S. Iacovelli, A. Sacchi, F. Moretti and S. Soddu, MDM2-regulated degradation of HIPK2 prevents p53Ser46 phosphorylation and DNA damage-induced apoptosis, *Mol. Cell*, 2007, **25**, 739–750.

37. Z. Jin and W. S. El-Deiry, Overview of cell death signaling pathways, *Cancer Biol. Ther.*, 2005, **4**, 139–163.

38. E. R. Flores, K. Y. Tsai, D. Crowley, S. Sengupta, A. Yang, F. McKeon and T. Jacks, p63 and p73 are required for p53-dependent apoptosis in response to DNA damage, *Nature*, 2002, **416**, 560–564.

39. A. E. Sayan, B. S. Sayan, V. Gogvadze, D. Dinsdale, U. Nyman, T. M. Hansen, B. Zhivotovsky, G. M. Cohen, R. A. Knight and G. Melino, P73 and caspase cleaved p73 fragments localize to mitochondria and augment TRAIL-induced apoptosis, *Oncogene*, 2008, **27**, 4363–4372.

40. Q. Zhang, Y. Yoshimatsu, J. Hildebrand, S. M. Frisch and R. H. Goodman, Homeodomain interacting protein kinase 2 promotes apoptosis by downregulating the transcriptional corepressor CtBP, *Cell*, 2003, **115**, 177–186.

41. T. G. Hofmann, N. Stollberg, M. L. Schmitz and H. Will, HIPK2 regulates transforming growth factor beta-induced c-Jun NH(2)-terminal kinase activation and apoptosis in human hepatoma cells, *Cancer Res.*, 2003, **63**, 8271–8277.

42. K. Lei and R. J. Davis, JNK phosphorylation of Bim-related members of the Bcl2 family induces Bax-dependent apoptosis, *Proc. Natl. Acad. Sci. U. S. A.*, 2003, **100**, 2432–2437.

43. S. Zhang, M. Ekman, N. Thakur, S. Bu, P. Davoodpour, S. Grimsby, S. Tagami, C. H. Heldin and M. Landstrom, TGF-beta1-induced activation of ATM and p53 mediates apoptosis in a Smad7-dependent manner, *Cell Cycle*, 2006, **5**, 2787–2795.

44. H. Li, S. K. Kolluri, J. Gu, M. I. Dawson, X. Cao, P. D. Hobbs, B. Lin, G. Chen, J. Lu, F. Lin, Z. Xie, J. A. Fontana, J. C. Reed and X. Zhang, Cytochrome *c* release and apoptosis induced by mitochondrial targeting of nuclear orphan receptor TR3, *Science*, 2000, **289**, 1159–1164.

45. L. de Le'se'leuc and F. Denis, Inhibition of apoptosis by Nur77 through NF-kB activity modulation, *Cell Death Differ.*, 2006, **13**, 293–300.

46. A. Tinel and J. Tschopp, The PIDDosome, a protein complex implicated in activation of caspase-2 in response to genotoxic stress, *Science*, 2004, **304**, 843–846.

47. M. Shi, C. J. Vivian, K. J. Lee, C. Ge, K. Morotomi-Yano, C. Manzl, F. Bock, S. Sato, C. Tomomori-Sato, R. Zhu, J. S. Haug, S. K. Swanson, M. P. Washburn, D. J. Chen, B. P. Chen, A. Villunger, L. Florens and C. Du, DNA-PKcs-PIDDosome: a nuclear caspase-2-activating complex with role in G2/M checkpoint maintenance, *Cell*, 2009, **136**, 508–520.

48. H. Vakifahmetoglu-Norberg and B. Zhivotovsky, The unpredictable caspase-2: what can it do? *Trends Cell Biol.*, 2010, **20**, 150–159.

49. H. Vakifahmetoglu, M. Olsson, C. Tamm, N. Heidari, S. Orrenius and B. Zhivotovsky, DNA damage induces two distinct modes of cell death in ovarian carcinomas, *Cell Death Differ.*, 2008, **15**, 555–566.

50. M. Castedo, J. L. Perfettini, T. Roumier, K. Yakushijin, D. Horne and R. Medema, The cell cycle checkpoint kinase Chk2 is a negative regulator of mitotic catastrophe, *Oncogene*, 2004, **23**, 4353–4361.

51. H. C. Reinhardt, A. S. Aslanian, J. A. Lees and M. B. Yaffe, p53-deficient cells rely on ATM- and ATR-mediated checkpoint signaling through the p38MAPK/MK2 pathway for survival after DNA damage, *Cancer Cell*, 2007, **11**, 175–189.

52. T. Waldman, C. Lengauer, K. W. Kinzler and B. Vogelstein, Uncoupling of S phase and mitosis induced by anticancer agents in cells lacking p21, *Nature*, 1996, **381**, 713–716.

53. M. Castedo, J. L. Perfettini, T. Roumier, K. Andreau, R. Medema and G. Kroemer, Cell death by mitotic catastrophe: a molecular definition, *Oncogene*, 2004, **236**, 2825–2837.

54. H. Vakifahmetoglu, M. Olsson and B. Zhivotovsky, Death through a tragedy: mitotic catastrophe, *Cell Death Differ.*, 2008, **15**, 1153–1162.

55. A. M. Cuervo, Autophagy: many paths to the same end, *Mol. Cell. Biochem.*, 2004, **263**, 55–72.

56. W. Bursch and K. Hochegger, T. Тиruk, B. Marian, A. Ellinger and R. Schulte Hermann, Autophagic and apoptotic types of programmed cell death exhibit different fates of cytoskeletal filaments, *J. Cell Sci.*, 2000, **113**, 1189–1198.

57. X. Zeng and T. J. Kinsella, BNIP3 is essential for mediating 6-thioguanine- and 5-fluorouracil-induced autophagy following DNA mismatch repair processing, *Cell Res.*, 2010, **20**, 665–675.

58. Z. Feng, H. Zhang, A. J. Levine and S. Jin, The coordinate regulation of the p53 and mTOR pathways in cells, *Proc. Natl. Acad. Sci. U. S. A.*, 2005, **102**, 8204–8209.

59. B. Levine and J. Abrams, p53: the Janus of autophagy? *Nat. Cell Biol.*, 2008, **10**, 637–639.
60. E. Tasdemir, M. C. Maiuri, L. Galluzzi, I. Vitale, M. Djavaheri-Mergny, M. D'Amelio, A. Criollo, E. Morselli, C. Zhu, F. Harper, U. Nannmark, C. Samara, P. Pinton, J.M. Vicencio, R. Carnuccio, U.M. Moll, F. Madeo, P. Paterlini-Brechot, R. Rizzuto, G. Szabadkai, G. Pierron, K. Blomgren, N. Tavernarakis, P. Codogno, F. Cecconi and G. Kroemer, Regulation of autophagy by cytoplasmic p53, *Nat. Cell Biol.*, 2008, **10**, 676–687.
61. A. Alexander, J. Kim and C. L. Walker, ATM engages the TSC2/mTORC1 signaling node to regulate autophagy, *Autophagy*, 2010, **6**, 672–673.
62. J. A. Muñoz-Gámez, J. M. Rodríguez-Vargas, R. Quiles-Pérez, R. Aguilar-Quesada, D. Martín-Oliva, G. de Murcia, J. Menissier de Murcia, A. Almendros, M. Ruiz de Almodóvar and F. J. Oliver, PARP-1 is involved in autophagy induced by DNA damage, *Autophagy*, 2009, **5**, 61–74.
63. C. Yang, V. Kaushal, S. V. Shah and G. P. Kaushal, Autophagy is associated with apoptosis in cisplatin injury to renal tubular epithelial cells, *Am. J. Physiol. Renal Physiol.*, 2008, **294**, 777–787.
64. J. Han, W. Hou, L. A. Goldstein, C. Lu, D. B. Stolz and X. M. Yin, Involvement of protective autophagy in TRAIL resistance of apoptosis-defective tumor cells, *J. Biol. Chem.*, 2008, **283**, 19665–19677.
65. M. J. Abedin, D. Wang, M. A. McDonnell, U. Lehmann and A. Kelekar, Autophagy delays apoptotic death in breast cancer cells following DNA damage, *Cell Death Differ.*, 2007, **14**, 500–510.
66. M. Gyrka, W. M. Daniewsk, B. Gajkowska, E. Lusakowska, M. M. Godlewski and T. Motyl, Autophagy is the dominant type of programmed cell death in breast cancer MCF-7 cells exposed to AGS 115 and EFDAC, new sesquiterpene analogs of paclitaxel, *Anticancer Drugs*, 2005, **16**, 777–788.
67. S. B. Lee, S. Y. Tong, J. J. Kim, S. J. Um and J. S. Park, Caspase-independent autophagic cytotoxicity in etoposide-treated CaSki cervical carcinoma cells, *DNA Cell Biol.*, 2007, **26**, 713–720.
68. S. Shimizu, T. Kanaseki, N. Mizushima, T. Mizuta, S. Arakawa-Kobayashi, C. B. Thompson and Y. Tsujimoto, Role of Bcl-2 family proteins in a nonapoptotic programmed cell death dependent on autophagy genes, *Nat. Cell Biol.*, 2004, **6**, 1221–1228.
69. K. S. Yee, S. Wilkinson, J. James, K. M. Ryan and K. H. Vousden, PUMA and Bax-induced autophagy contributes to apoptosis, *Cell Death Differ.*, 2009, **16**, 1135–1145.
70. S. Lorin, G. Pierron, K. M. Ryan, P. Codogno and M. Djavaheri-Mergny, Evidence for the interplay between JNK and p53-DRAM signalling pathways in the regulation of autophagy, *Autophagy*, 2010, **6**, 153–154.
71. N. A. Berger, Poly(ADP-ribose) in the cellular response to DNA damage, *Radiat. Res.*, 1985, **101**, 4–15.
72. L. Liaudet, F. G. Soriano, E. Szabó, L. Virág, J. G. Mabley, A. L. Salzman and C. Szabó, Protection against hemorrhagic shock in mice genetically deficient in poly(ADP-ribose)polymerase, *Proc. Natl. Acad. Sci. U. S. A.*, 2000, **97**, 10203–10208.

73. L. Virág, G. S. Scott, S. Cuzzocrea, D. Marmer, A. L. Salzman and C. Szabó, Peroxynitrite-induced thymocyte apoptosis: the role of caspases and poly (ADP-ribose) synthetase (PARS) activation, *Immunology*, 1998, **94**, 345–355.

74. S. H. Kaufmann, S. Desnoyers, Y. Ottaviano, N. E. Davidson and G. G. Poirier, Specific proteolytic cleavage of poly(ADP-ribose) polymerase: an early marker of chemotherapy-induced apoptosis, *Cancer Res.*, 1993, **53**, 3976–3985.

75. J. Menissier-de Murcia, C. Niedergang, C. Trucco, M. Ricoul, B. Dutrillaux, M. Mark and F. J. Oliver, Requirement of poly(ADP-ribose) polymerase in recovery from DNA damage in mice and in cells, *Proc. Natl. Acad. Sci. U. S. A.*, 1997, **94**, 7303–7307.

76. X. Lu and D. P. Lane, Differential induction of transcriptionally active p53 following UV or ionizing radiation: defects in chromosome instability syndromes? *Cell*, 1993, **75**, 765–778.

77. H. Otera, S. Ohsakaya, Z. Nagaura, N. Ishihara and K. Mihara, Export of mitochondrial AIF in response to proapoptotic stimuli depends on processing at the intermembrane space, *EMBO J.*, 2005, **24**, 1375–1386.

78. B. M. Polster, G. Basanez, A. Etxebarria, J. M. Hardwick and D. G. Nicholls, Calpain I induces cleavage and release of apoptosis-inducing factor from isolated mitochondria, *J. Biol. Chem.*, 2005, **280**, 6447–6454.

79. R. S. Moubarak, V. J. Yuste, C. Artus, A. Bouharrour, P. A. Greer and J. Menissier-de, Murcia and S. A. Susin, Sequential activation of poly(ADP-Ribose) polymerase 1, calpains, and Bax is essential in apoptosis-inducing factor-mediated programmed necrosis, *Mol. Cell Biol.*, 2007, **27**, 4844–4862.

80. H. Ye, C. Cande, N. C. Stephanou, S. Jiang, S. Gurbuxani, N. Larochette, E. Daugas and C. Garrido, G. Kroemer, H. Wu, DNA binding is required for the apoptogenic action of apoptosis inducing factor, *Nat. Struct. Biol.*, 2002, **9**, 680–684.

81. C. M. Simbulan-Rosenthal, D. S. Rosenthal, S. Iyer, A. H. Boulares and M. E. Smulson, Transient poly(ADP-ribosyl)ation of nuclear proteins and role of poly(ADP-ribose) polymerase in the early stages of apoptosis, *J. Biol. Chem*, 1998, **273**, 13703–13712.

82. T. Ruscetti, B. E. Lehnert, J. Halbrook, H. Le Trong, M. F. Hoekstra, D. J. Chen and S. R. Peterson, Stimulation of the DNA-dependent protein kinase by poly(ADP-ribose) polymerase, *J. Biol. Chem.*, 1998, **273**, 14461–14467.

83. S. A. Susin, H. K. Lorenzo, N. Zamzami, I. Marzo, B. E. Snow, G. M. Brothers, J. Mangion, E. Jacotot, P. Costantini, M. Loeffler, N. Larochette, D. R. Goodlett, R. Aebersold, D. P. Siderovski, J. M. Penninger and G. Kroemer, Molecular characterization of mitochondrial apoptosis-inducing factor, *Nature*, 1999, **397**, 441–446.

84. C. Artus, H. Boujrad, A. Bouharrour, M. N. Brunelle, S. Hoos, V. J. Yuste, P. Lenormand, J. C. Rousselle, A. Namane, P. England, H. K. Lorenzo and S. A. Susin, AIF promotes chromatinolysis and caspase-independent programmed necrosis by interacting with histone H2AX, *EMBO J.*, 2010, **29**, 1585–1599.

CHAPTER 4.3

Transcriptional Inhibition by DNA Damage as a Trigger for Cell Death

MATS LJUNGMAN

Division of Radiation and Cancer Biology, Department of Radiation
Oncology, University of Michigan Comprehensive Cancer Center and
Department of Environmental Health Sciences, School of Public Health,
University of Michigan, 4424C Med Sci I, 1301 Catherine Street,
Ann Arbor, MI 48109, USA
Email: ljungman@umich.edu

4.3.1 Introduction

Transcription is a fundamental process required for all life. By transcribing the
genetic information stored in DNA, RNA polymerase II generates mRNA that
instructs ribosomes how to build proteins. This flow of information is highly
regulated to ensure that cells express optimal levels of proteins for any given
time and situation. Events leading to the interruption of mRNA synthesis,
such as induction of DNA damage, may offset the balance of gene expression
and result in the preferential loss of unstable transcripts and proteins. If
transcription is not restored in a timely fashion, cells may die either by loss
of the ability to regenerate survival factors or by direct induction of stress
pathways leading to apoptosis. Furthermore, a slow accumulation of

Issues in Toxicology No. 13
The Cellular Response to the Genotoxic Insult: The Question of Threshold for
Genotoxic Carcinogens
Edited by Helmut Greim and Richard J. Albertini
© The Royal Society of Chemistry 2012
Published by the Royal Society of Chemistry, www.rsc.org

transcription-blocking DNA lesions over time leading to a sequential loss of gene expression may be an important underlying mechanism of aging. To circumvent the detrimental effects caused by transcription disruption following DNA damage, several cellular defense mechanisms have evolved and these are discussed in this chapter.

4.3.2 Transcription Factories

The process of transcription is commonly viewed as having the RNA polymerase translocate like a train down a stationary DNA template. This model of transcription would result in the wrapping of the nascent RNA molecule around the DNA template once every 10 bp transcribed. A more likely model is that the RNA polymerase and associated factors are anchored to the nuclear matrix while the DNA template is the moving part.[1] Having stationary RNA polymerases allows for the clustering of pre-assembled transcription 'factories'[2,3] to which genes selected to be expressed are directed.[4] The attachment and detachment of the DNA to transcription factors creates a dynamic self-organizing structure of DNA loops.[3] The estimated DNA loop sizes in human cells range between 5–200 kb.[5,6]

When transcription is inhibited by treatment with kinase inhibitors such as 5,6-dichloro-1-β-D-ribofuranosylbenzimidazole (DRB) (see below and Section 4.3.3.3.2) that inhibit the phosphorylation of the *C*-terminal domain (CTD) of RNA polymerase II, the initiated transcription machinery is prohibited from entering into the elongation mode. This eventually leads to the detachment of DNA from the transcription factories.[7] From photobleaching experiments using cells expressing RNA polymerase II tagged with green fluorescent protein (GFP), it has been shown that DRB treatment reduces the pool of RNA polymerase II being engaged on the DNA template.[8] In contrast, agents such as ultraviolet (UV) light and actinomycin D, which block transcription in the elongating mode, cause an increase in the pool of RNA polymerase II that is engaged on the DNA template.[8] These findings reflect the 'lock-down' of the RNA polymerase onto the DNA template after committing to elongation.[9,10] Inhibition of RNA polymerase II transcription has significant effects on the structure of chromatin and nuclear structure, suggesting that continuous transcription by RNA polymerase II shapes nuclear structure.[3,11,12] Inhibition of RNA polymerase II, but interestingly not RNA polymerase I, results in a dramatic disruption of the morphology of nucleoli.[11]

4.3.3 Transcription-blocking Agents

Inhibition of transcription by DNA damage triggers a cellular stress response that is aimed at arresting proliferating cells in the cell cycle, stimulating DNA repair and recovering mRNA synthesis.[13] If cells do not recover RNA synthesis in a timely fashion, cells will undergo apoptosis. In this manner, the transcription machinery acts as a cellular dosimeter for DNA damage and triggers the elimination of cells that have sustained sufficient damage to prohibit transcription recovery. Many genotoxic agents, such as UV light, alkylating agents and reactive oxygen species (ROS) can induce lesions that directly or indirectly block the elongation of RNA polymerase II.[14] Furthermore, many chemotherapeutic agents are at least partially associated with inhibition of either initiation or elongation of transcription.[15] The mechanisms by which some common genotoxic, chemotherapeutic and pharmacological agents inhibit transcription are discussed in Table 4.3.1.

Table 4.3.1   Mechanisms of transcription inhibition by common genotoxic, chemotherapeutic and pharmacological drugs.

Agent	Inhibition of initiation	Inhibition of transition into elongation	Inhibition of elongation
IR	Loss of DNA topology by DNA nicks (high doses)	no	no
Roscovitine/ flavopiridol	no	Inhibits phosphorylation of CTD of RNPII	no
DRB	no	Inhibits phosphorylation of CTD of RNPII	no
UV light	After depletion of initiating pool of RNAPII by trapping them in elongation mode	no	UV adducts trapped within active site
Alkylating agents	no	no	MMR-dependent
ROS	Loss of DNA topology by DNA nicks	no	MMR-dependent
Camptothecin	Increase in initiation due to accumulation of negative torsional tension at promoter	no	Blocked at trapped topo I, build-up of torsional tension
Cisplatin	no	no	Both intra and interstrand DNA C-L
Psoralen (ICL)	no	no	Interstrand cross-link (but not monoadduct)
Actinomycin D	no	Intercalates DNA preferentially in transcription bubble	Intercalates DNA preferentially in transcription bubble
α-Amanitin	no	Inhibits RNPII translocation	Inhibits RNPII translocation

4.3.3.1 Genotoxic Agents

4.3.3.1.1 UV Light

The most prevalent environmental agent inducing transcription-blocking lesions is UV light. That UV light causes inhibition of RNA synthesis was first discovered in *Escherichia coli*[16] and it is known that this inhibition is due to blockage of transcription elongation.[17,18] Localized UV-induced DNA lesions in cells cause blockage of transcription only in irradiated regions, suggesting that UV light does not cause a global shutdown of transcription.[19] However, some studies suggest that in addition to blockage of the elongation complex at sites of DNA lesions, UV light might trigger a caffeine-sensitive response leading to reduced initiation of transcription.[20,21]

Exonuclease footprinting studies of RNA polymerases stalled at a cyclo-butane pyrimidine dimer (CPD) show that the polymerase covers 35–40 nucleotides and is asymmetrically located over the lesion.[22,23] More detailed crystal structures of the stalled complex show that the CPD enters the active site of the RNA polymerase II complex where translocation stops due to uridine misincorpoartion into the mRNA.[24] The other major type of lesion induced by UV light, (6-4) photoproducts, also block RNA synthesis.[25]

Recovery of RNA synthesis following UV irradiation is dependent on transcription-coupled repair (TCR).[26,27] However, some studies have suggested that RNA synthesis may recover prior to the removal of the CPDs and (6-4) photoproducts from the DNA template.[28,29] The mechanism for this translesion RNA synthesis is not understood but it may contribute to transcriptional mutagenesis.[30–32] It has been shown that the largest subunit of RNA polymerase II becomes ubiquitylated and degraded following UV irradiation in a process dependent on the Cockayne syndrome factors A and B.[33–35] While this ubiquitin and proteasome-dependent degradation does not appear to be required for TCR,[36] it may be required for the recovery of RNA synthesis following UV irradiation.[35]

4.3.3.1.2 Ionizing Radiation (DNA Strand Breaks)

DNA in eukaryotic cells contains localized unconstrained torsional tension.[6,37,38] Negative torsional tension enhances the initiation step of transcription by lowering the energy required to unwind the DNA to form transcription initiation complexes.[39–42] DNA strand breaks induced by ionizing radiation effectively relax torsional tension, potentially affecting transcription of the genes in the nicked DNA loop.[6,37] A dose that induces, on average, a DNA strand break per 50 kb of cellular DNA completely relaxed all unconstrained torsional tension in the cell.[6]

It has been shown that DNA strand breaks induced by ionizing radiation or hydroxyl radicals reduce transcription in Tetrahymena[43,44] and mammalian cells.[45–48] Radiation-induced DNA strand breaks (mostly single-strand breaks) are rapidly repaired in mammalian cells with 50% of the breaks repaired in five minutes and 90% repaired within 20 minutes.[49] Although DNA strand breaks cause inhibition of transcription, the repair of strand breaks does not occur preferentially in active DNA but rather the repair rate is equally fast in mRNA and rRNA sequences as in the genome as a whole.[49] The lack of TCR of DNA strand breaks is probably because a strand break on either DNA strand anywhere within a topological domain would be predicted to result in the inhibition of transcription. To complete the repair event of DNA strand breaks, the topology of the DNA domain needs to be restored in order for transcription to recover. The mechanisms by which DNA topology is restored following strand break repair are not understood, but may involve the reversal of histone modifications such as phosphorylation of histone H2AX.[50]

4.3.3.1.3   Alkylating Agents (^{6}OMeG)

There is evidence *in vitro* that RNA polymerase can transcribe past the common alkylation adducts 3-methyladenine and 7-methylguanine[14,51] as well as O^6-methylguanine (O^6MeG).[52] However, *in vivo* O^6MeG adducts inhibit RNA polymerase II-mediated transcription efficiently.[53] This inhibition is dependent on the presence of functional mismatch repair (MMR) since no inhibition of transcription was observed in cells lacking MLH1. Transcription initiation may be reduced due to inhibition of transcription factor binding to promoter DNA containing O^6MeG adducts.[54]

Repair of 3-methyladenine and 7-methylguanine does not involve TCR,[55,56] while repair of O^6MeG adducts occurs preferentially in active genes.[57] The utilization of TCR is reflective of blocked transcription elongation, but an alternative explanation to the preferential repair of O^6MeG adducts in active chromatin is the finding that the major repair enzyme removing O^6MeG adducts, O^6MeG DNA methyltransferase, is preferentially associated with active transcription sites.[58]

Cancer cells defective in MMR have been shown to be resistance to alkylating chemotherapeutic agents inducing O^6MeG adducts.[59] This resistance to O^6MeG adducts correlates with lack of transcription inhibition and reduced subsequent DNA damage signaling.[53] Furthermore, lack of transcription inhibition will promote transcriptional mutagenesis at sites of O^6MeG adducts, which could affect the physiology of the treated cells.[60]

4.3.3.1.4   ROS

Reactive oxygen species (ROS) induce some of the most abundant adducts in cellular DNA such as 8-hydroxyguanine and thymine glycols. Purified RNA polymerase II partially arrests at 8-hydroxyguanine lesions *in vitro*.[61,62] However, if the transcription factor TFIIH is added, the purified RNA polymerase II is able to efficiently bypass the 8-hydroxyguanine lesions.[63] In cells, 8-hydroxyguanine lesions appear to block RNA polymerase II elongation and there is evidence of TCR of these lesions.[64,65] In addition, repair of 8-hydroxyguanine lesions is reduced in TCR-deficient Cockayne syndrome B cells.[66] Thymine glycols do not act as transcription blocks for purified RNA polymerase II *in vitro* but these lesions are repaired by transcription-coupled repair.[67] Thus, it is possible that cellular factors may bind to these lesions in cells and cause blockage of transcription.

It has been shown that high levels of oxidative stress to cells result in a general inhibition of RNA polymerase II-mediated transcription, perhaps due to suppression of histone acetylation.[68] In studies assessing the effect of hydrogen peroxide on mRNA synthesis in human HCT116 cells, which are defective in MMR, no inhibition of transcription was detected.[53] However, when the MMR defect was corrected, by either the addition of chromosome 3 or the MLH1 gene, transcription was significantly inhibited. This inhibition of mRNA synthesis correlated with an increased activation of apoptosis, but not p53, suggesting that MMR complexes may induce cell death by binding to ROS-induced DNA lesion and blocking RNA polymerase II elongation.[53]

4.3.3.2 Chemotherapeutic Agents

4.3.3.2.1 *Camptothecin*

Camptothecin is a powerful inhibitor of both mRNA and rRNA synthesis.[69–71] The drug target of camptothecin is DNA topoisomerase I (topo I), which gets trapped in a complex covalently bound to nicked DNA.[72,73] The major role of DNA topoisomerase I is to relax unconstrained positive and negative superhelicity generated during transcription elongation.[74] DNA topoisomerase I has been shown to be associated with the transcription elongation complex[75,76] and following targeting with camptothecin, both RNA polymerase I[77] and RNA polymerase II[78] elongation, but not initiation, are severely inhibited.

Treatment with camptothecin results in ubiquitin-dependent destruction of DNA topoisomerase I.[79] This degradation is dependent on transcription arrest at sites of camptothecin-induced topoisomerase I cleavable complexes.[80] In addition, the largest subunit of the arrested RNA polymerase II is targeted for ubiquitylation and the removal of both the polymerase and the topoisomerase is thought to promote the repair of the DNA nick.[80] The degradation of RNA polymerase II complexes following blockage at sites of DNA topoisomerase I cleavable complexes should prevent restart following the reversal of topoisomerase I cleavable complexes when camptothecin is removed. Indeed, assessing the recovery of RNA synthesis from the DHFR gene *in vivo* following camptothecin removal revealed that all new synthesis started at the 5′-end of the gene and moved as a wave through the gene.[78] Finally, the toxicity of camptothecin has long been attributed to its inhibitory effects on DNA replication.[81] However, its potent inhibitory effects on transcription elongation may also be considered in its cancer therapeutic activity.[13,82,83]

4.3.3.2.2   Cisplatin

Cisplatin is a commonly used chemotherapeutic agent that forms both intrastrand and interstrand DNA crosslinks in cells.[84-86] Although interstrand crosslinks are considered more toxic to cells than intrastrand crosslinks, only about 2% of the adducts formed by cisplatin are of this type. Thus, it is thought that the major toxicity exerted by cisplatin is due to intrastrand crosslinks.[84,87] Cisplatin-induced intrastrand crosslinks efficiently block RNA polymerase II elongation both *in vitro*[88-90] and *in vivo*.[91,92] In contrast to pyrimidine dimers, intrastrand DNA crosslinks are prevented from entering the active site of RNA polymerase II and thus the RNA polymerase stalls a few base pairs upstream of the adduct.[93] However, similarly to UV irradiation, cisplatin treatment results in the ubiquitylation and degradation of RNA polymerase II.[33,90,94] Thus, even though the RNA polymerase stalls upstream of the cisplatin lesion, it may not be able to resume synthesis following the removal of the blocking lesion but instead is targeted for degradation.

The intrastrand crosslinks are repaired by nucleotide excision repair (NER) while the interstrand crosslinks are dealt with by a combination of NER and homologous recombination (HR).[95-98] It has been shown that both the intrastrand and interstrand DNA crosslinks are repaired in a TCR-dependent manner, but the interstrand crosslink repair show, as expected, no strand specificity.[99,100] Cells defective in TCR are hypersensitive to cisplatin, suggesting that defective recovery of RNA synthesis following cisplatin triggers apoptosis.[92,101] Furthermore, loss of p53 function sensitized human fibroblast to cisplatin, suggesting that p53 protects these cells perhaps by promoting recovery of RNA synthesis in a similar way as after UV-irradiation.[101,102]

4.3.3.2.3   Psoralen

Psoralens have anti-proliferative activities and are used in the clinic to treat psoriasis.[103] Psoralens readily enter cells and intercalate DNA; following activation by UVA light, the intercalated psoralen molecules form covalent bonds with thymines resulting in either monoadducts or interstrand DNA crosslinks.[104] Psoralen monoadducts only marginally inhibit RNA polymerase II-mediated transcription while psoralen interstrand crosslinks, which induce a major conformational change of the DNA helix,[105] efficiently inhibit mRNA synthesis.[106] This preferential inhibition of transcription by the interstrand crosslinks correlates with induction of p53 and apoptosis in human fibroblast.[106] Furthermore, the psoralen interstrand crosslinks, but not the monoadducts, are preferentially removed from active genes by TCR.[107-110] TCR of psoralen interstrand crosslinks is efficient in RNA polymerase II transcribed genes, while very little repair occurs in ribosomal DNA sequences.[110]

4.3.3.2.4   CDK Inhibitors (Roscovitine and Flavopiridol)

The cyclin-dependent kinase (CDK) inhibitors roscovitine[111-113] and flavopiridol[114,115] are in clinical use to treat various malignancies. Both roscovitine[116] and flavopiridol[117] inhibit mRNA synthesis by blocking the phosphorylation of the *C*-terminal domain of RNA polymerase II, thereby prohibiting the initiated RNA polymerase to transition into elongation. Both drugs induce the nuclear accumulation of p53,[116,118,119] but this induction of p53 is not associated with stress-induced modification of p53 suggesting that transcription is inhibited prior to the transition into the elongation mode.[13,116,120,121] Both drugs induce apoptosis in cancer cells and it has been suggested that transcription inhibition leading to the loss of the short-lived anti-apoptosis factor MCL-1 triggers the induction of apoptosis.[122,123]

4.3.3.3 Mechanisms of Action of Commonly Used Transcription Inhibitors

4.3.3.3.1 *Actinomycin D*

Actinomycin D is a polypeptide antibiotic isolated from *Streptomyces* bacteria. It associates with duplex DNA and binds specifically to pre-melted DNA, such as DNA in 'transcription bubbles'.[124] Binding of actinomycin to DNA in the transcriptional complex immobilizes the RNA polymerase and inhibits transcription elongation. There is a strong correlation between the transcription unit size and the degree of inhibition exerted by actinomycin D, with large transcription units being more sensitive to inhibition.[125] Furthermore, transcription by RNA polymerase I or III is 50–100 times more sensitive to inhibition by actinomycin D than that by RNA polymerase II. Although the RNA polymerase I transcribed gene 45S RNA is very short compared with most RNA polymerase II-transcribed genes, the fact that it is arranged in 300–400 repeats transcribed together makes it a large target for actinomycin D. Therefore, 45S RNA synthesis is inhibited at lower doses of actinomycin D ($< 100 \, \text{ng mL}^{-1}$) compared with bulk mRNA synthesis, although RNA polymerase II transcription will also be affected proportionally to the transcription unit sizes.[125]

4.3.3.3.2 *DRB*

5,6-Dichloro-1-β-D-ribofuranosylbenzimidazole (DRB) is a nucleoside analogue that reversibly inhibits RNA polymerase II mediated transcription.[126] The mechanism of action of DRB involves the inhibition of the TFIIH-associated kinase[127] and the CDK9 kinase.[128,129] Both these kinases are responsible for the phosphorylation of the *C*-terminal domain (CTD) of RNA polymerase II.[130] By inhibiting the phosphorylation of the CTD of RNA polymerase II, DRB blocks the transition into the elongation mode resulting in transcription stalling shortly after initiation. Interestingly, some genes, such as the p53-iducible gene p21, appear to be insensitive to DRB suggesting that CTD phosphorylation is superfluous for transcription elongation for these genes.[131]

4.3.3.3.3 α-Amanitin

α-Amanitin is specific RNA polymerase II inhibitor isolated from the toxic mushroom *Amanita phalloides*.[132] Co-crystal structures of RNA polymerase II and α-amanitin show that the drug binds the bridge domain of RNA polymerase II near the nucleotide entry site, although nucleoside triphosphates can still enter into the active site of the polymerase and RNA chains can still be elongated.[133] However, the rate of chain elongation is reduced several thousand-fold and it is thought that α-amanitin binding to RNA polymerase II interferes with both nucleotide incorporation and polymerase translocation.[133–135]

4.3.4   The Transcription Stress Response

Transcription is a fundamental process for all cells. Interference with general transcription results in the preferential loss of unstable transcripts, which ultimately may lead to cell death. In order to rapidly respond to transcriptional arrest, cells activate a cellular stress response that involves the tumor suppressor p53. Cells respond differently depending on whether transcription is inhibited prior to the transition from initiation to elongation or whether transcription is blocked during elongation. When transcription is inhibited in the early phase of the cell cycle by inhibition of the phosphorylation of the CTD of RNA polymerase II, p53 accumulates in the nucleus without concomitant phosphorylation of the Ser15 site or acetylation on the Lys382 site of p53. However, if RNA polymerase II transcription is blocked during elongation, p53 is modified at both Ser15 and Lys382.[120,121,136–138] This distinction is probably very important since RNA polymerase is normally transiently arrested prior to elongation in promoter-proximal regions for many of genes.[139] Thus, an activation of a stress response due to such normal arrest would not be preferable. What then senses transcription elongation arrest specifically resulting in p53 modifications?

   Many of the agents used to block transcription elongation, such as UV light, actinomycin D, camptothecin, cisplatin and photoactivated psoralen, may trigger a stress response due to the processing of DNA adducts rather than from the transcription blockage. Indeed, induction of an ATR-mediated phosphorylation of the histone variant H2AX following UV irradiation occurs as a result of DNA repair-induced processing of UV lesions.[140–142]

   To directly test whether blockage of RNA polymerase II elongation is sufficient to induce p53 activation and phosphorylation in the absence of DNA damage, antibodies directed toward the CTD of RNA polymerase II were used to block transcription either in the early phase of transcription or during elongation.[121] Nuclear accumulation of p53 was observed with antibodies blocking either the early phase or the elongation phase of RNA polymerase II mediated transcription. Moreover, the nuclear accumulation of p53 was accompanied by phosphorylation of the Ser15 site only when elongating RNA polymerases were targeted. Thus, blockage of RNA polymerase II elongation is linked specifically to a pathway resulting in Ser15 phosphorylation of p53. This phosphorylation is dependent on the single-strand DNA-binding protein RPA (replication protein A) and the stress kinase ATR.[121,138] The mechanism responsible for the nuclear accumulation of p53 even when p53 is not phosphorylated at the Ser15 site may involve the general inhibition of nuclear export of p53 due to loss of mRNA export.[13,121,138,143,144] The model propose that p53 is normally kept at a low level in the nucleus due to ubiquitylation by MDM2 followed by the active nuclear export *via* the mRNA export machinery and proteasome-mediated degradation in the cytoplasm. When transcription is inhibited, mRNA export will cease and p53 is not exported even though it is fully ubiquitylated by MDM2.[144]

4.3.5 Post-transcriptional Effects

In response to genotoxic insults, cells activate a comprehensive DNA damage response (DDR) to ensure genome integrity and suppression of mutagenesis. An important mechanism for inducing DDR is to modify expression of specific genes involved in DNA repair, cell cycle checkpoints and apoptosis, and for the adaptation to potential future insults.[145,146] The induction of many genes following exposure to DNA-damaging agents can be transcriptionally induced by p53,[147,148] but gene expression may also be regulated post-transcriptionally through alternative splicing,[149] stabilization of specific mRNAs[150–153] and by preferential translation.[154–156] Some of these mRNAs may be regulated by microRNAs, which may be induced or repressed after exposure to DNA damage.[157–159] The mechanisms by which DNA damage affects mRNA stability are not well understood. It is likely that cells switch on post-transcriptional regulation of gene expression at times when the DNA template contains transcription-blocking DNA lesions.

The stability of a specific mRNA is regulated at many levels.[160,161] During mRNA biogenesis, the 5′-end of the transcript is capped to protect against 5′-3′ exonucleolytic degradation. Furthermore, the pre-mRNA is cleaved at the 3′-end by the 3′-end processing complex followed by 3′-end polyadenylation by the polyadenylation polymerase (PAP). The polyadenylation of the 3′-end protects the mRNA against 3′-5′ exonucleolytic degradation. Sequence elements present in the 3′UTR, such as AU-rich elements (AREs), can under certain circumstances confer instability to the transcript by attracting specific RNA-binding proteins. One such protein is tristetraproline (TTP), which recruits the poly(A)-specific exonucleases resulting in the removal of the 3′ poly(A) tail and subsequent degradation by the exosome, a multiprotein complex with 3′-5′ exonuclease activity. One potential mechanism for the stabilization of mRNAs following DNA damage-induced transcription blockage is that the p38 and MK2 kinases become activated and inhibit TTP binding to AREs while stimulating HuR binding.[153,162–169] Another potential mechanism is the activation of ATR in response to blocked transcription.[121,170] ATR has hundreds of substrates, some of which may be involved in regulating mRNA stability following exposure to transcription-blocking lesions.

4.3.6 Induction of Apoptosis by Transcription-blocking DNA lesions

What are the mechanisms responsible for triggering apoptosis following blockage of transcription? There are at least four different mechanisms that could be involved; these are depicted in Figure 4.3.1 and discussed below.

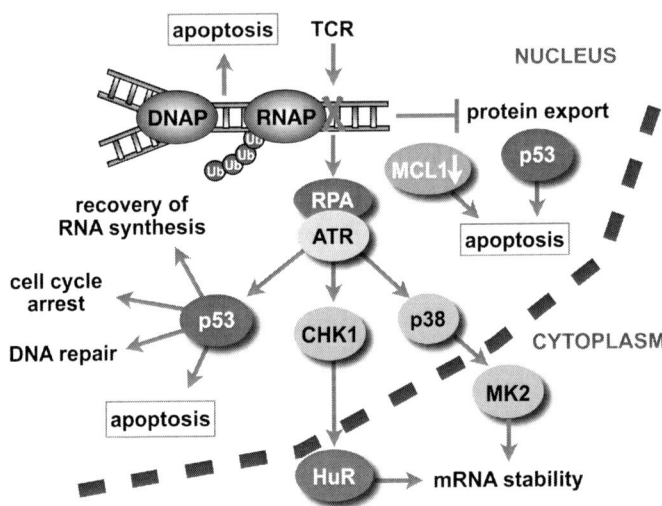

Figure 4.3.1 Cellular responses to transcription blocked at sites of DNA damage. Blockage of RNA polymerase II elongation triggers the activation of the ATR kinase in an RPA-dependent manner. Activated ATR has many hundred potential substrates including p53, CHK1 and p38. Activation of both the p38 and CHK1 kinases can lead to post-transcriptional regulation of certain transcripts. This is an important mechanism to regulate gene expression at times when transcription is blocked by DNA damage. ATR also activates p53, which in turn can promote recovery of RNA synthesis, cell cycle arrest, DNA repair and in some cases induce apoptosis. Apoptosis following transcription blockage occurs preferentially during S-phase, presumably as a consequence of complications between the replication fork and blocked RNA polymerase II complexes. Other mechanisms by which apoptosis may be induced following DNA damage-induced blockage of transcription include loss of survival factors such as MCL-1 and nuclear accumulation of nucleocytoplasmic shuttling proteins that are normally exported in a mRNA export-dependent manner. In order for cells to avoid undergoing apoptosis following transcription blockage, cells must recover RNA synthesis in a timely fashion and this is promoted by transcription-coupled repair, ubiquitylation-dependent degradation of stalled RNA polymerase II and by p53. See the text for more details.

4.3.6.1 Loss of Short-lived Survival Factors

When RNA synthesis is inhibited, cells need to rely on pre-existing mRNAs for the expression of essential proteins. With time cells will run out of essential factors as mRNA levels are diminished by degradation.[15] The ability of cells to swiftly recover RNA synthesis following UV irradiation is critical for the cells to avoid the induction of apoptosis.[102] Cells defective in transcription-coupled repair do not recover RNA synthesis efficiently and induce apoptosis at much lower doses of UV light than repair-proficient cells.[27,171] It is as if cells have an 'apoptotic timer' where cells can operate without transcription for a fixed amount of time before running out of essential factors. One of these essential factors is the cytoplasmic anti-apoptotic factor MCL-1. It has been shown that loss of MCL-1 expression is necessary for UV-induced apoptosis[172] and loss of MCL-1 expression correlates to apoptosis induced by other agents that inhibit transcription.[122,123,173] Although MCL-1 has a short half-life and is important for survival, there are many other survival factors that may also diminish in levels following transcription inhibition which may result in the tipping of the scale in favor of apoptosis.[15]

4.3.6.2 Role of p53

As discussed in Section 4.3.4, the tumor suppressor p53 accumulates in the nucleus in cells treated with agents that block transcription.[27,92,174] The role of activated p53 is to induce cell cycle arrest, promote DNA repair or to trigger cells to undergo apoptosis.[175–179] These are important tumor-suppressing functions aimed at reducing the mutagenic effect of DNA-damaging agents. It appears that different cell types have their unique settings determining whether the induction of p53 will lead to survival or cell death.[180] However, during times of transcription inhibition, the transcription-dependent functions of p53 will be blunted. Following UV-mediated transcription inhibition, expression of a particular gene will depend on the size of the transcribed gene and small genes will be preferentially induced by p53.[181] Moreover, it has been noted that p53-inducible pro-apoptotic genes are in general smaller than p53-inducible survival genes and thus, as the dose of UV light increases, the balance between apoptosis and survival factors will favor apoptosis.[181]

Despite this well-established role of p53 in promoting apoptosis, the presence of a functional p53 has actually been found to protect human fibroblasts against apoptosis induced by UV light[101,102,182] or cisplatin.[101] This p53-mediated protective effect of p53 correlates with a p53-dependent enhanced recovery of RNA synthesis following treatment.[102,182] The mechanism by which p53 stimulates the recovery of mRNA synthesis following exposure to UV-C light does not appear to involve stimulation of transcription-coupled repair since TCR is not influenced by the p53 status of the cells.[183,184] However, it has been found that p53 associates with the elongating RNA polymerase II complex[185,186] and plays a role in global chromatin relaxation.[187] It is possible that, as a component of the transcription factory and with its chromatin remodeling activity, p53 may aid in the post-repair restoration of the chromatin structure to promote re-initiation of transcription in gene domains that have undergone transcription-coupled repair.

4.3.6.3   Nuclear Accumulation of Nucleocytoplasmic Shuttling Proteins

The development of mechanisms allowing for the nuclear export of mRNA was essential for the evolution of eukaryotes.[188] As nascent RNA molecules are being processed in the nucleus, they become decorated with a variety of proteins that aid in their nuclear export.[189] These proteins include Aly, hnRNP A1, SRp20, HuR and ASF/SF2, and they are thought to act as adaptors between the mRNA and the nuclear export protein TAP.[190–196] The machinery involved in the nuclear export of proteins developed later in evolution and utilizes some of the same export mechanisms as the mRNA export machinery. In fact, nuclear microinjection studies using a GFP-NES (nuclear export signal) reporter protein revealed that if transcription is blocked by UV light, actinomycin D, DRB, H7 or microinjected anti-RNA polymerase II antibodies, the export of the NES reporter protein was severely attenuated.[143] Furthermore, when the nuclear export of mRNA was blocked by interference with components of the nuclear pore complex or the TAP protein, the microinjected NES reporter protein was prohibited from being exported to the cytoplasm. Thus, the process of nuclear export of proteins containing nuclear export signals (NES) requires ongoing transcription and is linked to the nuclear export of mRNA.[143]

One example of a NES-containing protein that becomes trapped in the nucleus following blockage of transcription or inhibition of mRNA export is p53.[121] Thus, the loss of mRNA export-linked transport of p53 to the cytoplasm explains why p53 accumulates in the nucleus of cells following transcription inhibition regardless of its phosphorylation or ubiquitylation status.[144] In addition to p53, other nucleocytoplasmic shuttling proteins such as HIF-α,[197] VHL[198] and the mitochondrial protein p32[199] have been shown to accumulate in the nucleus of cells treated with transcription inhibitors. Overall, there may possibly be hundreds of proteins that shuttle between the nucleus and the cytoplasm that would be predicted to accumulate in the nucleus following transcription blockage.[15] Whether accumulation of any of these proteins in the nucleus following transcription inhibition may trigger the induction of apoptosis will need further exploration.

4.3.6.4   Tug-of-war between Replication and Blocked Transcription Factories

When exploring the dose–response of UV-induced apoptosis in human fibroblasts derived from individuals with various DNA repair deficiencies, it was found that cells deficient in TCR induced more apoptosis at lower doses

$(5\,\mathrm{J\,m^{-2}})$ than at higher doses $(20\,\mathrm{J\,m^{-2}})$.[200] This surprising finding was explained by the observation that the cells receiving the higher dose were arrested in the cell cycle and did not enter S-phase. Indeed, if irradiated cells were blocked from entering S-phase by serum starvation or treatment with mimosine, less apoptosis was observed. Furthermore, when UV-irradiated cells were labeled with BrdU to mark cells that entered S-phase during the post-UV incubation period, it was found that the majority of cells that underwent apoptosis were labeled with BrdU. A similar correlation between transcription blockage, S-phase entry and apoptosis was found following treatment with photoactivated psoralen.[106] Thus, the observation that UV- and psoralen-induced apoptosis is linked to blocked transcription and the finding that treated cells undergo apoptosis preferential when entering S-phase suggest that apoptosis may be triggered as a consequence of complications arising when a replication fork encounters a blocked transcription complex.

Since transcription and replication factories are anchored to the nuclear matrix, it is not conceivable that transcription and replication can occur on the same piece of DNA at the same time. Indeed, sites of RNA and DNA synthesis are not found to co-localize in cells.[201,202] Thus, there must be mechanisms in place to clear replicons from transcription-complexes engaged in transcription elongation before origins of replication are fired.[203,204] Furthermore, the orientation of genes has been selected during evolution to preferentially be co-directional with replication so as to limit physical conflicts between transcription and replication.[203,205] In a situation when the elongating transcription machinery has encountered a DNA lesion and is not able to move out of the way of an approaching replication fork, the cell has to come up with measures to resolve the ensuing 'tug-of-war' between the transcription and replication machineries. DNA topoisomerase II and the high mobility group (HMG) protein HMO1 are proteins thought to prevent chromosome fragility at transcription sites during S-phase.[206] Homologous recombination is another potential mechanism by which cells can resolve conflicts occurring between replication and stalled transcription.[207,208]

4.3.7 Transcription Inhibition and Aging

Accumulation of DNA damage in the genome over time has been suggested to be contributing to organismal aging.[209–211] This claim is supported by the fact that deficiencies in DNA repair pathways are associated with premature aging phenotypes in mice and people.[212–214] While unrepaired DNA damage may result in mutations that inactivate critical genes, it is thought that it is the effect the DNA lesions have on gene expression that is associated with the aging process.[215] In fact, persistent transcription-blocking lesions are responsible for the inactivation of specific genes later in life.[216] These genes inactivated later in life by transcription-blocking lesions are involved in the somatic growth axis and include genes such as insulin-like growth factor-1 (IGF-1) and the growth hormone (GH) receptor genes. The gene expression profile of cells from older mice differed dramatically from that of cells from younger mice. However, when the cells from younger mice were irradiated with UV light, their gene expression profile changed into a profile resembling that of older cells. Thus, it appears that the DNA lesions accumulating during aging behave similarly to UV-induced lesions.[216] Since the inactivation of gene expression by UV light is proportional to the size of the gene,[217] it is possible that genes like IGF-1 and the GH receptor, which would be beneficial to inactivate later in life, reside in large transcription units that due to their large size would be selectively inactivated by stochastic transcription-blocking lesions.

4.3.8 Conclusions

To circumvent the detrimental effects caused by transcription disruption following DNA damage, several cellular defense mechanisms have evolved. First, stalled RNA polymerase II complexes trigger the activation of stress kinases, which induce stress responses leading to cell cycle arrest or apoptosis. Secondly, RNA polymerase II complexes stalled at DNA lesions recruit DNA-repair enzymes to remove blocking DNA lesions *via* transcription-coupled repair (TCR). Finally, post-transcriptional regulation of the existing pool of mRNA so as to compensate for the loss of nascent RNA may 'tide the cell over' until the transcription machinery can re-engage in mRNA synthesis. Future studies should be directed at using next generation sequencing technology to comprehensively map areas of the genome being transcribed and where DNA lesions are induced and removed in a genome-wide fashion. Furthermore, our knowledge about the mechanisms driving cells into apoptosis following transcription blockage should stimulate the development of novel cancer therapeutics.

References

1. D. A. Jackson, S. J. McReady and P. R. Cook, *Nature*, 1981, **292**, 552–555.
2. D. A. Jackson, A. B. Hassan, R. J. Errington and P. R. Cook, *EMBO J.*, 1993, **12**, 1059–1065.
3. P. R. Cook, *J. Mol. Biol.*, 2010, **395**, 1–10.
4. C. S. Osborne, L. Chakalova, K. E. Brown, D. Carter, A. Horton, E. Debrand, B. Goyenechea, J. A. Mitchell, S. Lopes, W. Reik and P. Fraser, *Nat. Genet.*, 2004, **36**, 1065–1071.
5. D. A. Jackson, P. Dickinson and P. R. Cook, *EMBO J.*, 1990, **9**, 567–571.
6. M. Ljungman and P. Hanawalt, *Nucleic Acids Res.*, 1995, **23**, 1782–1789.
7. J. A. Mitchell and P. Fraser, *Genes Dev.*, 2008, **22**, 20–25.
8. M. Fromaget and P. R. Cook, *Exp. Cell Res.*, 2007, **313**, 3026–3033.
9. P. Cramer, D. A. Bushnell, J. H. Fu, A. L. Gnatt, B. Maier-Davis, N. E. Thompson, R. R. Burgess, A. M. Edwards, P. R. David and R. D. Kornberg, *Science*, 2000, **288**, 640–649.
10. P. Cramer, K. J. Armache, S. Baumli, S. Benkert, F. Brueckner, C. Buchen, G. E. Damsma, S. Dengl, S. R. Geiger, A. J. Jasiak, A. Jawhari, S.

Jennebach, T. Kamenski, H. Kettenberger, C. D. Kuhn, E. Lehmann, K. Leike, J. F. Sydow and A. Vannini, *Annu. Rev. Biophys.*, 2008, **37**, 337–352.

11. T. Haaf and D. C. Ward, *Exp. Cell Res.*, 1996, **224**, 163–173.
12. A. Papantonis and P. R. Cook, *Curr. Opin. Cell Biol.*, 2010, **22**, 271–276.
13. M. Ljungman and D. P. Lane, *Nat. Rev. Cancer*, 2004, **4**, 727–737.
14. S. Tornaletti, *Cell Mol. Life Sci.*, 2009, **66**, 1010–1020.
15. F. A. Derheimer, C. W. Chang and M. Ljungman, *Eur. J. Cancer*, 2005, **41**, 2569–2576.
16. T. Kameyama and G. D. Novelli, *Arch. Biochem. Biophys.*, 1962, **97**, 529–537.
17. B. A. Donahue, S. Yin, J. S. Taylor, D. Reines and P. C. Hanawalt, *Proc. Natl. Acad. Sci. U.S.A.*, 1994, **91**, 8502–8506.
18. J. S. Mei Kwei, I. Kuraoka, K. Horibata, M. Ubukata, E. Kobatake, S. Iwai, H. Handa and K. Tanaka, *Biochem. Biophys. Res. Commun.*, 2004, **320**, 1133–1138.
19. M. J. Mone, M. Volker, O. Nikaido and L. H. F. Mullenders, A. A. van Zeeland, P. J. Verschure, E. M. M. Manders and R. van Driel, *EMBO Reports*, 2001, **2**, 1013–1017.
20. D. A. Rockx, R. Mason, A. van Hoffen, M. C. Barton, E. Citterio and D. B. Bregman, A. A. van Zeeland, H. Vrieling and L. H. Mullenders, *Proc. Natl. Acad. Sci. U.S.A.*, 2000, **10**, 10503–10508.
21. G. Napolitano, S. Amente, V. Castiglia, B. Gargano, V. Ruda, X. Darzacq, O. Bensaude, B. Majello and L. Lania, *PLoS One*, 2010, **5**, e11245.
22. S. Tornaletti, D. Reines and P. C. Hanawalt, *J. Biol. Chem.*, 1999, **274**, 24124–24130.
23. C. P. Selby, R. Drapkin, D. Reinberg and A. Sancar, *Nucleic Acids Res.*, 1997, **25**, 787–793.
24. F. Brueckner, U. Hennecke, T. Carell and P. Cramer, *Science*, 2007, **315**, 859–862.
25. C. Petit-Frère, P. Clingen, C. Arlett and M. Green, *Mutat. Res.*, 1996, **354**, 87–94.
26. I. Mellon, G. Spivak and P. C. Hanawalt, *Cell*, 1987, **51**, 241–249.
27. M. Ljungman and F. Zhang, *Oncogene*, 1996, **13**, 823–831.
28. S. Leadon and M. Snowden, *Mol. Cell. Biol.*, 1988, **8**, 5331–5338.
29. M. Ljungman, *Carcinogenesis*, 1999, **20**, 395–399.
30. P. W. Doetsch, *Mutat. Res.*, 2002, **510**, 131–140.
31. C. Marietta and P. J. Brooks, *EMBO Rep.*, 2007, **8**, 388–393.
32. D. Bregeon and P. W. Doetsch, *Nat. Rev. Cancer*, 2011, **11**, 218–227.
33. D. B. Bregman, R. Halaban, A. J. Vangool, K. A. Henning, E. C. Friedberg and S. L. Warren, *Proc. Natl. Acad. Sci. U.S.A.*, 1996, **93**, 11586–11590.
34. J. N. Ratner, B. Balasubramanian, J. Corden, S. L. Warren and D. B. Bregman, *J. Biol. Chem.*, 1998, **273**, 5184–5189.
35. B. C. McKay, F. Chen, S. T. Clarke, H. E. Wiggin, L. M. Harley and M. Ljungman, *Mutat. Res.*, 2001, **485**, 93–105.
36. L. Lommel, M. E. Bucheli. and K. S. Sweder, *Proc. Natl. Acad. Sci. U.S.A.*, 2000, **97**, 9088–9092.

37. M. Ljungman and P. C. Hanawalt, *Proc. Natl. Acad. Sci. USA*, 1992, **89**, 6055–6059.

38. E. R. Jupe, R. R. Sinden and I. L. Cartwright, *EMBO J.*, 1993, **12**, 1067–1075.

39. R. M. Harland, H. Weintraub and S. L. McKnight, *Nature*, 1983, **302**, 38–43.

40. H. Weintraub, P. F. Cheng and K. Conrad, *Cell*, 1986, **46**, 115–122.

41. J. D. Parvin and P. A. Sharp, *Cell*, 1993, **73**, 533–540.

42. M. Dunaway and E. A. Ostrander, *Nature*, 1993, **361**, 746–748.

43. S. G. Ernst, R. C. Rustad and N. L. Oleinick, *Int. J. Radiat. Biol. Relat. Stud. Phys. Chem. Med.*, 1975, **28**, 67–74.

44. R. E. Stephens, I. J. Paul and A. M. Zimmerman, *Int. J. Radiat. Biol. Relat. Stud. Phys. Chem. Med.*, 1976, **30**, 83–89.

45. E. V. Boudnitskaya, M. Brunfaut and M. Errera, *Biochim. Biophys. Acta*, 1964, **80**, 567–575.

46. A. N. Luchnik, T. A. Hisamutdinov and P. A. Georgiev, *Nucleic Acids Res.*, 1988, **16**, 5175–5190.

47. C. Rodi and W. Sauerbier, *J. Cell. Physiol.*, 1989, **141**, 346–352.

48. R. Ghosh, P. Amstad and P. Cerutti, *Mol. Cell. Biol.*, 1993, **13**, 6992–6999.

49. M. Ljungman, *Radiat. Res.*, 1999, **152**, 444–449.

50. M. Ljungman, *Mutat. Res.*, 2005, **577**, 203–217.

51. B. Plosky, L. Samson, B. P. Engelward, B. Gold, B. Schlaen, T. Millas, M. Magnotti, J. Schor and D. A. Scicchitano, *DNA Repair (Amst.)*, 2002, **1**, 683–696.

52. A. Dimitri, J. A. Burns, S. Broyde and D. A. Scicchitano, *Nucleic Acids Res.*, 2008, **36**, 6459–6471.

53. S. Yanamadala and M. Ljungman, *Mol. Cancer Res.*, 2003, **1**, 747–754.

54. M. Bonfanti, M. Broggini, C. Prontera and M. D'Incalci, *Nucleic Acids Res.*, 1991, **19**, 5739–5742.

55. D. A. Scicchitano and P. C. Hanawalt, *Proc. Natl. Acad. Sci. U.S.A.*, 1989, **86**, 3050–3054.

56. W. Wang, A. Sitaram and D. A. Scicchitano, *Biochemistry*, 1995, **34**, 1798–1804.

57. A. J. Ryan, M. A. Billett and P. J. O'Connor, *Carcinogenesis*, 1986, **7**, 1497–1503.

58. R. B. Ali, A. K. Teo, H. K. Oh, L. S. Chuang, T. C. Ayi and B. F. Li, *Mol. Cell Biol.*, 1998, **18**, 1660–1669.

59. P. Branch, G. Aquilina, M. Bignami and P. Karran, *Nature*, 1993, **362**, 652–654.

60. J. A. Burns, K. Dreij, L. Cartularo and D. A. Scicchitano, *Nucleic Acids Res.*, 2010, **38**, 8178–8187.

61. I. Kuraoka, M. Endou, Y. Yamaguchi, T. Wada, H. Handa and K. Tanaka, *J. Biol. Chem.*, 2003, **278**, 7294–7299.

62. S. Tornaletti, L. S. Maeda, R. D. Kolodner and P. C. Hanawalt, *DNA Repair (Amst.)*, 2004, **3**, 483–494.

63. N. Charlet-Berguerand, S. Feuerhahn, S. E. Kong, H. Ziserman, J. W. Conaway, R. Conaway and J. M. Egly, *EMBO J.*, 2006, **25**, 5481–5491.
64. F. Le Page, V. Randrianarison, D. Marot, J. Cabannes, M. Perricaudet, J. Feunteun and A. Sarasin, *Cancer Res.*, 2000, **60**, 5548–5552.
65. M. Pastoriza Gallego and A. Sarasin, *Biochimie*, 2003, **85**, 1073–1082.
66. G. Dianov, C. Bischoff, M. Sunesen and V. A. Bohr, *Nucleic Acids Res.*, 1999, **27**, 1365–1368.
67. S. Tornaletti, L. S. Maeda, D. R. Lloyd, D. Reines and P. C. Hanawalt, *J. Biol. Chem.*, 2001, **276**, 45367–45371.
68. M. Berthiaume, N. Boufaied, A. Moisan and L. Gaudreau, *DNA Cell Biol.*, 2006, **25**, 124–134.
69. R. Wu, A. Kumar and J. Warner, *Proc. Natl. Acad. Sci. U.S.A.*, 1971, **68**, 3009–3014.
70. H. Abelson and S. Penman, *Nature (new biology)*, 1972, **237**, 144–146.
71. D. Kessel, H. Bosmann and K. Lohr, *Biochim. Biophys. Acta*, 1972, **269**, 210–216.
72. Y.-H. Hsiang, R. Hertzberg, S. Hecht and L. Liu, *J. Biol. Chem.*, 1985, **260**, 14873–14878.
73. Y. Pommier, *Nat. Rev. Cancer*, 2006, **6**, 789–802.
74. L. F. Liu and J. C. Wang, *Proc. Natl. Acad. Sci. U.S.A.*, 1987, **84**, 7024–7027.
75. A. Merino, K. Madden, W. Lane, J. Champoux and D. Reinberg, *Nature*, 1993, **365**, 227–232.
76. M. Kretzschmar, M. Meisterernst and R. Roeder, *Proc. Natl. Acad. Sci. U.S.A.*, 1993, **90**, 11508–11512.
77. H. Zhang, J. Wang and L. Liu, *Proc. Natl. Acad. Sci. U.S.A.*, 1988, **85**, 1060–1064.
78. M. Ljungman and P. C. Hanawalt, *Carcinogenesis*, 1996, **17**, 31–35.
79. S. Desai, L. Liu, D. Vazquez-Abad and P. D'Arpa, *J. Biol. Chem.*, 1997, **272**, 24159–24164.
80. S. D. Desai, H. Zhang, A. Rodriguez-Bauman, J. M. Yang, X. Wu, M. K. Gounder, E. H. Rubin and L. F. Liu, *Mol. Cell. Biol.*, 2003, **23**, 2341–2350.
81. Y.-H. Hsiang, M. Lihou and L. Liu, *Cancer Res.*, 1989, **49**, 5077–5082.
82. G. Capranico, F. Ferri, M. V. Fogli, A. Russo, L. Lotito and L. Baranello, *Biochimie*, 2007, **89**, 482–489.
83. I. Collins, A. Weber and D. Levens, *Mol. Cell. Biol.*, 2001, **21**, 8437–8451.
84. A. L. Pinto and S. J. Lippard, *Biochim. Biophys. Acta*, 1985, **780**, 167–180.
85. D. Wang and S. J. Lippard, *Nat. Rev. Drug Discov.*, 2005, **4**, 307–320.
86. L. Kelland, *Nat. Rev. Cancer*, 2007, **7**, 573–584.
87. Z. H. Siddik, *Oncogene*, 2003, **22**, 7265–7279.
88. Y. Corda, M. F. Anin, M. Leng and D. Job, *Biochemistry*, 1993, **32**, 8582–8588.
89. S. Tornaletti, S. M. Patrick, J. J. Turchi and P. C. Hanawalt, *J. Biol. Chem.*, 2003, **278**, 35791–35797.
90. Y. Jung and S. J. Lippard, *J. Biol. Chem.*, 2006, **281**, 1361–1370.
91. J. A. Mello, S. J. Lippard and J. M. Essigmann, *Biochemistry*, 1995, **34**, 14783–14791.

92. M. Ljungman, F. F. Zhang and F. Chen, A. J. Rainbow and B. C. McKay, *Oncogene*, 1999, **18**, 583–592.

93. G. E. Damsma, A. Alt, F. Brueckner, T. Carell and P. Cramer, *Nat. Struct. Mol. Biol.*, 2007, **14**, 1127–1133.

94. K. B. Lee, D. Wang, S. J. Lippard and P. A. Sharp, *Proc. Natl. Acad. Sci. U.S.A.*, 2002, **99**, 4239–4244.

95. J. T. Sczepanski, A. C. Jacobs, B. Van Houten and M. M. Greenberg, *Biochemistry*, 2009, **48**, 7565–7567.

96. X. Wang, C. A. Peterson, H. Zheng, R. S. Nairn, R. J. Legerski and L. Li, *Mol. Cell. Biol.*, 2001, **21**, 713–720.

97. I. Kuraoka, W. R. Kobertz, R. R. Ariza, M. Biggerstaff, J. M. Essigmann and R. D. Wood, *J. Biol. Chem.*, 2000, **275**, 26632–26636.

98. T. Bessho, D. Mu and A. Sancar, *Mol. Cell. Biol.*, 1997, **17**, 6822–6830.

99. J. C. Jones, W. P. Zhen, E. Reed, R. J. Parker, A. Sancar and V. A. Bohr, *J. Biol. Chem.*, 1991, **266**, 7101–7107.

100. A. May, R. S. Nairn, D. S. Okumoto, K. Wassermann, T. Stevnsner, J. C. Jones and V. A. Bohr, *J. Biol. Chem.*, 1993, **268**, 1650–1657.

101. B. C. McKay, C. Becerril and M. Ljungman, *Oncogene*, 2001, **20**, 6805–6808.

102. B. McKay and M. Ljungman, *Neoplasia*, 1999, **1**, 276–284.

103. J. A. Parrish, T. B. Fitzpatrick, L. Tanenbaum and M. A. Pathak, *N. Engl. J. Med.*, 1974, **291**, 1207–1211.

104. G. D. Cimino, H. B. Gamper, S. T. Isaacs and J. E. Hearst, *Annu. Rev. Biochem.*, 1985, **54**, 1151–1193.

105. Y. B. Shi, J. Griffith, H. Gamper and J. E. Hearst, *Nucleic Acids Res.*, 1988, **16**, 8945–8952.

106. F. A. Derheimer, J. K. Hicks, M. T. Paulsen, C. E. Canman and M. Ljungman, *Mol. Pharmacol.*, 2009, **75**, 599–607.

107. A. L. Islas, F. J. Baker and P. C. Hanawalt, *Biochemistry*, 1994, **33**, 10794–10799.

108. A. L. Islas, J. M. Vos and P. C. Hanawalt, *Cancer Res.*, 1991, **51**, 2867–2873.

109. J. M. Vos and P. C. Hanawalt, *Cell*, 1987, **50**, 789–799.

110. J. M. Vos and E. L. Wauthier, *Mol. Cell. Biol.*, 1991, **11**, 2245–2252.

111. C. Le Tourneau, S. Faivre, V. Laurence, C. Delbaldo, K. Vera, V. Girre, J. Chiao, S. Armour, S. Frame, S. R. Green, A. Gianella-Borradori, V. Dieras and E. Raymond, *Eur. J. Cancer*, 2010, **46**, 3243–3250.

112. I. T. Aldoss, T. Tashi and A. K. Ganti, *Expert Opin. Investig. Drugs*, 2009, **18**, 1957–1965.

113. W. S. Hsieh, R. Soo, B. K. Peh, T. Loh, D. Dong, D. Soh, L. S. Wong, S. Green, J. Chiao, C. Y. Cui, Y. F. Lai, S. C. Lee, B. Mow, R. Soong, M. Salto-Tellez and B. C. Goh, *Clin. Cancer Res.*, 2009, **15**, 1435–1442.

114. L. M. Wang and D. M. Ren, *Mini Rev. Med. Chem.*, 2010, **10**, 1058–1070.

115. J. E. Karp, B. D. Smith, L. S. Resar, J. M. Greer, A. Blackford, M. Zhao, D. Moton-Nelson, K. Alino, M. J. Levis, S. D. Gore, B. Joseph, H. Carraway, M. A. McDevitt, L. Bagain, K. Mackey, J. Briel, L. A. Doyle, J. J. Wright and M. A. Rudek, *Blood*, 2011, **117**, 3302–3310.

116. M. Ljungman and M. T. Paulsen, *Mol. Pharm.*, 2001, **60**, 785–789.
117. M. V. Blagosklonny, *Cell Cycle*, 2004, **3**, 1537–1542.
118. T. David-Pfeuty, *Oncogene*, 1999, **18**, 7409–7422.
119. Z. N. Demidenko and M. V. Blagosklonny, *Cancer Res.*, 2004, **64**, 3653–3660.
120. M. Ljungman, H. M. O'Hagan and M. T. Paulsen, *Oncogene*, 2001, **20**, 5964–5971.
121. F. A. Derheimer, H. M. O'Hagan, H. H. M. Krueger, S. Hanasoge, M. T. Paulsen and M. Ljungman, *Proc. Natl. Acad. Sci. U.S.A.*, 2007, **104**, 12778–12783.
122. D. E. MacCallum, J. Melville, S. Frame, K. Watt, S. Anderson, A. Gianella-Borradori, D. P. Lane and S. R. Green, *Cancer Res.*, 2005, **65**, 5399–5407.
123. I. Gojo, B. Zhang and R. G. Fenton, *Clin. Cancer Res.*, 2002, **8**, 3527–3538.
124. H. M. Sobell, *Proc. Natl. Acad. Sci. U.S.A.*, 1985, **82**, 5328–5331.
125. R. Perry and D. Kelley, *J. Cell. Physiol.*, 1970, **76**, 127–140.
126. E. Egyhazi, *Nature*, 1976, **262**, 319–321.
127. K. Yankulov, K. Yamashita, R. Roy, J.-M. Egly and D. L. Bentley, *J. Biol. Chem.*, 1995, **270**, 23922–23925.
128. Y. K. Kim, C. F. Bourgeois, C. Isel, M. J. Churcher and J. Karn, *Mol. Cell. Biol.*, 2002, **22**, 4622–4637.
129. M. F. Dubois, V. T. Nguyen, S. Bellier and O. Bensaude, *J. Biol. Chem.*, 1994, **269**, 13331–13336.
130. S. Buratowski, *Mol. Cell*, 2009, **36**, 541–546.
131. N. P. Gomes, G. Bjerke, B. Llorente, S. A. Szostek, B. M. Emerson and J. M. Espinosa, *Genes Dev.*, 2006, **20**, 601–612.
132. T. Lindell, F. Weinberg, P. Morris, R. Roeder and W. Rutter, *Science*, 1970, **170**, 447–449.
133. D. A. Bushnell, P. Cramer and R. D. Kornberg, *Proc. Natl. Acad. Sci. U.S.A.*, 2002, **99**, 1218–1222.
134. C. D. Kaplan, K. M. Larsson and R. D. Kornberg, *Mol. Cell*, 2008, **30**, 547–556.
135. F. Brueckner and P. Cramer, *Nat. Struct. Mol. Biol.*, 2008, **15**, 811–818.
136. S. A. Klibanov, H. M. O'Hagan and M. Ljungman, *J. Cell Sci.*, 2001, **114**, 1867–1873.
137. H. M. O'Hagan and M. Ljungman, *Mutat. Res.*, 2004, **546**, 7–15.
138. M. Ljungman, *Cell Cycle*, 2007, **6**, 2252–2257.
139. G. W. Muse, D. A. Gilchrist, S. Nechaev, R. Shah, J. S. Parker, S. F. Grissom, J. Zeitlinger and K. Adelman, *Nat. Genet.*, 2007, **39**, 1507–1511.
140. S. Hanasoge and M. Ljungman, *Carcinogenesis*, 2007, **28**, 2298–304.
141. J. A. Marteijn, S. Bekker-Jensen, N. Mailand, H. Lans, P. Schwertman, A. M. Gourdin, N. P. Dantuma, J. Lukas and W. Vermeulen, *J. Cell. Biol.*, 2009, **186**, 835–847.
142. M. G. Vrouwe, A. Pines, R. M. Overmeer, K. Hanada and L. H. Mullenders, *J. Cell. Sci.*, 2011, **124**, 435–446.

143. H. M. O'Hagan and M. Ljungman, *Exp. Cell Res.*, 2004, **297**, 548–559.

144. H. M. O'Hagan and M. Ljungman, *Oncogene*, 2004, **23**, 5505–5512.

145. A. J. Fornace, Jr., I. Alamo, Jr. and M. C. Hollander, *Proc. Natl. Acad. Sci.*, 1988, **85**, 8800–8804.

146. P. Herrlich, H. Ponta and H. J. Rahmsdorf, *Rev. Physiol. Biochem. Pharmacol.*, 1992, **119**, 187–223.

147. W. El-Deiry, T. Tokino, V. Velculescu, D. Levy, R. Parsons, J. Trent, D. Lin, W. Mercer, K. Kinzler and B. Vogelstein, *Cell*, 1993, **75**, 817–825.

148. R. B. Zhao, K. Gish, M. Murphy, Y. X. Yin, D. Notterman, W. H. Hoffman, E. Tom, D. H. Mack and A. J. Levine, *Genes Dev.*, 2000, **14**, 981–993.

149. M. J. Munoz, M. S. Perez Santangelo, M. P. Paronetto, M. de la Mata, F. Pelisch, S. Boireau, K. Glover-Cutter, C. Ben-Dov, M. Blaustein, J. J. Lozano, G. Bird, D. Bentley, E. Bertrand and A. R. Kornblihtt, *Cell*, 2009, **137**, 708–720.

150. J. Jackman, I. Alamo, Jr. and A. J. Fornace, Jr., *Cancer Res.*, 1994, **54**, 5656–5662.

151. C. Blattner, P. Kannouche, M. Litfin, K. Bender, H. J. Rahmsdorf, J. F. Angulo and P. Herrlich, *Mol. Cell. Biol.*, 2000, **20**, 3616–3625.

152. W. Wang, H. Furneaux, H. Cheng, M. C. Caldwell, D. Hutter, Y. Liu, N. Holbrook and M. Gorospe, *Mol. Cell. Biol.*, 2000, **20**, 760–769.

153. H. C. Reinhardt, P. Hasskamp, I. Schmedding, S. Morandell, M. A. van Vugt, X. Wang, R. Linding, S. E. Ong, D. Weaver, S. A. Carr and M. B. Yaffe, *Mol. Cell.*, 2010, **40**, 34–49.

154. X. Lu, L. de la Pena, C. Barker, K. Camphausen and P. J. Tofilon, *Cancer Res.*, 2006, **66**, 1052–1061.

155. S. Kumaraswamy, P. Chinnaiyan, U. T. Shankavaram, X. Lu, K. Camphausen and P. J. Tofilon, *Cancer Res.*, 2008, **68**, 3819–3826.

156. S. Braunstein, M. L. Badura, Q. Xi, S. C. Formenti and R. J. Schneider, *Mol. Cell. Biol.*, 2009, **29**, 5645–5656.

157. J. Pothof, N. S. Verkaik, W. van Ijcken, E. A. Wiemer, V. T. Ta and G. T. van der Horst, N. G. Jaspers, D. C. van Gent, J. H. Hoeijmakers and S. P. Persengiev, *EMBO J.*, 2009, **28**, 2090–2099.

158. S. Shin, H. J. Cha, E. M. Lee, J. H. Jung, S. J. Lee, I. C. Park, Y. W. Jin and S. An, *Int. J. Oncol.*, 2009, **34**, 1645–1652.

159. N. L. Simone, B. P. Soule, D. Ly, A. D. Saleh, J. E. Savage, W. Degraff, J. Cook, C. C. Harris, D. Gius and J. B. Mitchell, *PLoS One*, 2009, **4**, e6377.

160. K. S. Khabar, *Cell. Mol. Life Sci.*, 2010, **67**, 2937–2955.

161. J. Houseley and D. Tollervey, *Cell*, 2009, **136**, 763–776.

162. P. Pandey, J. Raingeaud, M. Kaneki, R. Weichselbaum, R. J. Davis, D. Kufe and S. Kharbanda, *J. Biol. Chem.*, 1996, **271**, 23775–23779.

163. V. Lafarga, A. Cuadrado, I. Lopez de Silanes, R. Bengoechea, O. Fernandez-Capetillo and A. R. Nebreda, *Mol. Cell. Biol.*, 2009, **29**, 4341–4351.

164. J. L. Dean, M. Brook, A. R. Clark and J. Saklatvala, *J. Biol. Chem.*, 1999, **274**, 264–269.

165. K. R. Mahtani, M. Brook, J. L. Dean, G. Sully, J. Saklatvala and A. R. Clark, *Mol. Cell. Biol.*, 2001, **21**, 6461–6469.
166. M. A. Bachelor and G. T. Bowden, *J. Biol. Chem.*, 2004, **279**, 42658–42668.
167. T. G. Tessner, F. Muhale, S. Schloemann, S. M. Cohn, A. R. Morrison and W. F. Stenson, *Carcinogenesis*, 2004, **25**, 37–45.
168. B. Li, J. Si and J. W. DeWille, *J. Cell. Biochem.*, 2008, **103**, 1657–1669.
169. N. S. Fernau, D. Fugmann, M. Leyendecker, K. Reimann, S. Grether-Beck, S. Galban, N. Ale-Agha, J. Krutmann and L. O. Klotz, *J. Biol. Chem.*, 2009.
170. Z. Guo, A. Kumagai, S. X. Wang and W. G. Dunphy, *Genes Dev.*, 2000, **14**, 2745–2756.
171. B. C. McKay and M. Ljungman, and A. J. Rainbow, *Oncogene*, 1998, **17**, 545–555.
172. D. Nijhawan, M. Fang, E. Traer, Q. Zhong, W. Gao, F. Du and X. Wang, *Genes Dev.*, 2003, **17**, 1475–1486.
173. N. Raje, S. Kumar, T. Hideshima, A. Roccaro, K. Ishitsuka, H. Yasui, N. Shiraishi, D. Chauhan, N. C. Munshi, S. R. Green and K. C. Anderson, *Blood*, 2005, **106**, 1042–1047.
174. H. E. van Gijssel, L. H. Mullenders, M. F. van Oosterwijk and J. H. Meerman, *Life Sci.*, 2003, **73**, 1759–1771.
175. S. Lowe, E. Schmitt, S. Smith, B. Osborne and T. Jacks, *Nature*, 1993, **362**, 847–849.
176. K. Polyak, Y. Xia, J. L. Zweier, K. W. Kinzler and B. Vogelstein, *Nature*, 1997, **389**, 300–305.
177. M. Ljungman, *Neoplasia*, 2000, **2**, 208–225.
178. F. Murray-Zmijewski, E. A. Slee and X. Lu, *Nat. Rev. Mol. Cell. Biol.*, 2008, **9**, 702–712.
179. C. F. Cheok, C. S. Verma, J. Baselga and D. P. Lane, *Nat. Rev. Clin. Oncol.*, 2011, **8**, 25–37.
180. A. V. Gudkov and E. A. Komarova, *Nat. Rev. Cancer*, 2003, **3**, 117–129.
181. B. C. McKay, L. J. Stubbert, C. C. Fowler, J. M. Smith, R. A. Cardamore and J. C. Spronck, *Proc. Natl. Acad. Sci. U.S.A.*, 2004, **101**, 6582–6586.
182. B. C. McKay, F. Chen, C. R. Perumalswami, F. F. Zhang and M. Ljungman, *Mol. Biol. Cell*, 2000, **11**, 2543–2551.
183. J. M. Ford and P. C. Hanawalt, *Proc. Natl. Acad. Sci. U.S.A.*, 1995, **92**, 8876–8880.
184. G. Mathonnet, C. Leger, J. Desnoyers, R. Drouin, J. P. Therrien and E. A. Drobetsky, *Proc. Natl. Acad. Sci. U.S.A.*, 2003, **100**, 7219–7224.
185. C. P. Rubbi and J. Milner, *Oncogene*, 2000, **19**, 85–96.
186. S. K. Balakrishnan and D. S. Gross, *Oncogene*, 2008, **27**, 2661–2672.
187. C. P. Rubbi and J. Milner, *EMBO Journal*, 2003, **22**, 975–986.
188. M. Serpeloni, N. M. Vidal, S. Goldenberg, A. R. Avila and F. G. Hoffmann, *BMC Evol. Biol.*, 2011, **11**, 7.
189. S. M. Kelly and A. H. Corbett, *Traffic*, 2009, **10**, 1199–1208.
190. Y. Huang, R. Gattoni, J. Stevenin and J. A. Steitz, *Mol. Cell.*, 2003, **11**, 837–843.

191. J. P. Rodrigues, M. Rode, D. Gatfield, B. J. Blencowe, M. Carmo-Fonseca and E. Izaurralde, *Proc. Natl. Acad. Sci. U.S.A.*, 2001, **98**, 1030–1035.

192. I. E. Gallouzi and J. A. Steitz, *Science*, 2001, **294**, 1895–1901.

193. D. Gatfield and E. Izaurralde, *J. Cell Biol.*, 2002, **159**, 579–588.

194. M. L. Luo, Z. Zhou, K. Magni, C. Christoforides, J. Rappsilber, M. Mann and R. Reed, *Nature*, 2001, **413**, 644–647.

195. Z. Zhou, M. J. Luo, K. Straesser, J. Katahira, E. Hurt and R. Reed, *Nature*, 2000, **407**, 401–405.

196. W. M. Michael, M. Choi and G. Dreyfuss, *Cell*, 1995, **83**, 415–422.

197. I. Groulx and S. Lee, *Mol. Cell. Biol.*, 2002, **22**, 5319–5336.

198. S. Lee, M. Neumann, R. Stearman, R. Stauber, A. Pause, G. N. Pavlakis and R. D. Klausner, *Mol. Cell. Biol.*, 1999, **19**, 1486–1497.

199. K. A. Brokstad, K. H. Kalland, W. C. Russell and D. A. Matthews, *Biochem. Biophys. Res. Commun.*, 2001, **281**, 1161–1169.

200. B. McKay, C. Becerril, J. Spronck and M. Ljungman, *DNA Repair*, 2002, **1**, 811–820.

201. C. Bouniol-Baly, E. Nguyen, D. Besombes and P. Debey, *Exp. Cell Res.*, 1997, **236**, 201–211.

202. D. G. Wansink, E. E. Manders, I. van der Kraan, J. A. Aten, R. van Driel and L. de Jong, *J. Cell. Sci.*, 1994, **107** (Pt6), 1449–1456.

203. C. J. Rudolph, P. Dhillon, T. Moore and R. G. Lloyd, *DNA Repair (Amst.)*, 2007, **6**, 981–993.

204. M. Ljungman, *Environ. Mol. Mutagen.*, 2010, **51**, 879–889.

205. H. Merrikh, C. Machon, W. H. Grainger, A. D. Grossman and P. Soultanas, *Nature*, 2011, **470**, 554–557.

206. R. Bermejo, T. Capra, V. Gonzalez-Huici, D. Fachinetti, A. Cocito, G. Natoli, Y. Katou, H. Mori, K. Kurokawa, K. Shirahige and M. Foiani, *Cell*, 2009, **138**, 870–884.

207. P. Gottipati, T. N. Cassel, L. Savolainen and T. Helleday, *Mol. Cell. Biol.*, 2008, **28**, 154–164.

208. P. Gottipati and T. Helleday, *Mutagenesis*, 2009, **24**, 203–210.

209. J. H. Hoeijmakers, *N. Engl. J. Med.*, 2009, **361**, 1475–1485.

210. E. G. Seviour and S. Y. Lin, *Aging (Albany NY)*, 2010, **2**, 900–907.

211. G. A. Garinis, G. T. van der Horst, J. Vijg and J. H. Hoeijmakers, *Nat. Cell. Biol.*, 2008, **10**, 1241–1247.

212. J. de Boer, J. O. Andressoo, J. de Wit, J. Huijmans, R. B. Beems, H. van Steeg, G. Weeda, G. T. J. van der Horst, W. van Leeuwen, A. P. N. Themmen, M. Meradji and J. H. J. Hoeijmakers, *Science*, 2002, **296**, 1276–1279.

213. P. Hasty, J. Campisi, J. Hoeijmakers, H. van Steeg and J. Vijg, *Science*, 2003, **299**, 1355–1359.

214. L. J. Niederhofer, G. A. Garinis, A. Raams, A. S. Lalai, A. R. Robinson, E. Appeldoorn, H. Odijk, R. Oostendorp, A. Ahmad, W. van Leeuwen, A. F. Theil, W. Vermeulen, G. T. van der Horst, P. Meinecke, W. J. Kleijer and J. Vijg, N. G. Jaspers and J. H. Hoeijmakers, *Nature*, 2006, **444**, 1038–1043.

215. T. Lu, Y. Pan, S. Y. Kao, C. Li, I. Kohane, J. Chan and B. A. Yankner, *Nature*, 2004, **429**, 883–891.
216. G. A. Garinis, L. M. Uittenboogaard, H. Stachelscheid, M. Fousteri, W. van Ijcken, T. M. Breit, H. van Steeg, L. H. Mullenders, G. T. van der Horst, J. C. Bruning, C. M. Niessen, J. H. Hoeijmakers and B. Schumacher, *Nat. Cell. Biol.*, 2009, **11**, 604–615.
217. W. Sauerbier and K. Hercules, *Annu. Rev. Genet.*, 1978, **12**, 329–363.

5. Epigenetic Mechanisms

CHAPTER 5.1

The Interplay between Epigenetics and Gap Junctional Intercellular Communication

M. VINKEN,* V. ROGIERS AND T. VANHAECKE

Vrije Universiteit Brussel, Department of Toxicology, Laarbeeklaan 103, B-1090 Brussels, Belgium
*Email: mvinken@vub.ac.be

5.1.1 Introduction

Gap junctional intercellular communication (GJIC) is a key player in the control of the entire cellular life cycle. The production of connexin proteins, the gap junction building stones, is driven by a plethora of factors. At the pretranscriptional expression level, epigenetic mechanisms, including histone modifications and DNA methylation, are crucial regulators. These epigenetic determinants of GJIC are addressed in the first part of this chapter.

Inherent to its pivotal task in maintaining tissue homeostasis, GJIC is frequently disrupted upon impairment of this critical physiological balance, such as during toxicity and carcinogenicity. In fact, gap junctions are specifically targeted by epigenetic carcinogens and hence inhibition of GJIC is considered a

Issues in Toxicology No. 13
The Cellular Response to the Genotoxic Insult: The Question of Threshold for
Genotoxic Carcinogens
Edited by Helmut Greim and Richard J. Albertini
© The Royal Society of Chemistry 2012
Published by the Royal Society of Chemistry, www.rsc.org

suitable indicator for the detection of non-genotoxic carcinogenicity. The practical value of this concept is illustrated in the second part of this chapter.

5.1.2   Gap Junctional Intercellular Communication: Basic Concepts

5.1.2.1   *Structural Aspects*

Morphologically, gap junctions appear as plaques at the cell plasma membrane surface and arise from the docking of two hemichannels (connexons) of adjacent cells, which in turn are composed of six connexin (Cx) units. The connexin family comprises as many as 20 isoforms in mammals. They all share an identical molecular architecture, consisting of four membrane-spanning domains, two extracellular loops, one intracellular loop, one cytoplasmic *N*-terminal tail and one cytoplasmic *C*-terminal tail (Figure 5.1.1).

Connexins are named after their molecular weight and are expressed in a tissue-specific and even in a cell-specific manner. Thus, the most abundant connexin species in the human body has a predicted weight of 43 kDa and is therefore designated Cx43.[1–5] Connexins interact with a number of other cellular proteins including scaffolding proteins, junctional proteins, cytoskeletal proteins, trafficking regulators, posttranslational modifiers and growth regulators, all which may affect connexin metabolism and functionality.[1,6]

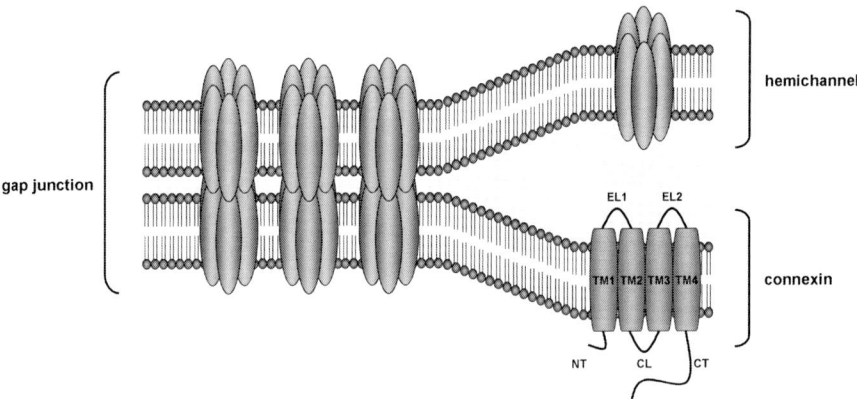

Figure 5.1.1   Molecular architecture of gap junctions. Gap junctions are grouped in plaques at the cell plasma membrane surface of two apposed cells and are composed of twelve connexin proteins, organized as two hexameric hemichannels. The connexin protein is organized as four membrane-spanning domains (TM1-4), two extracellular loops (EL1-2), one cytoplasmic loop (CL), one cytoplasmic aminotail (NT) and one cytoplasmic carboxytail (CT).

5.1.2.2 Functional Aspects

Gap junctions provide an essential pathway for the intercellular exchange of small and hydrophilic molecules including glucose, glutamate, glutathione, cyclic adenosine monophosphate (cAMP), adenosine triphosphate (ATP), inositol trisphosphate (IP$_3$) and ions such as calcium, sodium and potassium.[7,8] The biophysical permeation characteristics of these substances rely on the nature of the connexin species that forms the gap junction.[7,9] For instance, ATP passes significantly better through Cx43-based gap junctions compared with channels composed of Cx32.[10] Clearly, numerous, if not all, physiological processes are driven by substances that are conveyed *via* these channels, and GJIC is therefore considered a key mechanism in the maintenance of tissue homeostasis.[3–5,8]

In the last decade, it has become clear that hemichannels in non-junctional areas at the cell plasma membrane surface can also function as transmembrane channels. Connexons remain closed most of the time, but upon their opening, a pathway for communication between the intracellular compartment and the extracellular environment becomes available. The substances that travel through hemichannels are quite similar to those implied in GJIC, namely ATP, nicotinamide adenine dinucleotide (NAD$^+$), glutamate, glutathione and prostaglandins.[2,8,11–13]

5.1.2.3 Regulatory Aspects

A labyrinth of mechanisms underlies the regulation of the connexin life cycle and activity. Instant control (*i.e.* minute range) of connexin channel functionality through the process of channel gating is governed by a number of factors including transmembrane voltage, calcium ions and hydrogen ions.[3–5] Undoubtedly, phosphorylation has received most attention in this respect. All connexins, with the exception of Cx26, are phosphoproteins. The outcome of the phosphorylation event, mainly occurring at the *C*-terminal connexin tail, depends on both the identity of the connexin species and the kinase type.[14,15] Regulation of GJIC and hemichannel activity over the long-term (*i.e.* hour range) basically concerns peritranscriptional control of connexin expression.

The structure of most connexin genes is rather simple and consists of a first exon containing the 5′-untranslated region, which is separated by an intron from a second exon bearing the complete coding sequence and the 3′-untranslated region.[1,16,17] Connexin gene transcription as such is ruled by conventional *cis/trans* actions, involving both ubiquitous transcription factors such as specificity protein 1 (Sp1) and activator protein 1 (AP1), and tissue-specific transcription factors such as hepatocyte nuclear factor 1.[17] At the posttranscriptional level, connexin production is primarily controlled by microRNA species.[18–21] Pretranscriptional regulatory mechanisms of connexin production are discussed in detail in the following sections.

5.1.3   Epigenetic Determinants of Gap Junctional Intercellular Communication

5.1.3.1   Introduction

In the past decades, it has become very clear that transcriptional control of gene expression is not only controlled by regular *cis/trans* actions. Indeed, epigenetic mechanisms, including histone modifications and DNA methylation, are nowadays known to dominate the pretranscriptional platform and thus to act as crucial determinants of gene expression. With respect to the former, most attention has so far been paid to reversible histone acetylation. The primary sites for histone acetylation are ε-amino groups in the positively charged lysine residues in core histones H3 and H4. A number of histone acetyltransferase complexes mediate the addition of an acetyl group from acetyl coenzyme A, resulting in the neutralization of the positive charge and thus the loosening of histone-DNA contacts. In turn, this process promotes decondensation of the chromatin, thereby facilitating the accessibility of the transcriptional machinery to the DNA. The inverse reaction is catalyzed by histone deacetylase (HDAC) enzymes and is frequently associated with transcriptional repression.[22–25] At present, as many as eighteen mammalian HDAC enzymes have been identified. These have been assigned to four classes according to their homology to yeast HDAC enzymes, type of catalytic site and cellular localization.[25] The other prominent epigenetic mechanism, namely DNA methylation, includes the addition of a methyl group from *S*-adenosylmethionine to the carbon-5 position of cytosine and is controlled by DNA methyltransferase (DNMT) enzymes. *De novo* methylation is preferentially performed by DNMT3a and DNMT3b, whereas DNMT1 is crucial for maintaining established methylation patterns upon cell division.[24,26] DNA methylation plays a key role in maintaining gene silencing, which is necessary for tissue-specific and development-specific gene expression, but also underlies abrogation of gene expression patterns in pathophysiological conditions such as during cancer.[24,27] The negative correlation between the DNA methylation status and gene expression is, at least in part, mediated by methylated DNA-binding proteins, among which the methyl-CpG-binding protein 2 is known to interact with HDAC enzymes to trigger chromatin condensation and gene silencing.[24,28]

5.1.3.2   Histone Modifications

The majority of evidence that demonstrates the involvement of histone modifications, and reversible histone acetylation in particular, in the control of connexin expression comes from work with HDAC inhibitors (HDACi) including suberoylanilide hydroxamic acid (SAHA) and trichostatin A (TSA) (Table 5.1.1). Several reports describe the inductive effects of these epigenetic modifiers on GJIC in a broad variety of experimental settings[29–38] and, in most cases, this is directly linked to increased connexin production.[29,31–36] Their specific outcome, however, depends on the cellular context, the nature of the

Table 5.1.1 Effects of histone deacetylase inhibitors on connexin expression.

Cellular model	HDACi	Upregulation	Downregulation	No effect	Reference
Human neural progenitor cells	4-PB	Cx43			33
Human neural progenitor cells	TSA	Cx43			33
Human adipose tissue-derived stem cells	TSA	Cx43			142
Human dermal fibroblasts	NaB	Cx43			143
Human kB nasopharyngeal tumor cells	4-PB	Cx43			144
Human kB nasopharyngeal tumor cells	NaB	Cx43			144
Human kB nasopharyngeal tumor cells	TSA			Cx43	144
Human prostate carcinoma cell lines	TSA	Cx43			32
Human glioblastoma cells	NaB			Cx43	145
Human glioblastoma cells	4-PB	Cx43			31
Human peritoneal mesothelial cells	HMBA	Cx43			34, 35
Human peritoneal mesothelial cells	SAHA	Cx43			36
Human Huh7 liver cancer cells	TSA		Cx43	Cx32/Cx26	40
Human normal prostate epithelial cells	TSA	Cx43			32
Human Cx43-transfected HeLa cervix carcinoma cells	NaB	Cx43			143
Rat C6 glioma cells	NaB			Cx43	145
Rat C6 glioma cells	4-PB	Cx43			29
Rat CC531 colon cancer cells	NaB			Cx43	146
Rat WB-F344 cells (*ras*-transformed)	SAHA	Cx43			36
Rat primary hepatocytes	TSA	Cx43/Cx32	Cx26		38
Rat primary hepatocytes	4-Me$_2$N-BAVAH	Cx32	Cx43/Cx26		37
Mouse pancreas cell lines	TSA	Cx36			39
Mouse AtT20 pituitary corticotrophic cells	TSA			Cx36	39
Mouse SN56 neuronal cells	TSA			Cx36	39
Mouse LMTK⁻ fibroblasts	TSA			Cx36	39
Mouse embryonic stem cells	TSA		Cx43		51
Cardiac tissue from Hop-overexpressing mice (*in vivo*)	TSA	Cx40		Cx43	147
Cardiac tissue from Mdx mice (*in vivo*)	SAHA	Cx32/Cx37	Cx40	Cx43/Cx45	41

4-Me$_2$N-BAVAH = 5-(4-dimethylaminobenzoyl)aminovaleric acid hydroxamide; 4-PB = 4-phenylbutyrate; Cx = connexin; HDACi = histone deacetylase inhibitor; HMBA = hexamethylene bisacetamide; NaB = sodium butyrate; SAHA = suberoylanilide hydroxamic acid; TSA = trichostatin A.

connexin species and the epigenetic agent in question. For instance, TSA induces Cx36 expression in several mouse pancreatic cell lines, but not in mouse fibroblasts, neuronal cells and pituitary cells.[39] On the other hand, TSA down-regulates Cx43 production in human liver cancer cells, although it does not affect Cx26 and Cx32.[40] A single HDACi may even differentially affect different connexin species in one tissue. Thus, in cardiac tissue of dystrophin-lacking mice (an *in vivo* model of human muscular dystrophy) that are administered SAHA, Cx40 is negatively affected, whereas both Cx32 and Cx37 are up-regulated and Cx43 and Cx45 are unaltered.[41]

Very few studies have actually addressed the molecular mechanisms that underlie the effects of HDACi on connexin expression and activity. Hernandez and co-workers showed that the TSA-mediated induction of Cx43 in human prostate cancer cells depends on the recruitment of p300/cAMP responsive element binding protein, a transcriptional co-activator displaying histone acetyltransferase activity, and the transcription factors AP1 and Sp1 to the Cx43 gene promoter. This was accompanied by hyperacetylation of histone H4 in the AP1-responsive and Sp1-responsive elements.[32.] Similarly, SAHA caused accumulation of acetylated histones H3 and H4 in the Cx43 gene locus, which was associated with enhancement of Cx43 expression and GJIC between human peritoneal mesothelial cells.[35] Cx36 expression in pancreatic cells is controlled by the RE-1 silencing transcription factor, a transcriptional repressor that contains two independently acting HDAC-recruiting repression domains.[42,43] Active Cx36 production in these cells is evidenced by the presence of trimethylated lysine 4 residues in histone H4 in its gene promoter, an epigenetic marker of actively transcribed genes, and can be induced by TSA.[39] Such epigenetic markers that define the transcriptional status are also found in the promoter of other connexin genes. In human prostate cancer cells, low expression levels of Cx25, Cx26, Cx30, Cx30.3, Cx31, Cx31.1 and Cx32 are observed, corresponding with a high content of trimethylated lysine 27 in histone H3 in their gene promoters.[44] Also, selenomethione increases phosphorylation of serine 10 in histone H3 in the Cx26 gene promoter in human colon cancer cells, which results in its increased expression.[45]

Little is known about the identity of the specific HDAC enzymes involved in connexin expression. Upon transfection, the breast cancer metastasis suppressor 1 protein localizes in the cell nucleus and restores GJIC in human breast cancer cells[46–48] and melanoma cells.[49] In the former case, this coincides with elevated Cx43 mRNA levels and simultaneous decreased Cx32 gene transcription[46,48] Later on it was found that breast cancer metastasis suppressor 1 interacts with the large mSin3 HDAC complex, which contains both HDAC1 and HDAC2, but also forms smaller complexes with HDAC1.[50] Specific deletion of HDAC1 as well as exposure to TSA decreased Cx43 mRNA levels in mouse embryonic stem cells. Loss of HDAC1 drastically increased trimethylation of lysine 9 residues in histone H3 in the Cx43 gene promoter region, an epigenetic indicator of silenced genes, and only slightly reduced histone H3 and H4 acetylation. Thus, the Cx43 gene requires both HDAC1 presence and activity for its transcription, but histones H3 and H4 are merely minor targets in this regulatory process.[51]

Interestingly, HDACi can also elicit effects on connexin expression levels other than the transcriptional one. Research from our laboratory showed that TSA and its structural analogue 5-(4-dimethylaminobenzoyl)aminovaleric acid hydrox-amide both enhance GJIC between primary cultured hepatocytes, a finding which was associated with differential effects on the Cx26, Cx32 and Cx43 protein contents but not with alterations in the corresponding mRNA amounts.[37,38] The mechanisms that underpin these transcription-independent effects of HDACi on connexin production remain elusive, but may involve direct acetylation of connexin proteins. Acetylation sites have indeed been identified in the *N*-terminal tails of bovine Cx44 and Cx49, and are thought to play a direct role in the regulation of gating and permeability properties of the corresponding gap junctions.[52] Similarly, acetylation sites were observed at methionine and lysine residues in the *N*-terminal tail and in the intracellular loop of Cx26, which could determine protein stability.[53] Furthermore, HDACi also interfere with posttranslational connexin control, as they both increase[31,34–36] and decrease[33] the abundance of phosphorylated Cx43 isoforms. Resveratrol, an activator of class III HDAC enzymes, induces GJIC in human glioblastoma cells[54] as well as between rat liver epithelial cells pretreated with GJIC-reducing compounds.[55–57] This occurs independently of changes in Cx43 mRNA levels[56] and protein content,[54,56] but is allied with an altered Cx43 phosphorylation status.[54–56] Likewise, the HDACi sodium butyrate prevents tumor promoter-mediated inhibition GJIC *via* extracellular signal-regulated kinase 1/2 inactivation, whilst TSA restored experimentally elicited GJIC reduction and Cx43 hyperphosphorylation by preventing p38 mitogen-activated protein kinase in rat liver epithelial cells.[58]

5.1.3.3 DNA Methylation

DNA methylation, like other epigenetic mechanisms, is frequently studied in a clinical context. Hypermethylation in the promoter region of (tumor suppressor) genes is typically linked to transcriptional silencing.[24] This also holds true for connexins, which have repeatedly been demonstrated to possess potent anti-tumor properties by inducing cell cycle arrests, differentiation and apoptosis in neoplastic cells.[59–61] In fact, abrogation of connexin expression (*e.g.* Cx26, Cx32, Cx36 and Cx43) is associated with a high DNA methylation content in the connexin gene promoter in a plethora of malignant cells including human lung cancer cells,[62] human renal carcinoma cells,[63,64] human oesophageal cancer cells,[65] human breast cancer cells,[66] human nasopharyngeal cancer cells,[67] human colon cancer cells,[68] rat hepatoma cells[69–71] and rat lung cancer cells.[72] Shimizu and co-workers showed that disturbances in the DNA methylation patterns in the Cx26 gene are particularly observed in the early phase of hepatocarcinogenesis, induced by a choline-deficient L-amino acid defined diet in rats, and correlate with elevated DNMT1 mRNA levels.[70] Although exceptions exist,[73] the prototypical DNMT inhibitor decitabine indeed counteracts repression of connexin gene transcription[62,63,65,67,69,74–76] and triggers GJIC[63,67] in a variety of cancer cell lines, although this occurs in a cell type-dependent and connexin-specific fashion (Table 5.1.2).[63,65,69]

Table 5.1.2 Effects of DNA methyltransferase inhibitors on connexin expression.

Cellular model	DNMTi	Upregulation	Downregulation	No effect	Reference
Human Caki-2 renal carcinoma cells	decitabine	Cx32			63
Human Caki-1 renal carcinoma cells	decitabine	Cx32			74, 148
Human HK-2 renal tubular cells	decitabine			Cx32	63
Human oesophageal cancer cells	decitabine			Cx26/Cx43	65
Human MDA-MB-453 breast cancer cells	decitabine	Cx26			76
Human breast cancer cells	decitabine			Cx26	66
Human cervical adenocarcinoma cells	decitabine	Cx43			75
Human CNE-1 nasopharyngeal cancer cells	decitabine	Cx43			67
Human lung carcinoma cell lines	decitabine	Cx26			62
Human bone marrow mesenchymal stem cells	5-azacytidine	Cx32		Cx43	149, 150
Rat adipose tissue mesenchymal stem cells	5-azacytidine	Cx32		Cx43	151
Rat WB-F344 liver epithelial cells	decitabine	Cx43		Cx32	69
Mouse pancreas cell lines	5-azacytidine			Cx36	39
Mouse AtT20 pituitary corticotrophic cells	5-azacytidine			Cx36	39
Mouse SN56 neuronal cells	5-azacytidine			Cx36	39
Mouse LMTK⁻ fibroblasts	5-azacytidine			Cx36	39
Mouse preimplantation embryos	decitabine		Cx31/Cx43/Cx45		73

Cx = connexin; DNMTi = deoxyribonucleic acid methyltransferase inhibitor.

The functional link between DNA hypermethylation and impairment of connexin production is, however, not entirely clear. Aberrant binding of transcription factors to the methylated connexin gene promoter regions may be responsible for the poor connexin expression in cancer cells. In this respect, Chen and co-workers found that the decreased Cx43 gene transcription in human non-small cell lung cancer cells is accompanied by DNA methylation and is simultaneously related to reduced binding of AP1 to the Cx43 gene promoter.[77] Furthermore, hypermethylated DNA regions are preferentially located in Sp1 binding sites of the Cx26 gene promoter and the Cx32 gene promoter in human breast cancer cells[76] and rat liver cancer cells,[69] respectively.

5.1.4 Gap Junctional Intercellular Communication as a Target for Epigenetic Carcinogens

5.1.4.1 Introduction

From a mechanistic point of view, chemical carcinogens are usually classified as genotoxic (DNA-reactive) carcinogens or non-genotoxic (non-DNA-reactive) carcinogens. The latter are also referred to as epigenetic carcinogens.[78] However, this term should be interpreted in a broad sense in this context, as it implies a wide variety of mechanisms that are not solely restricted to the pre-transcriptional level. Indeed, epigenetic carcinogens perform quite diverse actions including disruption of the hormonal balance, induction of chronic inflammation, immunosuppression, induction of xenobiotic biotransformation enzymes, initiation of oxidative stress, induction of peroxisome proliferation and induction of sustained cytotoxicity.[79] Independent of their primary mode of action, epigenetic carcinogens share the ability to disturb the equilibrium between cell growth and cell death, ultimately leading to tumor formation.[5,79,80] Inherent to their pivotal task in maintaining tissue homeostasis, gap junctions are frequently involved in this process. Although several exceptions exist, non-genotoxic carcinogens decrease GJIC, whereas their genotoxic counterparts usually leave gap junctions unaffected. The deleterious actions of this type of carcinogens are thereby mostly targeted towards translational and posttranslational control, but do not involve the most upper regulatory levels of connexin expression. The subsequent suppression of GJIC occurs in a tissue-specific, species-specific and gender-specific manner, which is a typical feature of non-genotoxic carcinogenicity.[5,80] These concepts are illustrated in the following sections by discussing a number of prototypical epigenetic carcinogens that affect the liver, being the main site of xenobiotic biotransformation and thereby a major target for toxicity in the body.

5.1.4.2 Tetradecanoyl Phorbol Acetate

Phorbol esters are tetracyclic diterpenoids derived from the seed oil of the *Croton tiglium* plant. The most common phorbol ester is tetradecanoyl phorbol acetate (TPA), also called phorbol-12-myristate-13-acetate, which is frequently used as a tumor promoter in models of hepatocarcinogenesis. TPA mimics the action of diacylglycerol, an activator of protein kinase C, which regulates different signal transduction pathways and other cellular metabolic activities.[81,82] It has been demonstrated on many occasions that TPA down-regulates GJIC in liver-based *in vitro* models, both primary hepatocytes and hepatic cell lines. With some exceptions,[83,84] hepatic connexin mRNA levels remained unaffected,[56,85–88] whereas both increased[89] and decreased[83,90,91] protein contents have been observed in the presence of TPA. The principal mode of TPA action is actually located at the posttranslational level. It has been repeatedly described that TPA induces Cx43 hyperphosphorylation, which burgeons into the loss of GJIC.[56,58,84,86–89,92–109] Besides reducing the unphosphorylated Cx43 signal, TPA was also reported to induce the appearance of a particular phosphorylated Cx43 variant during immunoblot analysis.[84,85,88] Both protein kinase C[88,92,101,102,105–108,110] and mitogen-activated protein kinases[58,104,106–108] have been shown to mediate TPA-induced Cx43 hyperphosphorylation. In fact, TPA induces oxidative stress and the activation of different protein kinase C isoforms.[96,97,101] Protein kinase C then directly phosphorylates Cx43 in its *C*-terminal region,[103,108] followed by its internalization and degradation.[88,96,97]

5.1.4.3 Phenobarbital

Phenobarbital is a widely used anti-epileptic drug that also has sedative and hypnotic properties. It is frequently applied as a model tumor promoter in rodent liver,[111,112] whereby the expression of a broad set of genes is altered, of which genes related to cytochrome P450 (CYP450)-dependent xenobiotic biotransformation have gained particular attention.[113] The presence of functional gap junctions consisting of Cx32, but not of Cx26, is a prerequisite for the promotional activity of phenobarbital, since Cx32 knock-out mice,[111,112] unlike Cx26 knock-out animals,[114] are resistant to promotion of hepatocarcinogenesis by the barbiturate. Furthermore, a subset of genes is differentially affected by phenobarbital in the liver of Cx32-deficient mice compared with their wild-type counterparts,[113] thus further pointing to a critical role for GJIC in phenobarbital-mediated tumor promotion.

It has been shown by several groups that gap junction activity becomes reduced upon administration of phenobarbital to rodents.[115–120] This was associated with abnormal frequency and size of gap junctions on the hepatocyte membrane surface,[121] decreased Cx32 immunoreactivity[115,118,119,122] and aberrant Cx32 localization,[117] whereas Cx26 expression was not affected.[115,118,119] Both unchanged[118,120] and decreased[123,124] hepatic Cx32 mRNA levels were reported in phenobarbital-treated rodents. Interestingly, phenobarbital specifically reduced Cx32 protein production in centrolobular hepatocytes of male rodent liver,[115,118,119] which is the acinar area where the phenobarbital-induced expression of CYP450BII1/2 is mostly manifested. These co-localized modifications in Cx32 production and CYP450BII1/2 expression are believed to be physiologically important for the effective bio-transformation of xenobiotics, *in casu* phenobarbital, by limiting the cytoplasmic diffusion of toxic reactive intermediates.[118] As shown in rodent models both *in vivo*[120] and *in vitro*,[88,125] the reduction of GJIC by phenobarbital occurs in a strain-specific way. Furthermore, the inhibitory effect of phenobarbital on GJIC between primary cultured mouse hepatocytes depends on xenobiotic biotransformation capacity, as it was abolished by a CYP450 inhibitor.[126]

5.1.4.4   Dichlorodiphenyltrichloroethane

The prototypical pesticide 1,1,1-trichloro-2,2 bis(4-chlorophenyl)ethane, commonly known as dichlorodiphenyltrichloroethane (DDT), is a non-systemic insecticide which persists strongly in the environment and accumulates in animal fats.[127] DDT is neurotoxic, causes endocrine disruption and acts as a tumor promoter, whereby the liver is a principal target.[128]

Early freeze-fracture studies showed that the size of gap junction plaques on hepatocytes from rats exposed to DDT is reduced.[121] Subsequent animal experiments demonstrated that DDT provokes gap junction closure in male rat liver, which is associated with decreased Cx32 immunoreactivity and/or protein levels, aberrant Cx32 localization, but not with changes in the Cx32 phosphorylation status or mRNA content.[117,129–132] DDT did not affect the overall Cx26 protein and mRNA levels,[129,132] but enhanced Cx26 protein production by centrolobular hepatocytes[117] and negatively affected Cx26 immunoreactivity in the perilobular region.[132] Furthermore, DDT also promoted the appearance of Cx43 in the cytoplasm of hepatocytes when administered to male rats.[117] *In vitro* studies, carried out in primary hepatocyte cultures from rat and mouse, showed that the inhibition of GJIC caused by DDT was likely to be caused by the formation of radical intermediates following lipid peroxidation.[133] In rat liver epithelial cell line models, DDT also reduced GJIC,[56,88,89,134] which was allied with decreased Cx43 protein levels,[89] an altered Cx43 phosphorylation status,[88,89,134] increased endocytosis of gap junctions and lysosomal degradation of phosphorylated Cx43,[94] whereas changes in Cx43 gene transcription remained absent.[134]

5.1.5   Conclusions and Perspectives

Virtually all aspects of the cellular life cycle are regulated by chemical signals that are intercellularly exchanged *via* gap junctions. Accordingly, GJIC is considered a central control platform in the maintenance of the homeostatic balance.[4,5,60,135,136] A strict and well-coordinated regulation is compulsory for appropriate GJIC functioning. Management of GJIC over the long-term occurs mainly at the level of connexin expression. Many efforts have yet been focused on the elucidation of the *cis/trans* machinery that drives connexin gene transcription.[17] Increasing evidence also points to the critical involvement of epigenetic phenomena in this process. As discussed in this chapter, classical epigenetic mechanisms, like histone acetylation and DNA methylation, are essential determinants of connexin expression. These mechanisms are directly connected with less acknowledged components of the epigenetic circuit such as ATP-dependent chromatin remodelling systems,[24,137] but also with post-transcriptional regulatory entities, including microRNA species. In fact, a major challenge lies ahead in fully elucidating the involvement of these molecular mechanisms in connexin expression. A number of tools are now available that will allow clarification of this peritranscriptional crosstalk in GJIC control, including isotype-specific epigenetic modifiers,[138] small interfering RNA duplexes[139] and microRNA inhibitors.[140]

At the same time, the use of GJIC suppression as a biomarker for the *in vitro* detection of epigenetic carcinogenicity should be further explored. Indeed, in contrast to genotoxic carcinogens, for which a classical *in vitro* testing battery is applied, no validated *in vitro* assay is currently available for the testing of non-genotoxic carcinogenicity.[5] In this chapter, specific attention was paid to hepatic gap junctions in this respect, which are negatively affected by prototypical non-genotoxic carcinogens such as TPA, phenobarbital and DDT. When developing such an *in vitro* assay, care should be taken while selecting the cellular system. Thus, to allow reliable detection of non-genotoxic hepatocarcinogens, the liver-based *in vitro* model must exhibit the *in vivo*-like hepatocellular connexin expression pattern including high amounts of Cx32 and low, and even absent, production of Cx43.[136] In practice, this means that hepatic cell lines should be avoided, as they are frequently derived from cancerous tissue, and that cultures of primary hepatocytes, isolated from healthy liver, should be preferred.[79] The latter also display sufficient metabolic capacity, at least during short-term culture regimes.[79,141] This is of crucial importance, since several carcinogenic compounds rely on CYP450-dependent biotransformation in order to perform their harmful effects, including GJIC suppression.[5] Further efforts should be additionally focussed on the optimization and standardization of test conditions before a solid GJIC inhibition assay for routine use can be delivered. It can be anticipated that the resulting validated *in vitro* GJIC inhibition test will be a valuable tool for the evaluation of the hazardous potential of chemical compounds during the process of risk assessment.

Acknowledgement

This work was supported by the grants from the Fund of Scientific Research Flanders (FWO-Vlaanderen), Belgium, the Research Council of the Vrije Universiteit Brussel (OZR-VUB), and the European Union (FP6 project carcinoGENOMICS).

References

1. H. A. Dbouk, R. M. Mroue, M. E. El-Sabban and R. S. Talhouk, *Cell Commun., Signaling*, 2009, **7**, 4.
2. D. A. Goodenough and D. L. Paul, *Cold Spring Harbor Perspect., Biol.*, 2009, **1**, a002576.

3. M. Rackauskas, V. Neverauskas and V. A. Skeberdis, *Medicina*, 2010, **46**, 1–12.
4. M. Vinken, E. De Rop, E. Decrock, E. De Vuyst, L. Leybaert, T. Vanhaecke and V. Rogiers, *Biochim. Biophys. Acta*, 2009, **1795**, 53–61.
5. M. Vinken, T. Doktorova, E. Decrock, L. Leybaert, T. Vanhaecke and V. Rogiers, *Crit. Rev. Biochem. Mol. Biol.*, 2009, **44**, 201–222.
6. D. W. Laird, *Trends Cell Biol.*, 2010, **20**, 92–101.
7. D. B. Alexander and G. S. Goldberg, *Curr. Med. Chem.*, 2003, **10**, 2045–2058.
8. E. Decrock, M. Vinken, E. De Vuyst, D. V. Krysko, K. D'Herde, T. Vanhaecke, P. Vandenabeele, V. Rogiers and L. Leybaert, *Cell Death Differ.*, 2009, **16**, 524–536.
9. G. T. Cottrell and J. M. Burt, *Biochim. Biophys. Acta*, 2005, **1711**, 126–141.
10. G. S. Goldberg, A. P. Moreno and P. D. Lampe, *J. Biol. Chem.*, 2002, **277**, 36725–36730.
11. C. D'Hondt, R. Ponsaerts, H. De Smedt, G. Bultynck and B. Himpens, *Bioessays*, 2009, **31**, 953–974.
12. W. H. Evans, E. De Vuyst and L. Leybaert, *Biochem. J.*, 2006, **397**, 1–14.
13. K. A. Schalper, N. Palacios-Prado, J. A. Orellana and J. C. Saez, *Cell Commun. Adhes.*, 2008, **15**, 207–218.
14. A. P. Moreno and A. F. Lau, *Prog. Biophys. Mol. Biol.*, 2007, **94**, 107–119.
15. J. L. Solan and P. D. Lampe, *Biochem. J.*, 2009, **419**, 261–272.
16. G. Sohl and K. Willecke, *Cardiovasc. Res.*, 2004, **62**, 228–232.
17. M. Oyamada, Y. Oyamada and T. Takamatsu, *Biochim. Biophys. Acta*, 2005, **1719**, 6–23.
18. C. Anderson, H. Catoe and R. Werner, *Nucleic Acids Res.*, 2006, **34**, 5863–5871.
19. H. Inose, H. Ochi, A. Kimura, K. Fujita, R. Xu, S. Sato, M. Iwasaki, S. Sunamura, Y. Takeuchi, S. Fukumoto, K. Saito, T. Nakamura, H. Siomi, H. Ito, Y. Arai, K. I. Shinomiya and S. Takeda, *Proc. Natl. Acad. Sci. U. S. A.*, 2009, **106**, 20794–20799.
20. H. K. Kim, Y. S. Lee, U. Sivaprasad, A. Malhotra and A. Dutta, *J. Cell Biol.*, 2006, **174**, 677–687.
21. B. Yang, H. Lin, J. Xiao, Y. Lu, X. Luo, B. Li, Y. Zhang, C. Xu, Y. Bai, H. Wang, G. Chen and Z. Wang, *Nat. Med.*, 2007, **13**, 486–491.
22. T. Kouzarides, *Cell*, 2007, **128**, 693–705.
23. S. M. Reamon-Buettner and J. Borlak, *Reprod. Toxicol.*, 2007, **24**, 20–30.
24. M. Szyf, *Toxicol. Sci.*, 2007, **100**, 7–23.
25. X. J. Yang and E. Seto, *Oncogene*, 2007, **26**, 5310–5318.
26. P. Siedlecki and P. Zielenkiewicz, *Acta Biochim. Pol.*, 2006, **53**, 245–256.
27. K. Gronbaek, C. Hother and P. A. Jones, *Apmis*, 2007, **115**, 1039–1059.
28. M. Szyf, *Ageing Res. Rev.*, 2003, **2**, 299–328.
29. O. Ammerpohl, D. Thormeyer, Z. Khan, I. B. Appelskog, Z. Gojkovic, P. M. Almqvist and T. J. Ekstrom, *Biochem. Biophys. Res. Commun.*, 2004, **324**, 8–14.

30. O. Ammerpohl, A. Trauzold, B. Schniewind, U. Griep, C. Pilarsky, R. Grutzmann, H. D. Saeger, O. Janssen, B. Sipos, G. Kloppel and H. Kalthoff, *Br. J. Cancer*, 2007, **96**, 73–81.

31. T. Asklund, I. B. Appelskog, O. Ammerpohl, T. J. Ekstrom and P. M. Almqvist, *Eur. J. Cancer*, 2004, **40**, 1073–1081.

32. M. Hernandez, Q. Shao, X. J. Yang, S. P. Luh, M. Kandouz, G. Batist, D. W. Laird and M. A. Alaoui-Jamali, *Prostate*, 2006, **66**, 1151–1161.

33. Z. Khan, M. Akhtar, T. Asklund, B. Juliusson, P. M. Almqvist and T. J. Ekstrom, *Exp. Cell Res.*, 2007, **313**, 2958–2967.

34. T. Ogawa, T. Hayashi, S. Kyoizumi, T. Ito, J. E. Trosko and N. Yorioka, *Lab. Invest.*, 1999, **79**, 1511–1520.

35. T. Ogawa, T. Hayashi, M. Tokunou, K. Nakachi, J. E. Trosko, C. C. Chang and N. Yorioka, *Cancer Res.*, 2005, **65**, 9771–9778.

36. T. Ogawa, T. Hayashi, N. Yorioka, S. Kyoizumi and J. E. Trosko, *Kidney Int.*, 2001, **60**, 996–1008.

37. M. Vinken, T. Henkens, S. Snykers, A. Lukaszuk, D. Tourwe, V. Rogiers and T. Vanhaecke, *Toxicology*, 2007, **236**, 92–102.

38. M. Vinken, T. Henkens, T. Vanhaecke, P. Papeleu, A. Geerts, E. Van Rossen, J. K. Chipman, P. Meda and V. Rogiers, *Toxicol. Sci.*, 2006, **91**, 484–492.

39. M. Hohl and G. Thiel, *Eur. J. Neurosci.*, 2005, **22**, 2216–2230.

40. Y. Yamashita, M. Shimada, N. Harimoto, S. Tanaka, K. Shirabe, H. Ijima, K. Nakazawa, J. Fukuda, K. Funatsu and Y. Maehara, *Cell Transplant.*, 2004, **13**, 793–799.

41. C. Colussi, R. Berni, J. Rosati, S. Straino, S. Vitale, F. Spallotta, S. Baruffi, L. Bocchi, F. Delucchi, S. Rossi, M. Savi, D. Rotili, F. Quaini, E. Macchi, D. Stilli, E. Musso, A. Mai, C. Gaetano and M. C. Capogrossi, *Cardiovasc. Res.*, 2010, **87**, 73–82.

42. D. Martin, F. Allagnat, G. Chaffard, D. Caille, M. Fukuda, R. Regazzi, A. Abderrahmani, G. Waeber, P. Meda, P. Maechler and J. A. Haefliger, *Diabetologia*, 2008, **51**, 1429–1439.

43. D. Martin, T. Tawadros, L. Meylan, A. Abderrahmani, D. F. Condorelli, G. Waeber and J. A. Haefliger, *J. Biol. Chem.*, 2003, **278**, 53082–53089.

44. X. S. Ke, Y. Qu, K. Rostad, W. C. Li, B. Lin, O. J. Halvorsen, S. A. Haukaas, I. Jonassen, K. Petersen and N. Goldfinger, V. Rotter, L. A. Akslen, A. M. Oyan and K. H. Kalland, *PloS One*, 2009, **4**, e4687.

45. A. C. Goulet, G. Watts, J. L. Lord and M. A. Nelson, *Cancer Biol. Ther.*, 2007, **6**, 494–503.

46. P. Kapoor, M. M. Saunders, Z. Li, Z. Zhou, N. Sheaffer, E. L. Kunze, R. S. Samant, D. R. Welch and H. J. Donahue, *Int. J. Cancer*, 2004, **111**, 693–697.

47. R. S. Samant, M. J. Seraj, M. M. Saunders, T. S. Sakamaki, L. A. Shevde, J. F. Harms, T. O. Leonard, S. F. Goldberg, L. Budgeon, W. J. Meehan, C. R. Winter, N. D. Christensen, M. F. Verderame, H. J. Donahue and D. R. Welch, *Clin. Exp. Metastasis*, 2000, **18**, 683–693.

48. M. M. Saunders, M. J. Seraj, Z. Li, Z. Zhou, C. R. Winter, D. R. Welch and H. J. Donahue, *Cancer Res.*, 2001, **61**, 1765–1767.

49. L. A. Shevde, R. S. Samant, S. F. Goldberg, T. Sikaneta, A. Alessandrini, H. J. Donahue, D. T. Mauger and D. R. Welch, *Exp. Cell Res.*, 2002, **273**, 229–239.
50. W. J. Meehan, R. S. Samant, J. E. Hopper, M. J. Carrozza, L. A. Shevde, J. L. Workman, K. A. Eckert, M. F. Verderame and D. R. Welch, *J. Biol. Chem.*, 2004, **279**, 1562–1569.
51. G. Zupkovitz, J. Tischler, M. Posch, I. Sadzak, K. Ramsauer, G. Egger, R. Grausenburger, N. Schweifer, S. Chiocca, T. Decker and C. Seiser, *Mol. Cell. Biol.*, 2006, **26**, 7913–7928.
52. D. Shearer, W. Ens, K. Standing and G. Valdimarsson, *Invest. Ophthalmol. Visualization Sci.*, 2008, **49**, 1553–1562.
53. D. Locke, S. Bian, H. Li and A. L. Harris, *Biochem. J.*, 2009, **424**, 385–398.
54. S. Leone, M. Fiore, M. G. Lauro, S. Pino, T. Cornetta and R. Cozzi, *Mol. Carcinog.*, 2008, **47**, 587–598.
55. J. H. Kim, B. K. Lee, K. W. Lee and H. J. Lee, *J. Nutr. Biochem.*, 2009, **20**, 149–154.
56. M. Nielsen, R. J. Ruch and O. Vang, *Biochem. Biophys. Res. Commun.*, 2000, **275**, 804–809.
57. B. L. Upham, M. Guzvic, J. Scott, J. M. Carbone, L. Blaha, C. Coe, L. L. Li, A. M. Rummel and J. E. Trosko, *Nutr. Cancer*, 2007, **57**, 38–47.
58. J. W. Jung, S. D. Cho, N. S. Ahn, S. R. Yang, J. S. Park, E. H. Jo, J. W. Hwang, O. I. Aruoma, Y. S. Lee and K. S. Kang, *Cancer Lett.*, 2006, **241**, 301–308.
59. J. Czyz, *Cell. Mol. Biol. Lett.*, 2008, **13**, 92–102.
60. G. Pointis, C. Fiorini, J. Gilleron, D. Carette and D. Segretain, *Curr. Med. Chem.*, 2007, **14**, 2288–2303.
61. M. Vinken, T. Vanhaecke, P. Papeleu, S. Snykers, T. Henkens and V. Rogiers, *Cell. Signalling*, 2006, **18**, 592–600.
62. Y. Chen, D. Huhn, T. Knosel, M. Pacyna-Gengelbach, N. Deutschmann and I. Petersen, *Int. J. Cancer*, 2005, **113**, 14–21.
63. A. Hirai, T. Yano, K. Nishikawa, K. Suzuki, R. Asano, H. Satoh, K. Hagiwara and H. Yamasaki, *Am. J. Nephrol.*, 2003, **23**, 172–177.
64. T. Yano, F. Ito, K. Kobayashi, Y. Yonezawa, K. Suzuki, R. Asano, K. Hagiwara, H. Nakazawa, H. Toma and H. Yamasaki, *Cancer Lett.*, 2004, **208**, 137–142.
65. J. Loncarek, H. Yamasaki, P. Levillain, S. Milinkevitch and M. Mesnil, *Mol. Carcinog.*, 2003, **36**, 74–81.
66. R. Singal, Z. J. Tu, J. M. Vanwert, G. D. Ginder and D. T. Kiang, *Anticancer Res.*, 2000, **20**, 59–64.
67. Z. C. Yi, H. Wang, G. Y. Zhang and B. Xia, *Oral Oncol.*, 2007, **43**, 898–904.
68. S. C. Borinstein, M. Conerly, S. Dzieciatkowski, S. Biswas, M. K. Washington, P. Trobridge, S. Henikoff and W. M. Grady, *Mol. Carcinog.*, 2010, **49**, 94–103.
69. M. P. Piechocki, R. D. Burk and R. J. Ruch, *Carcinogenesis*, 1999, **20**, 401–406.

70. K. Shimizu, M. Onishi, E. Sugata, Y. Sokuza, C. Mori, T. Nishikawa, K. Honoki and T. Tsujiuchi, *Cancer Sci.*, 2007, **98**, 1318–1322.
71. T. Tsujiuchi, K. Shimizu, Y. Itsuzaki, M. Onishi, E. Sugata, H. Fujii and K. Honoki, *Mol. Carcinog.*, 2007, **46**, 269–274.
72. K. Shimizu, Y. Shimoichi, D. Hinotsume, Y. Itsuzaki, H. Fujii, K. Honoki and T. Tsujiuchi, *Mol. Carcinog.*, 2006, **45**, 710–714.
73. J. N. Yu, C. Y. Xue, X. G. Wang, F. Lin, C. Y. Liu, F. Z. Lu and H. L. Liu, *Zygote*, 2009, **17**, 137–145.
74. H. Hagiwara, H. Sato, Y. Ohde, Y. Takano, T. Seki, T. Ariga, N. Hokaiwado, M. Asamoto, T. Shirai, Y. Nagashima and T. Yano, *Br. J. Pharmacol.*, 2008, **153**, 1373–1381.
75. T. J. King, L. H. Fukushima, T. A. Donlon, A. D. Hieber, K. A. Shimabukuro and J. S. Bertram, *Carcinogenesis*, 2000, **21**, 311–315.
76. L. W. Tan, T. Bianco and A. Dobrovic, *Carcinogenesis*, 2002, **23**, 231–236.
77. J. T. Chen, Y. W. Cheng, M. C. Chou, T. Sen-Lin, W. W. Lai, W. L. Ho and H. Lee, *Clin. Cancer Res.*, 2003, **9**, 4200–4204.
78. M. Vinken, T. Doktorova, H. Ellinger-Ziegelbauer, H. J. Ahr, E. Lock, P. Carmichael, E. Roggen, J. van Delft, J. Kleinjans, J. Castell, R. Bort, T. Donato, M. Ryan, R. Corvi, H. Keun, T. Ebbels, T. Athersuch, S. A. Sansone, P. Rocca-Serra, R. Stierum, P. Jennings, W. Pfaller, H. Gmuender, T. Vanhaecke and V. Rogiers, *Mutat. Res.*, 2008, **659**, 202–210.
79. M. Vinken, P. Papeleu, S. Snykers, E. De Rop, T. Henkens, J. K. Chipman, V. Rogiers and T. Vanhaecke, *Crit. Rev. Toxicol.*, 2006, **36**, 299–318.
80. J. K. Chipman, A. Mally and G. O. Edwards, *Toxicol. Sci.*, 2003, **71**, 146–153.
81. G. Goel, H. P. Makkar, G. Francis and K. Becker, *Int. J. Toxicol.*, 2007, **26**, 279–288.
82. E. M. Griner and M. G. Kazanietz, *Nat. Rev.*, 2007, **7**, 281–294.
83. M. Asamoto, M. Oyamada, A. el Aoumari, D. Gros and H. Yamasaki, *Mol. Carcinog.*, 1991, **4**, 322–327.
84. D. F. Matesic, H. L. Rupp, W. J. Bonney, R. J. Ruch and J. E. Trosko, *Mol. Carcinog.*, 1994, **10**, 226–236.
85. X. Guan, W. J. Bonney and R. J. Ruch, *Toxicol. Appl. Pharmacol.*, 1995, **130**, 79–86.
86. K. S. Kang, J. W. Yun, B. Yoon, Y. K. Lim and Y. S. Lee, *Cancer Lett.*, 2001, **166**, 147–153.
87. P. D. Lampe, *J. Cell Biol.*, 1994, **127**, 1895–1905.
88. P. Ren, P. P. Mehta and R. J. Ruch, *Carcinogenesis*, 1998, **19**, 169–175.
89. I. V. Budunova, G. M. Williams and D. C. Spray, *Arch. Toxicol.*, 1993, **67**, 565–572.
90. K. Kenne, R. Fransson-Steen, S. Honkasalo and L. Warngard, *Carcinogenesis*, 1994, **15**, 1161–1165.
91. E. Rivedal, H. Yamasaki and T. Sanner, *Carcinogenesis*, 1994, **15**, 689–694.

92. V. M. Berthoud, M. B. Rook, O. Traub, E. L. Hertzberg and J. C. Saez, *Eur. J. Cell Biol.*, 1993, **62**, 384–396.
93. C. Chaumontet, C. Droumaguet, V. Bex, C. Heberden, I. Gaillard-Sanchez and P. Martel, *Cancer Lett.*, 1997, **114**, 207–210.
94. X. Guan and R. J. Ruch, *Carcinogenesis*, 1996, **17**, 1791–1798.
95. C. S. Hill, S. Y. Oh, S. A. Schmidt, K. J. Clark and A. W. Murray, *Biochem. J.*, 1994, **303**, 475–479.
96. J. Hu and I. A. Cotgreave, *Chem.-Biol. Interact.*, 1995, **95**, 291–307.
97. J. Hu, L. Engman and I. A. Cotgreave, *Carcinogenesis*, 1995, **16**, 1815–1824.
98. M. Y. Kanemitsu and A. F. Lau, *Mol. Biol. Cell*, 1993, **4**, 837–848.
99. Y. Kato and K. Kenne, *Pharmacol. Toxicol.*, 1996, **79**, 23–28.
100. L. O. Klotz, P. Patak, N. Ale-Agha, D. P. Buchczyk, K. Abdelmohsen, P. A. Gerber, C. von Montfort and H. Sies, *Cancer Res.*, 2002, **62**, 4922–4928.
101. E. Leithe, V. Cruciani, T. Sanner, S. O. Mikalsen and E. Rivedal, *Carcinogenesis*, 2003, **24**, 1239–1245.
102. E. Leithe and E. Rivedal, *J. Biol. Chem.*, 2004, **279**, 50089–50096.
103. R. Loch-Caruso, M. M. Galvez, K. Brant and D. Chung, *Cell Biol. Toxicol.*, 2004, **20**, 147–169.
104. J. R. Park, J. S. Park, E. H. Jo, J. W. Hwang, S. J. Kim, J. C. Ra, O. I. Aruoma, Y. S. Lee and K. S. Kang, *Biofactors*, 2006, **27**, 147–155.
105. E. Rivedal and E. Leithe, *Exp. Cell Res.*, 2005, **302**, 143–152.
106. E. Rivedal and H. Opsahl, *Carcinogenesis*, 2001, **22**, 1543–1550.
107. R. J. Ruch, J. E. Trosko and B. V. Madhukar, *J. Cell. Biochem.*, 2001, **83**, 163–169.
108. S. Sirnes, A. Kjenseth, E. Leithe and E. Rivedal, *Biochem. Biophys. Res. Commun.*, 2009, **382**, 41–45.
109. B. L. Upham, K. S. Kang, H. Y. Cho and J. E. Trosko, *Carcinogenesis*, 1997, **18**, 37–42.
110. S. Y. Oh, B. V. Madhukar and J. E. Trosko, *Carcinogenesis*, 1988, **9**, 135–139.
111. E. G. Luebeck, A. Buchmann, D. Schneider, S. H. Moolgavkar and M. Schwarz, *Mutat. Res.*, 2005, **570**, 33–47.
112. O. Moennikes, A. Buchmann, A. Romualdi, T. Ott, J. Werringloer, K. Willecke and M. Schwarz, *Cancer Res.*, 2000, **60**, 5087–5091.
113. S. Stahl, C. Ittrich, P. Marx-Stoelting, C. Kohle, T. Ott, A. Buchmann and M. Schwarz, *Int. J. Cancer*, 2005, **115**, 861–869.
114. P. Marx-Stoelting, J. Mahr, T. Knorpp, S. Schreiber, M. F. Templin, T. Ott, A. Buchmann and M. Schwarz, *Toxicol. Sci.*, 2008, **103**, 260–267.
115. S. Ito, M. Tsuda, A. Yoshitake, T. Yanai and T. Masegi, *Toxicol. Pathol.*, 1998, **26**, 253–259.
116. S. H. Jeong, S. S. Habeebu and C. D. Klaassen, *Toxicol. Sci.*, 2000, **57**, 156–166.
117. V. A. Krutovskikh, G. Mesnil, G. Mazzoleni and H. Yamasaki, *Lab. Invest.*, 1995, **72**, 571–577.

118. M. J. Neveu, K. L. Babcock, E. L. Hertzberg, D. L. Paul, B. J. Nicholson and H. C. Pitot, *Cancer Res.*, 1994, **54**, 3145–3152.

119. M. J. Neveu, J. R. Hully, D. L. Paul and H. C. Pitot, *Cancer Commun.*, 1990, **2**, 21–31.

120. K. A. Warner, M. J. Fernstrom and R. J. Ruch, *Toxicol. Sci.*, 2003, **71**, 190–197.

121. S. Sugie, H. Mori and M. Takahashi, *Carcinogenesis*, 1987, **8**, 45–51.

122. H. Okamiya, K. Mitsumori, H. Onodera, S. Ito, T. Imazawa, K. Yasuhara and M. Takahashi, *Arch. Toxicol.*, 1998, **72**, 744–750.

123. D. G. Beer, M. J. Neveu, D. L. Paul, U. R. Rapp and H. C. Pitot, *Cancer Res.*, 1988, **48**, 1610–1617.

124. M. Mesnil, D. J. Fitzgerald and H. Yamasaki, *Mol. Carcinog.*, 1988, **1**, 79–81.

125. J. E. Klaunig and R. J. Ruch, *Cancer Lett.*, 1987, **36**, 161–168.

126. J. E. Klaunig, R. J. Ruch and C. M. Weghorst, *Toxicol. Appl. Pharmacol.*, 1990, **102**, 553–563.

127. M. Maroni, C. Colosio, A. Ferioli and A. Fait, *Toxicology*, 2000, **143**, 1–118.

128. J. Beard, *Sci. Total Environ.*, 2006, **355**, 78–89.

129. C. Cowles, A. Mally and J. K. Chipman, *Toxicology*, 2007, **238**, 49–59.

130. T. Harada, S. Yamaguchi, R. Ohtsuka, M. Takeda, H. Fujisawa, T. Yoshida, A. Enomoto, Y. Chiba, J. Fukumori, S. Kojima, N. Tomiyama, M. Saka, M. Ozaki and K. Maita, *Toxicol. Pathol.*, 2003, **31**, 87–98.

131. S. Ito, C. Tateno, M. Tanaka and A. Yoshitake, *Cell Biol. Toxicol.*, 1993, **9**, 189–196.

132. C. Tateno, S. Ito, M. Tanaka, M. Oyamada and A. Yoshitake, *Carcinogenesis*, 1994, **15**, 517–521.

133. E. Leibold and L. R. Schwarz, *Carcinogenesis*, 1993, **14**, 2377–2382.

134. R. J. Ruch, W. J. Bonney, K. Sigler, X. Guan, D. Matesic, L. D. Schafer, E. Dupont and J. E. Trosko, *Carcinogenesis*, 1994, **15**, 301–306.

135. E. Kardami, X. Dang, D. A. Iacobas, B. E. Nickel, M. Jeyaraman, W. Srisakuldee, J. Makazan, S. Tanguy and D. C. Spray, *Prog. Biophys. Mol. Biol.*, 2007, **94**, 245–264.

136. M. Vinken, T. Henkens, E. De Rop, J. Fraczek, T. Vanhaecke and V. Rogiers, *Hepatology*, 2008, **47**, 1077–1088.

137. L. Lafon-Hughes, M. V. Di Tomaso, L. Mendez-Acuna and W. Martinez-Lopez, *Mutat. Res.*, 2008, **658**, 191–214.

138. Y. Itoh, T. Suzuki and N. Miyata, *Curr. Pharm. Des.*, 2008, **14**, 529–544.

139. N. K. Sahu, G. Shilakari, A. Nayak and D. V. Kohli, *Curr. Pharm. Biotechnol.*, 2007, **8**, 291–304.

140. D. Catalucci, M. V. Latronico and G. Condorelli, *Ann. N. Y. Acad. Sci.*, 2008, **1123**, 20–29.

141. G. Elaut, T. Henkens, P. Papeleu, S. Snykers, M. Vinken, T. Vanhaecke and V. Rogiers, *Curr. Drug Metab.*, 2006, **7**, 629–660.

142. Y. S. Choi, G. J. Dusting, S. Stubbs, S. Arunothayaraj, X. L. Han, P. Collas, W. A. Morrison and R. J. Dilley, *J. Cell. Mol. Med.*, 2010, **14**, 878–889.

143. S. R. Johnstone, A. K. Best, C. S. Wright, B. E. Isakson, R. J. Errington and P. E. Martin, *J. Cell. Biochem.*, 2010, **110**, 772–782.
144. Y. Hattori, M. Fukushima and Y. Maitani, *Int. J. Oncol.*, 2007, **30**, 1427–1439.
145. P. A. Robe, O. Jolois, M. N'Guyen, F. Princen, B. Malgrange, M. P. Merville and V. Bours, *Int. J. Oncol.*, 2004, **25**, 187–192.
146. A. Germann, S. Dihlmann, M. Hergenhahn, M. K. Doeberitz and R. Koesters, *Int. J. Cancer*, 2003, **106**, 187–197.
147. F. Liu, M. D. Levin, N. B. Petrenko, M. M. Lu, T. Wang, L. J. Yuan, A. L. Stout, J. A. Epstein and V. V. Patel, *J. Mol. Cell. Cardiol.*, 2008, **45**, 715–723.
148. Y. Takano, H. Iwata, Y. Yano, M. Miyazawa, N. Virgona, H. Sato, K. Ueno and T. Yano, *Biochem. Pharmacol.*, 2010, **80**, 463–470.
149. I. Aurich, L. P. Mueller, H. Aurich, J. Luetzkendorf, K. Tisljar, M. M. Dollinger, W. Schormann, J. Walldorf, J. G. Hengstler, W. E. Fleig and B. Christ, *Gut*, 2007, **56**, 405–415.
150. P. Stock, M. S. Staege, L. P. Müller, M. Sgodda, A. Völker, I. Volkmer, J. Lützkendorf and B. Christ, *Transplant. Proc.*, 2008, **40**, 620–623.
151. M. Sgodda, H. Aurich, S. Kleist, I. Aurich, S. König, M. M. Dollinger, W. E. Fleig and B. Christ, *Exp. Cell Res.*, 2007, **313**, 2875–2886.

Subject Index